AF598035

METHODS IN MOLECULAR BIOLOGY™

Series Editor
John M. Walker
School of Life Sciences
University of Hertfordshire
Hatfield, Hertfordshire, AL10 9AB, UK

For further volumes:
http://www.springer.com/series/7651

Membrane Biogenesis

Methods and Protocols

Edited by

Doron Rapaport

Interfaculty Institute of Biochemistry, University of Tuebingen, Tuebingen, Germany

Johannes M. Herrmann

Cell Biology, University of Kaiserslautern, Kaiserslautern, Germany

Editors
Doron Rapaport
Interfaculty Institute of Biochemistry
University of Tuebingen
Tuebingen, Germany

Johannes M. Herrmann
Cell Biology
University of Kaiserslautern
Kaiserslautern, Germany

ISSN 1064-3745 ISSN 1940-6029 (electronic)
ISBN 978-1-62703-486-9 ISBN 978-1-62703-487-6 (eBook)
DOI 10.1007/978-1-62703-487-6
Springer New York Heidelberg Dordrecht London

Library of Congress Control Number: 2013946909

Printed on acid-free paper

Humana Press is a brand of Springer
Springer is part of Springer Science+Business Media (www.springer.com)

Preface

Membranes are active interfaces between different cells or between different cellular compartments. Cellular membranes are composed of phospholipids, proteins, and sugars. Differences in the relative ratio and composition of these compounds give each subcellular compartment its distinctive character. In recent years, the essentially static picture of an inert, mainly homogeneous membrane has all but vanished. Instead we now perceive biological membranes as highly dynamic structures that contain a remarkable diversity of microdomains and a large variety of distinct local structures. Interestingly, recent studies suggest that not only the shape but also the composition of membranes is dynamic.

About a third of all cellular proteins are membrane proteins which are involved in a large variety of activities. The importance of membrane proteins is further reflected by the fact that most drug targets are membrane proteins, particularly those that are involved in the transport of ions or metabolites or which contribute to inter- and intracellular signaling processes. However, despite their broad significance in biology and medicine, it is still poorly understood how the different cellular membrane systems are put together and maintained, how they are organized in living cells, and how they fulfill their physiological tasks. This lack of knowledge is mainly due to the technical difficulties associated with working with membrane components. For example, in order to study membrane proteins in vitro, the membranes in which they reside need to be solubilized with detergents so that their natural environment, the membrane, is removed. This often has considerable influences for their structures and functions. Initially, membrane lipids were mainly regarded as the insolating grease around membrane proteins. However, recent studies clearly demonstrated that lipids play a decisive and active role in many membrane processes and interact in various ways with membrane proteins. Thus, membrane proteins and membrane lipids form complex interactive systems that are highly dynamic. These systems can be studied only by combinations of different in vivo and in vitro techniques.

The idea for this book was to present a broad collection of methods to study the biogenesis and function of cellular membranes. These methods were grouped in five parts (see Figure). The first two parts (I and II) describe how membrane lipids or membrane proteins, respectively, can be studied. In Part III, different procedures to investigate the interaction of membrane proteins among each other or with membrane lipids are described. Methods to study the biogenesis of membrane proteins and the dynamics of organelles are presented in Part IV. Finally, Part V provides protocols for the analyses of the functions or complex organization of membrane proteins. Our aim was to provide readers with a comprehensive but still concise collection of methods to study membrane biogenesis including both basic protocols of rather general application and more specialized methods for specific and novel techniques that were not presented before.

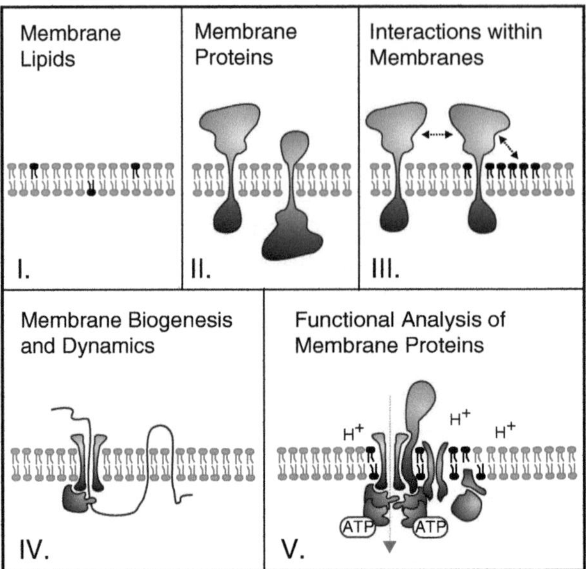

We would like to thank all the authors for their concise and clear contributions which made working with them a great pleasure. We also want to express our gratitude to the series editor, John Walker, for his constant support.

Tuebingen, Germany — *Doron Rapaport*
Kaiserslautern, Germany — *Johannes M. Herrmann*

Contents

Contributors

NATHAN N. ALDER • *Department of Molecular and Cell Biology, University of Connecticut, Storrs, CT, USA*

AVRAHAM ASHKENAZI • *Department of Biological Chemistry, The Weizmann Institute of Science, Rehovot, Israel*

ABDUSSALAM AZEM • *Department of Biochemistry and Molecular Biology, Tel Aviv University, Tel Aviv, Israel*

PHILIPP BARTSCH • *Biophysics, Department of Biology/Chemistry, University of Osnabrueck, Osnabrueck, Germany*

THOMAS BAUMGARTEN • *Center for Biomembrane Research, Department of Biochemistry and Biophysics, Stockholm University, Stockholm, Sweden*

LEA BLEIER • *Molecular Bioenergetics, Medical School, Goethe-University, Frankfurt, Germany*

STEFAN BÖCKLER • *Institut für Zellbiologie, Universität Bayreuth, Bayreuth, Germany*

RAINER A. BÖCKMANN • *Computational Biology, University of Erlangen-Nürnberg, Erlangen, Germany*

BRITTA BRÜGGER • *Heidelberg University Biochemistry Center (BZH), Heidelberg, Germany*

IAN COLLINSON • *School of Biochemistry, University of Bristol, Bristol, UK*

FLORIAN CSINTALAN • *Institute of Biochemistry, Goethe-University, Frankfurt, Germany*

GÜNTHER DAUM • *Institute of Biochemistry, Graz University of Technology, Graz, Austria*

JAN-WILLEM DE GIER • *Center for Biomembrane Research, Department of Biochemistry and Biophysics, Stockholm University, Stockholm, Sweden*

JEANINE DE KEYZER • *Department of Molecular Microbiology, University of Groningen, Groningen, The Netherlands*

RANIA M. DERANIEH • *Department of Biological Sciences, Wayne State University, Detroit, MI, USA*

DAVID DREW • *Center for Biomembrane Research, Department of Biochemistry and Biophysics, Stockholm University, Stockholm, Sweden*

ARNOLD J. M. DRIESSEN • *Department of Molecular Microbiology, University of Groningen, Groningen, The Netherlands*

JOHANNA DUDEK • *Medical Biochemistry and Molecular Biology, Saarland University, Homburg, Germany*

CHRISTINE EBEL • *Institut de Biologie Structurale, CEA, Grenoble, France*

TOSHIYA ENDO • *Department of Chemistry, Nagoya University, Nagoya, Japan*

JOHANNES FÖRTSCH • *Institut für Zellbiologie, Universität Bayreuth, Bayreuth, Germany*

ANA J. GARCÍA-SÁEZ • *German Cancer Research Center, Heidelberg, Germany*

MIRIAM L. GREENBERG • *Department of Biological Sciences, Wayne State University, Detroit, MI, USA*

MARKUS GREINER • *Medical Biochemistry and Molecular Biology, Saarland University, Homburg, Germany*
LUCIA E. GROß • *Cluster of Excellence Frankfurt, Center for Membrane Proteomics, Department of Biosciences, Molecular Cell Biology, Goethe University, Frankfurt, Germany*
MARTINA GSELL • *Institute of Biochemistry, Graz University of Technology, Graz, Austria*
ANKE HARSMAN • *Biophysics, Department of Biology/Chemistry, University of Osnabrueck, Osnabrueck, Germany*
JULIANA HEIDLER • *Department of Molecular Hematology, University of Frankfurt Medical School, Frankfurt, Germany*
JOHANNES M. HERRMANN • *Cell Biology, University of Kaiserslautern, Kaiserslautern, Germany*
DILEM HIZLAN • *Istanbul Sehir University, Istanbul, Turkey*
ANNA HJELM • *Center for Biomembrane Research, Department of Biochemistry and Biophysics, Stockholm University, Stockholm, Sweden*
JEAN-MICHEL JAULT • *Institut de Biologie Structurale, CEA, Grenoble, France*
AMIT S. JOSHI • *Department of Biological Sciences, Wayne State University, Detroit, MI, USA*
SANDRO KELLER • *Molecular Biophysics, University of Kaiserslautern, Kaiserslautern, Germany*
TILL KLECKER • *Institut für Zellbiologie, Universität Bayreuth, Bayreuth, Germany*
MIRJAM KLEPSCH • *Center for Biomembrane Research, Department of Biochemistry and Biophysics, Stockholm University, Stockholm, Sweden*
JOHANNES KLINGLER • *Molecular Biophysics, University of Kaiserslautern, Kaiserslautern, Germany*
KATRIN KRUMPE • *Interfaculty Institute of Biochemistry, University of Tübingen, Tübingen, Germany*
ILJA KUSTERS • *Department of Molecular Microbiology, University of Groningen, Groningen, The Netherlands*
C. ROY D. LANCASTER • *Department of Structural Biology, Saarland University, Homburg, Germany; Center of Human and Molecular Biology (ZHMB), Saarland University, Homburg, Germany*
SVEN LANG • *Medical Biochemistry and Molecular Biology, Saarland University, Homburg, Germany*
ALINE LE ROY • *Institut de Biologie Structurale, CEA, Grenoble, France*
JOHANNES LINXWEILER • *Medical Biochemistry and Molecular Biology, Saarland University, Homburg, Germany*
MILIT MAROM • *Department of Biochemistry and Molecular Biology, Tel Aviv University, Tel Aviv, Israel*
FLORIAN G. MÜLLER • *Department of Structural Biology, Saarland University, Homburg, Germany; Center of Human and Molecular Biology (ZHMB), Saarland University, Homburg, Germany*
IRIS NASIE • *Department of Biological Chemistry, Hebrew University of Jerusalem, Jerusalem, Israel*
SHUH-ICHI NISHIKAWA • *Department of Chemistry, Nagoya University, Nagoya, Japan*
HUGUES NURY • *Institut de Biologie Structurale, CEA, Grenoble, France*
CAGAKAN ÖZBALCI • *Heidelberg University Biochemistry Center (BZH), Heidelberg, Germany*
KRISTYNA PLUHACKOVA • *Computational Biology, University of Erlangen-Nürnberg, Erlangen, Germany*

DORON RAPAPORT • *Interfaculty Institute of Biochemistry, University of Tübingen, Tübingen, Germany*
RON SAAR-DOVER • *Department of Biological Chemistry, The Weizmann Institute of Science, Rehovot, Israel*
TIMO SACHSENHEIMER • *Heidelberg University Biochemistry Center (BZH), Heidelberg, Germany*
JONATHAN SARWAN • *Institut de Biologie Structurale, CEA, Grenoble, France*
SUSAN SCHLEGEL • *Center for Biomembrane Research, Department of Biochemistry and Biophysics, Stockholm University, Stockholm, Sweden*
ENRICO SCHLEIFF • *Cluster of Excellence Frankfurt, Center for Membrane Proteomics, Department of Biosciences, Molecular Cell Biology, Goethe University, Frankfurt, Germany*
DIRK SCHOLZ • *Institut für Zellbiologie, Universität Bayreuth, Bayreuth, Germany*
STEFAN SCHORR • *Medical Biochemistry and Molecular Biology, Saarland University, Homburg, Germany*
SHIMON SCHULDINER • *Department of Biological Chemistry, Hebrew University of Jerusalem, Jerusalem, Israel*
CHRISTINE T. SCHWALL • *Department of Molecular and Cell Biology, University of Connecticut, Storrs, CT, USA*
YECHIEL SHAI • *Department of Biological Chemistry, The Weizmann Institute of Science, Rehovot, Israel*
TAKUYA SHIOTA • *Department of Chemistry, Nagoya University, Nagoya, Japan*
MAIK S. SOMMER • *Cluster of Excellence Frankfurt, Center for Membrane Proteomics, Department of Biosciences, Molecular Cell Biology, Goethe University, Frankfurt, Germany*
MANUEL SOMMER • *Cluster of Excellence Frankfurt, Center for Membrane Proteomics, Department of Biosciences, Molecular Cell Biology, Goethe University, Frankfurt, Germany*
SONIA STEINER-MORDOCH • *Department of Biological Chemistry, Hebrew University of Jerusalem, Jerusalem, Israel*
VALENTINA STRECKER • *Molecular Bioenergetics, Medical School, Goethe-University, Frankfurt, Germany*
JOSEPH D. UNSAY • *German Cancer Research Center, Heidelberg, Germany*
CHRIS VAN DER DOES • *Max Planck Institute for Terrestrial Microbiology, Marburg, Germany*
MARTIN VAN DER LAAN • *Institut für Biochemie and Molekularbiologie, ZMBZ, Universität Freiburg, Freiburg, Germany*
CAROLYN VARGAS • *Molecular Biophysics, University of Kaiserslautern, Kaiserslautern, Germany*
JANET VONCK • *Department of Structural Biology, Max Planck Institute of Biophysics, Frankfurt, Germany*
RICHARD WAGNER • *Biophysics, Department of Biology/Chemistry, University of Osnabrueck, Osnabrueck, Germany*
TSJERK A. WASSENAAR • *Department of Biological Sciences and Institute for Biocomplexity and Informatics, University of Calgary, Calgary, AL, Canada*
DANIEL WECKBECKER • *Cell Biology, University of Kaiserslautern, Kaiserslautern, Germany*
BENEDIKT WESTERMANN • *Institut für Zellbiologie, Universität Bayreuth, Bayreuth, Germany*
DAVID WICKSTRÖM • *Xbrane Bioscience AB, Arrhenius Laboratories, Stockholm University, Stockholm, Sweden*
KATHARINA WIESEMANN • *Cluster of Excellence Frankfurt, Center for Membrane Proteomics, Department of Biosciences, Molecular Cell Biology, Goethe University, Frankfurt, Germany*

BENJAMIN WISEMAN • *Institut de Biologie Structurale, CEA, Grenoble, France*
ILKA WITTIG • *Molecular Bioenergetics, Medical School, Goethe-University, Frankfurt, Germany*
ZHT CHENG WU • *Department of Molecular Microbiology, University of Groningen, Groningen, The Netherlands*
RALF M. ZERBES • *Institut für Biochemie and Molekularbiologie, ZMBZ, Universität Freiburg, Freiburg, Germany*
RICHARD ZIMMERMANN • *Medical Biochemistry and Molecular Biology, Saarland University, Homburg, Germany*

Part I

Investigation of Membrane Lipids

Chapter 1

Quantitative Analysis of Cellular Lipids by Nano-Electrospray Ionization Mass Spectrometry

Cagakan Özbalci, Timo Sachsenheimer, and Britta Brügger

Abstract

Lipid analysis performed by nano-electrospray ionization mass spectrometry is a highly sensitive method for quantification of lipids including all lipid species of a given lipid class. Various instrumental setups are used for quantitative lipid analysis, including different modes of ionization, separation, and detection. Here we describe a work-flow for the rapid and quantitative analysis of lipid species from cellular membranes by direct infusion of lipid extracts to a nano-electrospray ionization triple quadrupole/linear ion trap mass spectrometer.

Key words Lipidomics, Nano-electrospray ionization mass spectrometry, QTRAP 5500, LipidView, Quantitative lipid analysis, Lipid standards

1 Introduction

Recent advances in the field of mass spectrometry have contributed to a rapid development of the newly emerged research area lipidomics. Lipidomics aims at elucidating lipid distribution, lipid metabolism, and roles of lipids in complex networks such as cell signaling. Introduction of electrospray ionization together with novel generations of mass spectrometers has greatly facilitated the analysis of lipids on a quantitative basis, including not only lipid classes but also individual molecular lipid species of a given lipid class [1–6].

Besides advances in instrumental setups, adapted protocols for lipid extractions and a continuously growing number of commercially available lipid standards are important factors towards comprehensive quantitative lipid analyses. Independent of the mass spectrometer used, there are mainly two different strategies used: direct injection versus liquid chromatography. Lipid analysis by direct injection of total lipid extracts relies on the different physico-chemical properties of lipids. This approach was defined as shotgun lipidomics [7, 8]. Various approaches are used for shotgun lipidomics, one of them represents a bottom-up strategy:

Doron Rapaport and Johannes M. Herrmann (eds.), *Membrane Biogenesis: Methods and Protocols*,
Methods in Molecular Biology, vol. 1033, DOI 10.1007/978-1-62703-487-6_1,

Individual lipid classes are selectively monitored based on unique fragmentation ions, which are generated by collision-induced dissociation. Which scan procedure is used depends on the nature of the fragment ion produced, e.g., charged fragment ions are the basis for precursor ion scanning, while neutral fragments are monitored by neutral loss scanning (Fig. 1). Multiple reaction monitoring can utilize neutral or charged fragments, but in contrast to precursor or neutral loss scanning selected parental (precursor) ions are fragmented, resulting in a higher sensitivity. Lipids are preferentially positively or negatively charged and are thus monitored in positive or negative ion mode, respectively. Acidic or basic condition as well as the use of ion adducts is employed to shift the charged state of lipids. As an example, addition of ammonium acetate allows analysis of phosphatidylcholine either as $[M+H]^+$ ion in positive ion mode or as $[M+OAc]^-$ in negative ion mode. While in many cases analysis in positive ion mode results in higher sensitivity, analysis in negative ion mode can provide structural information such as fatty acid composition.

Interpretation of mass spectra and data evaluation includes corrections for isotope distribution and mass dependency of ion intensities. While in the past most laboratories relied on self-made software tools for data evaluation, an increasing number of non-commercial and commercial software is available [9–12], and these were recently reviewed in [13]. Especially attractive are software tools that aim at providing a platform-independent approach, such as the LipidXplorer developed by Shevchenko and colleagues [14, 15]. Unfortunately, in some cases mass spectrometry data formats are still difficult to convert to allow accessibility by these open-source software solutions.

Here we describe a robust work-flow for quantitation of cellular lipids by nano-electrospray ionization mass spectrometry, including lipid extraction, mass spectrometric analysis by direct injection into a triple quadrupole/linear ion trap mass spectrometer, and data evaluation.

2 Materials

2.1 Chemicals

All chemicals used should be of highest chemical grade. Chloroform, methanol, HCl, H_2O, ammonium carbonate, TCA, EDTA, ammonium acetate, KH_2PO_4, perchloric acid, ammonium heptamolybdate, ascorbic acid, 2-ethoxy-1-ethoxycarbonyl-1,2-dihydroquino line, toluene, diethyl ether, serine, ethanolamine, glycerol, inositol, sodium acetate, calcium chloride, *n*-Octyl-β-D-glucopyranoside.

2.2 Lipids

Phosphatidylinositol-(4)-phosphate (PI4P) 17:0/20:4, Phosphatidylinositol-(4,5)-bisphosphate ($PI4,5P_2$) 17:0/20:4, diacylglycerol (DAG) 14:0/14:0, DAG 16:0/16:0, DAG 16:0/18:1, DAG

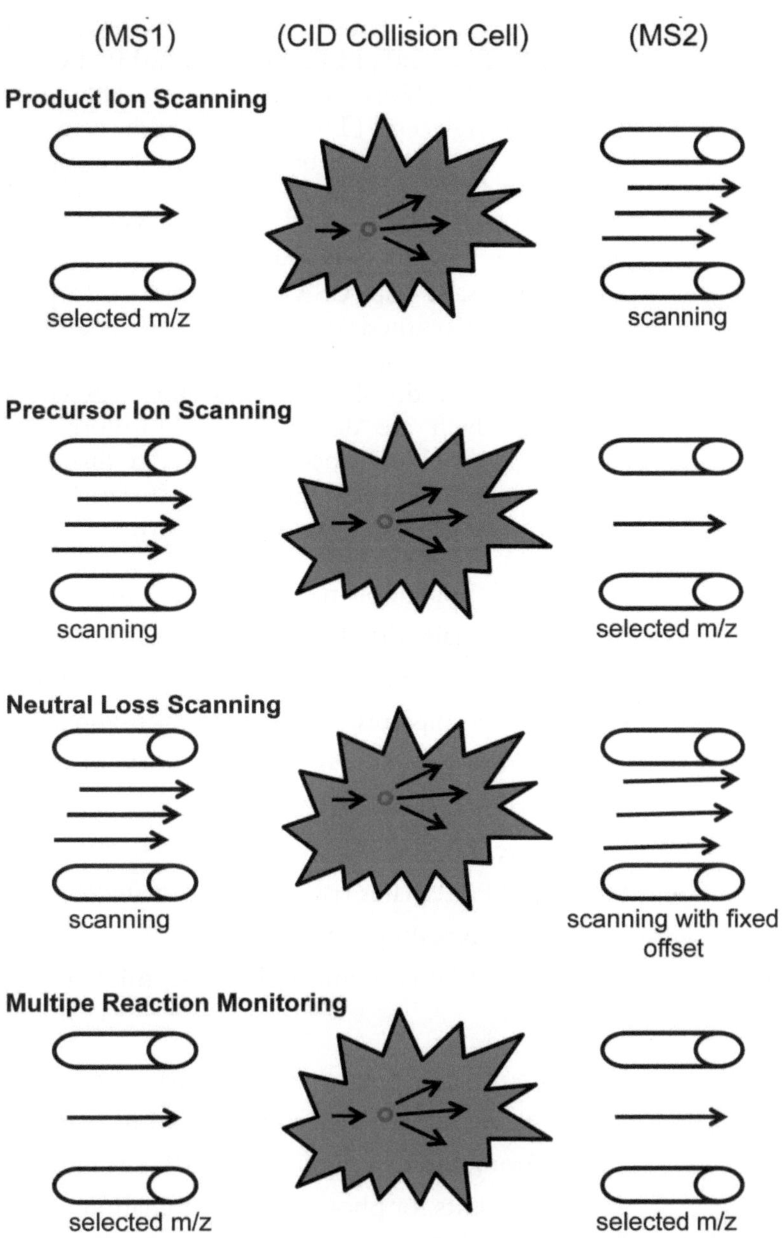

Fig. 1 Scan experiments on a triple quadrupole instrument. In product ion scan experiments, MS1 (the first quadrupole) is operating as a mass filter, allowing the passage to the collision chamber (the second quadrupole) of a selected *m/z* ion only. Following collision-induced dissociation (CID), fragment ions are transferred to MS2 (the third quadrupole), which operates in scanning mode, i.e., molecular ions are separated according to their *m/z* ratio. The resulting spectrum shows all fragment ions arising from fragmentation of the respective precursor ion. In precursor ion scan experiments, MS1 is operating in scanning mode and transfers all molecular ions into the collision cell. Upon collision-induced dissociation, fragment ions are transferred to MS2, operating as mass filter, allowing the passage to the collision chamber of a selected *m/z* ion only. The resulting spectrum shows all precursor ions that are produced in the collision cell from the fragment ion selected in MS2. In neutral loss scan experiments, both MS1 and MS2 are in scanning mode, with MS2 scanning with a fixed offset (with an *m/z* value corresponding to the neutral fragment to be monitored). In multiple reaction monitoring, both MS1 and MS2 are set to a selected mass, resulting in increased sensitivity

18:1/18:1, DAG 18:0/20:4, phosphatidylcholine (PC) 13:0/13:0, PC 14:0/14:0, PC 20:0/20:0, PC 21:0/21:0, PC 14:1/14:1, 20:1/20:1, PC 22:1/22:1, Sphingosylphosphorylcholine, triacylglycerol (TAG), phosphatidic acid (PA), phosphatidylinositol (PI), hemi-bis(monoacylglycero)phosphate, and cardiolipin (CL) are from Avanti Polar Lipids (Alabaster, USA). DAG 17:0/17:0 is from Larodan (Malmö, Sweden). Lysosphingolipids (except for sphingosylphoshorylcholine) is from Matreya (Pleasant Gap, USA). Fatty acids (C14:0, C17:0, C19:0, and C25:0) for synthesis of sphingolipid standards are from Fluka. D_6-cholesterol is from Cambridge Isotope Laboratories (Andover, USA). Lipids are dissolved in chloroform, except for PIP, PIP_2, and PIP_3, which are dissolved in chloroform/methanol/H_2O (1:1:0.2, v/v/v). Lipid stocks are diluted to final concentrations of 1–10 μM and stored over argon at −80 °C (*see* **Note 1**). PIP_2 mixtures were kept at concentrations >30 μM.

2.3 Instruments and Accessories

1. Safe lock Eppendorf tubes, 1.5 and 2 mL.
2. Wheaton vials with Teflon lids, 10 mL.
3. Evaporator (combined gasing unit and thermostat).
4. Glass micropipettes (10–100 μL, Drummond).
5. Hamilton syringes, gas tight (10–2,500 μL).
6. ART tips (aerosol resistant) (Molecular Bio Products).
7. Silica gel 60, RP 18.
8. 96-well PCR plates (twin.tec PCR Plate 96, Eppendorf).
9. Aluminum sealing tape (Corning).
10. QTRAP5500 (Thermo Fisher Scientific), Triversa Nanomate (Advion).

2.4 Solutions

1. Components for extractions and mass spec analysis:
   (a) Ammonium carbonate: 155 mM in water.
   (b) Ammonium acetate: 10 and 20 mM in methanol.
2. Components for phosphate determination:
   (a) Ammonium heptamolybdate: 1.25 % (w/v) in water.
   (b) Ascorbic acid: 5 % (w/v) in water.
3. Components for lipid synthesis:
   (a) Phospholipase D from *streptomyces sp.* (Sigma).
   (b) Phospholipase D buffer: 0.1 mM sodium acetate pH 5.5, 0.1 mM calcium chloride, 2 % (w/v) octyl-β-D-glucopyranoside, and either 2 % (v/v) ethanolamine, 5 g (w/v) l-serine, 1 % (v/v) glycerol, or 300 mM myo-inositol.

(c) Prepare buffer freshly from stock solutions:
   - 2 M sodium acetate in water.
   - 1 M $CaCl_2$ in water.
   - 10 % (w/v) octyl-β-D-glucopyranoside in water.

3 Methods

The lipid extraction method of choice depends on the lipids of interest and the sample itself. Two step extractions are recommended for comprehensive lipid analyses, including minor lipid classes (such as phosphoinositides), and allow efficient and simultaneous extraction of both neutral and charged lipids. While some lipids (e.g., phosphoinositides or sulfatides) are recovered with only low efficiencies by some one-step extraction protocols, their recovery by two step extractions with neutral and acidic phase is greatly enhanced. With the first neutral extraction apolar lipids are efficiently recovered, while negatively charged lipids preferentially partition into the acidic phase (e.g., acidification of extraction solvents facilitates protonation of the head group phosphates of phosphoinositides, which in turn facilitates disruption of ionic interactions with proteins and supports partitioning into the organic phase to increase the recovery of low abundant PIPs [16]).

As additives such as salt, detergents, and density gradient materials (e.g. sucrose) are only tolerated in limited amounts in mass spectrometric analysis, it is recommended—whenever possible—to wash samples prior to lipid extractions. Here ammonium carbonate buffers (with the compound being a good leaving group during evaporation) were shown to be suitable for this purpose. As most lipid extractions involve separation of an organic and an aqueous phase, extraction protocols were established, where the organic phase is on top, diminishing contaminations with aqueous phase or interphase during transfer of the organic phase [17].

Here we describe various extractions protocols. Two step extractions with a combination of neutral and acidic conditions are recommended for the simultaneous recovery of neutral and acidic lipids, while neutral two step extractions improve the recovery of glycosphingolipids [18]. Single step extractions are suitable for quantification of major lipids; in case of acid-labile plasmalogen lipids neutral extractions are recommended. In all cases it is recommended to run with the extractions a blank extraction (i.e., only standards). Sample extractions should always be in technical duplicates or triplicates, including different amounts of sample to ensure analysis within the linear range. To this end, it is recommended to perform the analysis in the range of 1–4 nmol of total lipid (with final total concentrations around 10 μM), depending on the lipid composition of sample analyzed. Higher concentrations might exceed the lipid species-dependent dynamic range. In addition, it was reported that high concentrations lead to an increase in response of unsaturated fatty acids [19].

Lipid extractions are performed in the presence of internal standards, i.e., standards are added to the organic phase prior to addition of the aqueous sample. Typically, 5–50 pmol of each lipid standard is added. To account for a loss in signal intensity at higher *m/z* values [20], it is recommended to add more than one standard to cover the whole mass range of endogenous lipids: select lipid standard with masses that are close to the lowest and highest *m/z* values, and one in between (Fig. 2). Mass-dependent changes of signal intensities (Fig. 2) that can vary in between measurements are accounted for in the data evaluation process.

3.1 Neutral/Acidic Extraction from Pellets [21]

1. Transfer cells (confluent 10 cm dish) to ice. Remove medium. Wash cells two times with 10 mL cold PBS. Add 1.2 mL cold 0.5 M TCA and scrape off cells immediately. Transfer cells to an Eppendorf tube and vortex for 10 s followed by 5 min incubation on ice. Centrifuge sample for 20 min at 4 °C and 13,000 × g_{av}. Discard supernatant and wash pellet two times with 1 mL cold 5 % TCA/1 mM EDTA.
2. For neutral extraction, resuspend pellet in 750 μL chloroform/methanol (1:2, v/v). Incubate for 3 min at RT, then vortex 30 s. Repeat this step three times. Centrifuge sample for 2 min at 4 °C and 13,000 × g_{av}. Transfer the supernatant to a new Eppendorf tube. Process the pellet as described below (*see* **step 3**). Add 250 μL chloroform and 450 μL water to the supernatant, vortex, and centrifuge for 2 min at 6,500 × g_{av} and 4 °C. Transfer the lower *neutral* phase to a new tube (*see* **Note 2**), remove 150 μL for cholesterol determination (*see* Subheading 3.6), and gently evaporate the solvent of the remaining sample (*see* **Note 3**). For mass spectrometric analysis dissolve lipids present in the film obtained after evaporation in 10 mM ammonium acetate in methanol (*see* Subheading 3.4).
3. For acidic extraction, resuspend pellet in 750 μL chloroform/methanol/37 % HCl (40:80:1, v/v/v) and incubate for 15 min at RT, while vortexing the sample for 30 s every 5 min. Transfer the tube to ice, then add 250 μL cold chloroform and 450 μL cold 0.1 N HCl, followed by 1 min vortexing and centrifugation (6,500 × g_{av}, 2 min at 4 °C). Transfer the bottom organic phase to a new Eppendorf tube (*see* **Note 2**) and keep on ice until further processing. For mass spectrometric analysis dilute an aliquot (10 μL) of the *acidic* organic phase 1:2 with 50 mM ammonium acetate in chloroform/methanol (1:1, v/v) (*see* Subheading 3.4).

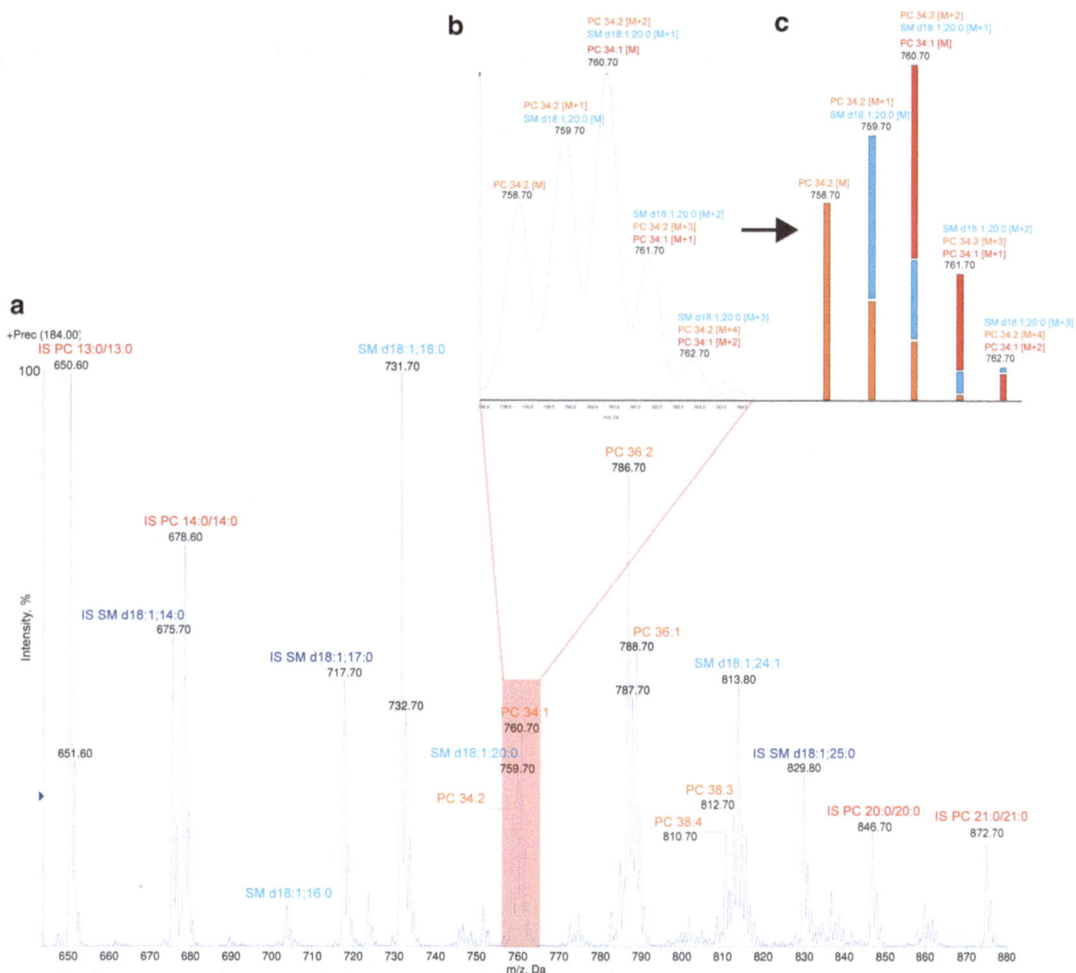

Fig. 2 Isotope and mass correction of peak intensities. (**a**) Precursor ion scan experiment selecting for fragment ion *m*/*z* 184 Da to monitor PC and SM species. Prior to lipid extraction the sample was spiked with equimolar amounts of three sphingomyelin standards (IS SM d18:1; 14:0, IS SM d18:1; 17:0, IS SM d18:1; 25:0) and four phosphatidylcholine standards (IS PC 13:0/13:0, IS PC 14:0/14:0, IS PC 20:0/20:0, IS PC 21:0/21:0). Monoisotopic masses are labeled. The mass-dependent decrease of signal intensity for PC and SM species depends on both mass spectrometer-dependent and independent factors. Here an exponential function describes the mass-dependent changes in signal intensities, and should be used for peak intensity corrections. (**b**) Range of *m*/*z* 758–762 Da of spectrum shown in (**a**). Peak labels include isotopic masses (up to M + 4) of each lipid species. (**c**) Correction of peak intensities for isotope distributions to determine contribution of different lipid species to a given peak. *IS* internal standard. Lipid nomenclature, e.g.,: SM d18:1; 14:0, sphingomyelin with a C18 sphingosine backbone and *N*-acylated C14 fatty acid with zero double bonds; PC 34:1, phosphatidylcholine with a total number of 34 C atoms in both fatty acids and one double bond

3.2 Neutral/Acidic Extraction from Solutions [21]

1. Transfer 990 μL chloroform/methanol (17:1) to a 2 mL Eppendorf test tube. If standards are added adjust the volumes and ratios of chloroform and methanol to end up with the indicated volume and solvent composition. Add the sample in a total volume of 200 μL 155 mM ammonium carbonate buffer. Incubate sample for 30 min at RT on a shaker.

Centrifuge sample for 2 min at 9,000 × g_{av}. Transfer the lower *neutral* chloroform phase to a new 1.5 mL Eppendorf test tube (*see* **Note 2**). Keep the upper phase for further treatment as described in **step 2**. Remove 150 μL for cholesterol determination (*see* Subheading 3.1.6). Following evaporation (*see* **Note 3**), subject remaining sample to mass spectrometric analysis (*see* Subheading 3.4).

2. Add 10 μL 2 N HCl, 375 μL chloroform/methanol/HCl (40:80:1, v/v/v), and 125 μL cold chloroform to the upper phase. Vortex after addition of each component. Keep on ice for a total of 15 min, vortex every 5 min in between. Vortex 1 min, centrifuge for 2 min at 9,000 × g_{av} and 4 °C. Transfer the lower *acidic* chloroform phase to a new 1.5 mL Eppendorf test tube (*see* **Note 2**). For analysis of phosphoinositides directly subject one aliquot to mass spectrometric analysis, for all other lipids evaporate organic solvent and dissolve lipids present in the film obtained after evaporation in an appropriate solvent (*see* **Note 3**).

3.3 Neutral/Neutral Extraction A [22]

1. Transfer 990 μL chloroform/methanol (17:1, v/v) to a 2 mL Eppendorf test tube. If standards are added adjust the volumes and ratios of chloroform and methanol to end up with the indicated volume and solvent composition. Add the sample in a total volume of 200 μL 155 mM ammonium carbonate buffer. Incubate sample for 30 min at RT on a shaker. Centrifuge sample for 2 min at 9,000 × g_{av}. Transfer the lower chloroform phase to a new 1.5 mL Eppendorf test tube (*see* **Note 2**). Keep the upper phase for further treatment as described in **step 2**. Remove 150 μL for cholesterol determination (*see* Subheading 3.1.6). Evaporate solvent or remaining sample (*see* **Note 3**), dissolve lipids present in the film obtained after evaporation in 10 mM ammonium acetate and subject an aliquot to mass spectrometric analysis.
2. Add 990 μL chloroform/methanol (2:1, v/v) to the upper phase obtained in **step 1** and incubate for 30 min on a shaker at RT. Centrifuge for 2 min at 9,000 × g_{av} and transfer the chloroform phase to a new 1.5 mL Eppendorf test tube (*see* **Note 2**). Evaporate solvent (*see* **Note 2**), dissolve lipids present in the film obtained after evaporation in 10 mM ammonium acetate, and subject an aliquot to mass spectrometric analysis.

3.4 Neutral/Neutral Extraction B [18]

1. Transfer 1 mL chloroform/methanol (10:1, v/v) to a 2 mL Eppendorf test tube. If standards are added adjust the volumes and ratios of chloroform and methanol to end up with the indicated volume and solvent composition. Add the sample in a total volume of 200 μL 150 mM ammonium carbonate buffer. Incubate sample for 60 min at 600 rpm and RT on a thermoshaker. Centrifuge sample for 5 min at 13,000 × g_{av}. Transfer

the lower chloroform phase to a new 1.5 mL Eppendorf test tube ("A") (*see* **Note 2**). Remove 150 μL for cholesterol determination (*see* Subheading 3.1.6). Keep the upper phase for further treatment as described in **step 2**.

2. Add 1 mL chloroform/methanol (2:1, v/v) to the upper phase and incubate for 60 min on a shaker at 600 rpm and RT. Centrifuge for 5 min at 13,000 × g_{av} and transfer the chloroform phase to a new 1.5 mL Eppendorf test tube ("B") (*see* **Note 2**).
3. Evaporate solvents in tubes ("A") and ("B") (*see* **Note 3**), dissolve lipids present in the film obtained after evaporation in an appropriate solvent, and subject to mass spectrometric analysis.

3.5 One-Step Extraction (Modified from [23])

1. Transfer 1.875 mL chloroform/methanol/37 % HCl (5:10:0.15, v/v/v) (acidic) or 1.875 mL chloroform/methanol (5:10, v/v) (neutral) to a 10 mL Wheaton vial with Teflon-screw cap (I). If standards are added, adjust the volumes and ratios of chloroform/methanol/37 % HCl to end up with the indicated volume and solvent composition.
2. Add sample in a total volume of 500 μL aqueous solution, add 500 μL chloroform, add 500 μL water, vortex 5 min after each step. Centrifuge sample for 5 min at 500 × g_{av} and 4 °C. Transfer lower chloroform phase to a new 10 mL Wheaton vial (II). Keep the upper phase for further treatment as described in **step 3**.
3. Re-extract upper phase with 500 μL chloroform in (I), wash lower chloroform phase with 500 μL water in (II). Vortex each for 5 min, centrifuge as above. Transfer lower organic phase from II to a new Wheaton vial (III), and from (I) into (II). Repeat vortexing and centrifugation for vial II, then transfer lower organic phase into (III).
4. Carefully evaporate combined chloroform phases in (III) (*see* **Note 3**), and resuspend lipid film in appropriate solvent for mass spectrometry. Once mass spectrometric analysis is done, evaporate solvent and subject remaining lipids to cholesterol determination (*see* Subheading 3.1.6).

3.6 Cholesterol

Quantification of cholesterol is performed following a single step chemical derivatization to either cholesterol sulfate [24] or cholesterol acetate [25]. Here we describe quantification of cholesterol by acetylation, which is a rapid one-step derivatization. Both derivatizations can also be used for the analysis of ergosterol and some other sterols.

1. Evaporate solvent of samples to be subjected to cholesterol determination.
2. Add 50 μL acetylchoride/chloroform (1:5, v/v, freshly prepared) to a dried lipid film, incubate with open lid for 2 h at

RT under the fume hood. If necessary, evaporate remaining solvent (*see* **Note 3**) and resuspend lipids present in the film obtained after evaporation in 50 μL 10 mM ammonium acetate in methanol. For mass spectrometric analysis *see* Subheading 3.10.

3.7 Phosphate Determination

This assay is based on the colorimetric determination of inorganic phosphate [26]. To this end, lipid extracts are subjected to oxidation by perchloric acid, releasing inorganic phosphate from phospholipids. Reaction to phosphomolybdate and reduction of ascorbic acid converts the colorless component to a blue colored component. Presence of sucrose or other density gradient materials can interfere with the measurement; it is recommended to run buffer only controls together with the samples.

1. Prepare a calibration curve from 2 to 40 nmol inorganic phosphate by transferring from a 0.4 mM KH_2PO_4 stock solution 0/5/12.5/25/50/75/100 μL into Teflon-sealed Wheaton vials. Add water to a final volume of 500 μL.
2. Perform a lipid extraction of sample of interest. Dry down lipid extract in a 10 mL Teflon-sealed Wheaton vial (*see* **Note 3**).
3. Add 150 μL 70 % perchloric acid to each vial (calibration and test samples) and vortex.
4. Incubate test tubes (calibration tubes do not need to be heated) for 40 min at 180 °C in a heating block. Do not close the test tubes firmly. After cooling down, add 500 μL water and vortex.
5. Add 400 μL of a freshly prepared mix of 1.25 % ammonium heptamolybdate/5 % ascorbic acid (1:1, v/v) to samples and calibration tubes, vortex.
6. Incubate for 5 min at 100 °C.
7. After cooling down, vortex and transfer samples to 1 mL plastic cuvettes. Read absorbance at 797 nm against the blank of the calibration curve (*see* **Note 4**).

3.8 Synthesis of Sphingolipid Standards

Currently, not many sphingolipid standards suitable for mass spectrometry are available. Here we describe a robust protocol for the synthesis of sphingolipids with defined *N*-acylated fatty acids (such as C14, C15, C17, C19, and C25), starting from lyso-sphingolipids (e.g., sphingosine, lyso-SM, lyso-glucosylceramide, lyso-hexosylceramide, lyso-lactosylceramide or lyso-gangliosides).

1. Dissolve 10.8 μmol lyso-sphingolipid (*see* **Note 5**) in 4 mL ethanol, and add to 21.6 μmol fatty acid in a Schlenk flask.
2. Add 50 μmol 2-ethoxy-1-ethoxycarbonyl-1,2-dihydroquinoline (EEDQ) and stir reaction mixture for 2 h at 45 °C. Add 50 μmol EEDQ a second time and stir for additional 12 h at RT.

3. Concentrate the reaction mixture under reduced pressure using a rotary evaporator, and subject the crude product to flash chromatography (silica gel, 11.5 × 2 cm column, eluent: toluene/diethyl ether/methanol/water, 3:3:3.5:0.7, v/v/v/v). Collect 1–2 mL fractions. Combine and concentrate product-containing fractions under reduced pressure (as monitored by TLC: silica gel, running solvent: toluene/ diethyl ether/methanol/water, 3:3:4:1, v/v/v/v).
4. Dissolve product in 2 mL chloroform/methanol (1:2, v/v) and subject to a second flash chromatography (RP-18 silica gel, 10 × 1 cm column). Pre-equilibrate column with methanol/water (1:1, v/v). Load product and wash column six times with 5 mL water, then six times with 5 mL methanol/ water (1:1, v/v). Elute stepwise with 5 mL methanol, 15 mL chloroform/methanol (1:1, v/v), 15 mL chloroform/methanol (2:1, v/v), and 10 mL chloroform. Collect 1–2 mL fractions. Dry product-containing fractions (monitored by TLC) under a gentle stream of nitrogen (*see* **Note 3**), dissolve in 100 mL chloroform/methanol (1:2, v/v), and store under argon at –20 °C.
5. Subject lipid product to phosphate determination as described. Products are verified by mass spectrometric and NMR analysis. Typically, a yield of 50–60 % is achieved.

3.9 Synthesis of Phospholipid Standards

Glycerophospholipids with mono-unsaturated fatty acids that are not commercially available were synthesized by enzymatically catalyzed head group exchange as described by the group of Penti Somerharju [19]. To this end, phosphatidylcholine 14:1/14:1, phosphatidylcholine 20:1/20:1, and phosphatidylcholine 22:1/22:1 (or any other appropriate lipid) were subjected to phospholipase D (PLD)-catalyzed transesterification in the presence of excess of the respective headgroup (e.g., serine, ethanolamine, inositol, or glycerol), resulting in an enzymatically catalyzed exchange of the headgroup. Educt and product are separated by HPLC as described.

1. Transfer 1 μmol of glycerophospholipid (e.g., PC 14:1/14:1) into a 10 mL Wheaton vial and evaporate organic solvent (*see* **Note 3**). No traces of organic solvent should remain.
2. Add 1 mL of PLD buffer containing the respective headgroup component (e.g., ethanolamine), vortex (remove one aliquot for mass spec and TLC analysis).
3. Add 5 U PLD (20 μL) and incubate for 5 h under rigorous shaking at 45 °C.
4. Subject reaction to lipid extraction by adding 200 μL 0.5 M EDTA and 4.3 volumes of chloroform/methanol (5:8, v/v). Incubate for 30 min on a rotating wheel at 4 °C. Add 1 volume

Table 1
Instrument settings QTRAP5500

Curtain gas	10
CAD gas	5
Operating pressure (torr)	1.6×10^{-5}
Interface heater temperature (°C)	60
Entrance potential	7
Detector (CEM)	2.1
Step size (Da)	0.1
Settling time (ms)	0–5
Scan rate (Da/s)	200
Pause between mass ranges (ms)	5
Scan mode	Profile
Synchronization mode	LC syn
Quadrupole resolution	Unit or high (Q1 and Q3)

Typically used scan settings are listed. *See* also Table 2

H_2O, vortex, add 3.7 volumes chloroform, vortex, and centrifuge for 10 min at $750 \times g_{av}$. Transfer lower organic phase to a new vial and evaporate the organic solvent (*see* **Note 3**). Resuspend the dried lipid film and subject the sample to HPLC separation as described [27]. Peak fractions are analyzed by TLC and mass spectrometry. Phosphate content is determined as described in Subheading 3.2.

3.10 Mass Spectrometric Analysis

Prior to quantitative analysis of lipids by mass spectrometry some basic parameters should be considered, including choice of lipid standards (i.e., lipid standards do not interfere with endogenous lipids), optimization of scan settings (e.g., collision energy), and quality of lipid extractions (e.g., background signals in blank extractions). Depending on the mass spectrometer used, sphingolipids undergo a variable degree of dehydration. Here, instrument parameters should be optimized to minimize this reaction.

Prior to mass spectrometric analysis, lipid extracts should be clarified by centrifugation ($13{,}800 \times g_{av}$, 5 min at 4 °C). Lipid infusion and ionization is performed via a Triversa Nanomate (Advion Biosciences), with the following settings: sample infusion volume: 10 μL, volume of air to aspirate after sample: 1 μL, air gap before chip: enabled, aspiration delay: 0 s, pre-piercing: with mandrel, spray sensing: enabled, cooling temperature: 12 °C, gas pressure: 0.5 psi. The ionization voltage is typically set to 1.3–1.6 kV, for other instrument settings see Tables 1 and 2. For infusion in 10 mM

Table 2
Scan settings QTRAP5500

Lipid	Polarity	Scan type	m/z (Da)	Mass range	DP	EP	CE	CXP	Cycles
PC	Positive	Precursor	184	644–880	100	7	35	10	300
SM	Positive	Precursor	184	644–880	100	7	35	10	300
PE	Positive	Neutral loss	141	620–860	100	7	25	19	300
16:0-pl-PE	Positive	Precursor	364.2	650–850	100	10	25	16	300
18:1-pl-PE	Positive	Precursor	390.3	650–850	100	10	27	16	300
18:0-pl-PE	Positive	Precursor	392.3	650–850	100	10	27	16	300
PS	Positive	Neutral loss	185	670–910	100	7	25	19	300
PG/LBPA	Positive	Neutral loss	189	670–910	100	7	25	19	300
PI	Positive	Neutral loss	277	800–950	100	11	19	19	300
PA	Positive	Neutral loss	115	600–900	100	10	25	14	300
PIP	Positive	Neutral loss	357	880–1,050	100	7	25	19	400
PIP2	Positive	Neutral loss	437	950–1,150	100	7	35	19	400
PIP3	Positive	Neutral loss	517	1,000–1,250	100	7	35	19	400
Cer	Positive	Precursor	264.3	500–670	100	10	35	14	300
HexCer	Positive	Precursor	264.3	670–860	100	10	40	14	300
LacCer	Positive	Precursor	264.3	830–1,020	100	10	45	14	300
Sulfatides	Negative	Precursor	97	700–1,000	-240	-10	-100	-11	150
Cholesterolesters	Positive	Precursor	369	500–800	100	10	15	14	271
Cholesterolacetate	Positive	Neutral loss	77	442–456	60	7	15	14	85
DAG	Positive	Neutral loss	217.3	500–750	100	7	25	14	300
	Positive	Neutral loss	245.3	500–750	100	7	25	14	300
	Positive	Neutral loss	259.3	500–750	100	7	25	14	300
	Positive	Neutral loss	271.3	500–750	100	7	25	14	300
	Positive	Neutral loss	273.3	500–750	100	7	25	14	300
	Positive	Neutral loss	285.3	500–750	100	7	25	14	300
	Positive	Neutral loss	287.3	500–750	100	7	25	14	300
	Positive	Neutral loss	295.3	500–750	100	7	25	14	300
	Positive	Neutral loss	297.3	500–750	100	7	25	14	300
	Positive	Neutral loss	299.3	500–750	100	7	25	14	300
	Positive	Neutral loss	301.3	500–750	100	7	25	14	300
	Positive	Neutral loss	315.3	500–750	100	7	25	14	300
	Positive	Neutral loss	319.3	500–750	100	7	25	14	300
	Positive	Neutral loss	321.3	500–750	100	7	25	14	300
	Positive	Neutral loss	325.3	500–750	100	7	25	14	300
	Positive	Neutral loss	327.3	500–750	100	7	25	14	300
	Positive	Neutral loss	329.3	500–750	100	7	25	14	300
	Positive	Neutral loss	345.3	500–750	100	7	25	14	300

(continued)

Table 2 (continued)

Lipid	Polarity	Scan type	*m/z* (Da)	Mass range	DP	EP	CE	CXP	Cycles
TAG	Positive	Neutral loss	217.3	750–1,000	100	7	40	14	300
	Positive	Neutral loss	245.3	750–1,000	100	7	40	14	300
	Positive	Neutral loss	259.3	750–1,000	100	7	40	14	300
	Positive	Neutral loss	271.3	750–1,000	100	7	40	14	300
	Positive	Neutral loss	273.3	750–1,000	100	7	40	14	300
	Positive	Neutral loss	285.3	750–1,000	100	7	40	14	300
	Positive	Neutral loss	287.3	750–1,000	100	7	40	14	300
	Positive	Neutral loss	295.3	750–1,000	100	7	40	14	300
	Positive	Neutral loss	297.3	750–1,000	100	7	40	14	300
	Positive	Neutral loss	299.3	750–1,000	100	7	40	14	300
	Positive	Neutral loss	301.3	750–1,000	100	7	40	14	300
	Positive	Neutral loss	315.3	750–1,000	100	7	40	14	300
	Positive	Neutral loss	319.3	750–1,000	100	7	40	14	300
	Positive	Neutral loss	321.3	750–1,000	100	7	40	14	300
	Positive	Neutral loss	325.3	750–1,000	100	7	40	14	300
	Positive	Neutral loss	327.3	750–1,000	100	7	40	14	300
	Positive	Neutral loss	329.3	750–1,000	100	7	40	14	300
	Positive	Neutral loss	345.3	750–1,000	100	7	40	14	300

Scan settings to cover major mammalian lipids. Depending on the sample analyzed, mass ranges and MRM settings (here exemplified for DAG and TAG) vary. Lyso-lipids are analyzed using the same scan types as corresponding di-acylated lipids, with the mass range shifted to lower masses. For lipid classes with species with a molecular weight above 1,250 Da (e.g., cardiolipin or gangliosides, only species with more than one charge/ molecule are accessible to QTRAP 5500 measurements)

ammonium acetate in methanol vent headspace is enabled and pre-wetting is done once. For infusion in 25 mM ammonium acetate in chloroform/methanol/H_2O (analysis of phosphoinositides), vent headspace is disabled and pre-wetting is performed two times.

1. Dissolve lipid extracts in 50 μL methanol (vortex and ultrasonicate). Add 10 μL of sample to 10 μL 20 mM (or 50 mM in case of phosphoinositides) ammonium acetate in methanol (positive ion mode) or 0.01 % piperidine in methanol (negative ion mode) in a 96 well plate. Mix the solvent in the well plate using a 20 μL pipette. Seal the wells tightly with sealing foil to avoid evaporation, and place the 96 wells plate carefully into the NanoMate plate holder.
2. Open ChipSoft Manager. Open a file with .meth extension under "File," and select a spray method. For automatic spraying select "Sequence" under "View." Open a new sequence and enter your parameters (number of the samples, injection per sample, and the initial point of the samples, e.g., "A01"). Save sequence.

3. Open Analyst 1.5.1 software. Go to "Project Navigator." Select "Build Acquisition Batch" in order to create your automatic scanning list. Name your sample set (e.g., date_sample). Select "Add set" and "Add samples" to specify sample number according to the 96 well plate. Choose "Acquisition Method" from "Acquisition" panel. Select "submit." Select "Sample Queue" under "View" to view listed samples with the selected method. Select "Ready" and then "Start Sample" under "Acquire." Start measurement via ChipSoft Software. Once the QTRAP 5500 gets a signal from the NanoMate, data acquisition is initiated. Once the measurements are completed the data are sorted as *.wiff format.

3.11 Identification and Quantification of Lipid Species

Lipid species identification and data evaluation can be performed using commercial or non-commercial software tools. Data evaluation includes isotope correction and correction for mass-dependent changes in response factors (Fig. 2) [20, 28, 29]. Data analysis should include both absolute and relative quantification, taking into account background signals (biological and technical). Here we describe data evaluation using LipidView, a commercially available software.

1. Open LipidView™ software by selecting the shortcut under the "Companion software" option under "Project Navigation." Select "Load Data" in order to choose your sample set. Select file to be processed and select samples of interest (confirm with click on "ok"). Select "Processing Setup" under "Workflow." Select method of choice for the measured samples under "Selected analysis method." Set up a target method by editing an existing method. Select "Edit Methods" and adjust parameters under "spectrum/data." Select "Processing Method" and select lipids to be quantified. Set an expected maximum of "double bonds." Select "Details" for advanced options. Once the target method is completed, select "save and close" (or (preferably) save new method under "save as"), and start analysis by selecting "run." Assign samples by selecting "Assign Method to X samples" using the selected method. Select "Find Lipids," then "Start" under "Workflow." Select "Data Spectra" choose "Scan type," select "Display" to open the corresponding spectra. Select "Report," then close the spectra window. Select "Advanced Data" and "View Mode." Choose "corrected intensities" or "areas" for isotope correction. Select "Profile Tests" to choose lipid species of interest (including internal standards). Save this target method, which can be used for other samples. Close the "Result view" window.
2. Select "New Session" under "File." Select "Load data" and choose "Mass Spectrometer Data" and samples of interest. Choose appropriate target method under "Selected Analysis

Method." Select "Use Target Method" under "Processing Method." Select "Use target method" and "Set" to select chosen target method. Save and close. Select "Find lipids" and "start" (do not to overwrite data, change name of new file). After processing, select "Advanced Data." From "View Mode" select "corrected intensities" or "areas." Select "Export" and save data as text file. The data can now be processed in an appropriate software spreadsheet for data calculation and graphing tools (e.g., excel). Data processing includes (if necessary) correction of molar amounts by response factors and statistics of technical and biological replicates.

3.12 Data Evaluation and Presentation

Quantitative lipid data are typically presented as relative values, with the sum of all measured lipids set to 100 mol%. Lipid species distributions within individual lipid classes (for nomenclature *see* ref. 30) can be either presented as mol% of total lipid or with all species of the given lipid class set to 100 %. With comprehensive lipidomics approaches complex data sets are generated. As an example, in human plasma samples around 200–500 different lipid species distributed among more than ten different lipid categories were assessed [31, 32]. To allow for rapid sample evaluations with respect to distributions based on lipid categories, lipid species fatty acid chain length, and degree of unsaturation, it is recommended to manage lipid data in data base formats [33].

4 Notes

1. Lipids should be stored under argon in glass vials with Teflon-sealed lids. Do not use Parafilm, in case sealing is required, Teflon tapes are recommended. All chemicals and materials should be used only for lipid extractions and mass spectrometric analysis.
2. Test tubes should be resistant to organic solvents; polypropylene tubes are recommended.
3. Evaporation should be performed using a gentle stream of nitrogen or argon and should be facilitated by keeping samples during evaporation at 37 °C.
4. Perchloric acid is a strong acid, wear lab coat, gloves, and glasses.
5. Prior to sphingolipid standard synthesis lyso-lipids should be analyzed for purity by mass spectrometry. Only lyso-lipids with a single defined sphingosine backbone are suitable as educts in standard synthesis.

Acknowledgment

This work was supported by a grant of the Deutsche Forschungsgemeinschaft (SFB/TRR83). B.B. is an investigator of the CellNetworks Cluster of Excellence (EXC81).

References

1. Han X, Yang K, Gross RW (2012) Multidimensional mass spectrometry-based shotgun lipidomics and novel strategies for lipidomic analyses. Mass Spectrom Rev 31:134–178
2. Ivanova PT, Milne SB, Myers DS et al (2009) Lipidomics: a mass spectrometry based systems level analysis of cellular lipids. Curr Opin Chem Biol 13:526–531
3. Merrill AH Jr, Sullards MC, Allegood JC et al (2005) Sphingolipidomics: high-throughput, structure-specific, and quantitative analysis of sphingolipids by liquid chromatography tandem mass spectrometry. Methods 36: 207–224
4. Murphy RC, Gaskell SJ (2011) New applications of mass spectrometry in lipid analysis. J Biol Chem 286:25427–25433
5. Shevchenko A, Simons K (2010) Lipidomics: coming to grips with lipid diversity. Nat Rev Mol Cell Biol 11:593–598
6. Wenk MR (2010) Lipidomics: new tools and applications. Cell 143:888–895
7. Han X, Gross RW (2003) Global analyses of cellular lipidomes directly from crude extracts of biological samples by ESI mass spectrometry: a bridge to lipidomics. J Lipid Res 44: 1071–1079
8. Han X, Gross RW (2005) Shotgun lipidomics: electrospray ionization mass spectrometric analysis and quantitation of cellular lipidomes directly from crude extracts of biological samples. Mass Spectrom Rev 24:367–412
9. Ejsing CS, Duchoslav E, Sampaio J et al (2006) Automated identification and quantification of glycerophospholipid molecular species by multiple precursor ion scanning. Anal Chem 78:6202–6214
10. Hartler J, Trotzmuller M, Chitraju C et al (2011) Lipid Data Analyzer: unattended identification and quantitation of lipids in LC-MS data. Bioinformatics 27:572–577
11. Pietilainen KH, Sysi-Aho M, Rissanen A et al (2007) Acquired obesity is associated with changes in the serum lipidomic profile independent of genetic effects—a monozygotic twin study. PLoS One 2:e218
12. Schwudke D, Oegema J, Burton L et al (2006) Lipid profiling by multiple precursor and neutral loss scanning driven by the data-dependent acquisition. Anal Chem 78:585–595
13. Köfeler HC, Fauland A, Rechberger GN et al (2012) Mass spectrometry based lipidomics: an overview of technological platforms. Metabolites 2:19–38
14. Herzog R, Schuhmann K, Schwudke D et al (2012) LipidXplorer: a software for consensual cross-platform lipidomics. PLoS One 7:e29851
15. Herzog R, Schwudke D, Schuhmann K et al (2011) A novel informatics concept for high-throughput shotgun lipidomics based on the molecular fragmentation query language. Genome Biol 12:R8
16. Haag M, Schmidt A, Sachsenheimer T et al (2012) Quantification of signaling lipids by nano-ESI MS/MS. Metabolites 2:57–76
17. Matyash V, Liebisch G, Kurzchalia TV et al (2008) Lipid extraction by methyl-tert-butyl ether for high-throughput lipidomics. J Lipid Res 49:1137–1146
18. Sampaio JL, Gerl MJ, Klose C et al (2011) Membrane lipidome of an epithelial cell line. Proc Natl Acad Sci USA 108:1903–1907
19. Koivusalo M, Haimi P, Heikinheimo L et al (2001) Quantitative determination of phospholipid compositions by ESI-MS: effects of acyl chain length, unsaturation, and lipid concentration on instrument response. J Lipid Res 42:663–672
20. Brügger B, Erben G, Sandhoff R et al (1997) Quantitative analysis of biological membrane lipids at the low picomole level by nano-electrospray ionization tandem mass spectrometry. Proc Natl Acad Sci USA 94:2339–2344
21. Gray A, Olsson H, Batty IH et al (2003) Nonradioactive methods for the assay of phosphoinositide 3-kinases and phosphoinositide phosphatases and selective detection of signaling lipids in cell and tissue extracts. Anal Biochem 313:234–245
22. Ejsing CS, Sampaio JL, Surendranath V et al (2009) Global analysis of the yeast lipidome by quantitative shotgun mass spectrometry. Proc Natl Acad Sci USA 106:2136–2141
23. Bligh EG, Dyer WJ (1959) A rapid method of total lipid extraction and purification. Can J Biochem Physiol 37:911–917

24. Sandhoff R, Brügger B, Jeckel D et al (1999) Determination of cholesterol at the low picomole level by nano-electrospray ionization tandem mass spectrometry. J Lipid Res 40: 126–132
25. Liebisch G, Binder M, Schifferer R et al (2006) High throughput quantification of cholesterol and cholesteryl ester by electrospray ionization tandem mass spectrometry (ESI-MS/MS). Biochim Biophys Acta 1761:121–128
26. Rouser G, Fkeischer S, Yamamoto A (1970) Two dimensional then layer chromatographic separation of polar lipids and determination of phospholipids by phosphorus analysis of spots. Lipids 5:494–496
27. Silversand C, Haux C (1997) Improved high-performance liquid chromatographic method for the separation and quantification of lipid classes: application to fish lipids. J Chromatogr B Biomed Sci Appl 703:7–14
28. Han X, Gross RW (2001) Quantitative analysis and molecular species fingerprinting of triacylglyceride molecular species directly from lipid extracts of biological samples by electrospray ionization tandem mass spectrometry. Anal Biochem 295:88–100
29. Han X, Yang J, Cheng H et al (2004) Toward fingerprinting cellular lipidomes directly from biological samples by two-dimensional electrospray ionization mass spectrometry. Anal Biochem 330:317–331
30. Fahy E, Subramaniam S, Murphy RC et al (2009) Update of the LIPID MAPS comprehensive classification system for lipids. J Lipid Res 50:S9–S14
31. Quehenberger O, Armando AM, Brown AH et al (2010) Lipidomics reveals a remarkable diversity of lipids in human plasma. J Lipid Res 51:3299–3305
32. Schuhmann K, Almeida R, Baumert M et al (2012) Shotgun lipidomics on a LTQ Orbitrap mass spectrometer by successive switching between acquisition polarity modes. J Mass Spectrom 47:96–104
33. Gerl MJ, Sampaio JL, Urban S et al (2012) Quantitative analysis of the lipidomes of the influenza virus envelope and MDCK cell apical membrane. J Cell Biol 196:213–221

Chapter 2

Thin-Layer Chromatography of Phospholipids

Rania M. Deranieh, Amit S. Joshi, and Miriam L. Greenberg

Abstract

Thin-layer chromatography (TLC) is a technique that has been routinely used for the separation and identification of lipids. Here we describe an optimized protocol for the steady state labeling, separation, and quantification of yeast phospholipids using 1-D TLC.

Key words Thin-layer chromatography, Phospholipids, Extraction, Orthophosphate ($^{32}P_i$), Phosphatidic acid, Phosphatidylserine, Phosphatidylethanolamine, Phosphatidylcholine, Phosphatidylglycerol, Cardiolipin, Phosphatidylinositol, Zymolyase, *Saccharomyces cerevisiae*

1 Introduction

Phospholipids constitute a major component of cell membranes, where they play essential roles in vital cellular processes. They are derived from the precursor phosphatidic acid (PA) and include phosphatidylserine (PS), phosphatidylethanolamine (PE), phosphatidylcholine (PC), phosphatidylglycerol (PG), cardiolipin (CL), and phosphatidylinositol (PI), which is further phosphorylated to form a variety of phosphoinositides.

Increasingly, diseases are found to be associated with alterations in composition and levels of phospholipids. For example, Barth syndrome results from defects in cardiolipin levels and species [1, 2]; Charcot-Marie-Tooth disease is linked to altered levels of the phosphoinositide PI3,5P_2 [3]; and Alzheimer's disease and cystic fibrosis are both associated with altered levels of PE and PC [4, 5]. The finding that altered phospholipids are implicated in many diseases necessitates quick and relatively easy methods for the detection and quantification of specific phospholipids.

Thin-layer chromatography (TLC) was first described in 1938 [6]. Since then, TLC has been used for the routine separation and identification of lipids. Its use offers several advantages in terms of cost, convenience, and simplicity. The method depends on the use of: (1) A stationary phase, most commonly silica gel, with particle

Doron Rapaport and Johannes M. Herrmann (eds.), *Membrane Biogenesis: Methods and Protocols*, Methods in Molecular Biology, vol. 1033, DOI 10.1007/978-1-62703-487-6_2,

size ranging from 10 to 50 μm for standard TLC, and 5 μm for high-performance TLC (HPTLC), and (2) a mobile phase, generally consisting of a combination of organic solvents. The composition varies in different protocols.

Phospholipids can be separated by 1-D or 2-D chromatography. 2-D TLC results in more sensitive separation; however, the number of samples is limited to one per plate. While 1-D TLC is somewhat less sensitive, multiple samples can be analyzed simultaneously on a single plate and in a shorter time than in 2-D TLC.

Here, we demonstrate the use of an optimized protocol for 1-D TLC for the detection and quantification of phospholipids from *Saccharomyces cerevisiae*.

2 Materials

2.1 Growth and Labeling of Phospholipids

1. Culture media: (a) Complex medium (yeast peptone dextrose, YPD): 1 % (w/v) yeast extract, 2 % (w/v) bactopeptone, 2 % (w/v) glucose, and (2 % (w/v) agar for plates only). (b) Synthetic medium (SM): To prepare 1 L of SM, use 0.69 g of vitamin-free yeast base mix, 2.01 g of ammonium sulfate, 20 g of glucose, vitamins, inositol, and amino acids as needed. The vitamin-free yeast base mix is based on the Difco recipe and contains the following: potassium phosphate (monobasic): 0.5 g; magnesium sulfate (anhydrous): 0.25 g; sodium chloride: 0.05 g; calcium chloride: 0.05 g; boric acid: 0.25 mg; copper sulfate (cupric sulfate anhydrous) 0.02 mg; potassium iodide: 0.05 mg; ferric chloride: 0.1 mg; manganese sulfate (monohydrate): 0.2 mg; sodium molybdate (molybdic acid, sodium salt): 0.1 mg; zinc sulfate: 0.2 mg.
2. Zymolyase 20-T (*see* **Note 1**).
3. Zymolyase buffer (50 mM Tris–SO_4 (pH 7.5), 1.2 M glycerol, 100 mM sodium thioglycolate): Weigh 0.6 g Tris base and transfer to a beaker. Add 60 mL dH_2O. Add 22.10 g of 50 % glycerol and 1.14 g of sodium thioglycolate. Mix and adjust pH to 7.5 using 1 N H_2SO_4. Make up to 100 mL with dH_2O. Store at 4 °C (*see* **Note 2**).
4. [$^{32}P_i$] orthophosphate: 10 μCi/mL for each sample used for steady state labeling (*see* **Note 3**).

2.2 Extraction

1. Extraction solvent: Chloroform/methanol 2:1 (v/v).
2. Microcentrifuge.
3. Agitator/Stirrer.

2.3 Chromatography

1. TLC plates (Whatman LK5 Silica gel 150Å).
2. Prewash solution: Chloroform/methanol 1:1 (v/v). Prepare 150 mL of prewash solution by mixing 75 mL of chloroform with 75 mL of methanol.

3. Boric acid solution 1.8 % (w/v in 100 % ethanol): Weigh 9.0 g boric acid and add to 500 mL ethanol (*see* **Note 4**).
4. TLC tank lined with filter paper: Cut an appropriate size of filter paper to completely line all sides of the tank.
5. Developing solution (mobile phase): Chloroform/ethanol/water/tri-ethylamine 30:35:7:35 (v/v/v/v). Make a total of about 160 mL. Mix together 45 mL chloroform, 52.5 mL ethanol, 10.5 mL water, and 52.5 mL tri-ethylamine. Keep in an air-tight container until ready to use.

3 Methods

3.1 Steady State Labeling of Phospholipids

1. Streak the desired yeast strain on a fresh YPD plate and grow until colonies are about 2 mm in diameter.
2. Prepare an overnight culture by picking one colony and transferring it to 5 mL of the desired medium (YPD or SM).
3. Inoculate each of three tubes containing 2 mL of the desired media with cells from the overnight culture to a starting A_{550} of 0.05 (*see* **Note 5**).
4. To two tubes only, add 20 μCi $^{32}P_i$ (10 μCi/mL culture). Use the third tube (with no $^{32}P_i$ added) to monitor the absorbance.
5. To achieve steady state labeling, incubate all three tubes in a rotary shaker at the desired temperature for 5–6 generations (*see* **Note 6**).
6. Harvest the cells by centrifuging at 5,500 × *g* for 5 min, and wash once with 1 mL distilled water. Remove the supernatant. *In this and all subsequent steps, discard radioactive materials properly.*
7. To facilitate the extraction of phospholipids, digest the cell wall by adding 1 mL zymolyase solution to the cell pellets (*see* **Note 7**). Mix and keep on stirrer for 15 min at room temperature.
8. Harvest the spheroplasts (the cells without the cell wall) by centrifuging at 5,500 × *g* for 5 min. Discard the supernatant, which contains the zymolyase and all debris.

3.2 Extraction of the Phospholipids

1. To lyse the spheroplasts and extract the phospholipids, add 250 μL of chloroform/methanol 2:1 (v/v), resuspend and incubate on stirrer at room temperature for 45 min to 1 h.
2. To separate the organic phase containing the phospholipids from the aqueous phase containing non-lipid cellular material, add 50 μL of water, vortex, and centrifuge at 1,000 × *g* for 1 min.

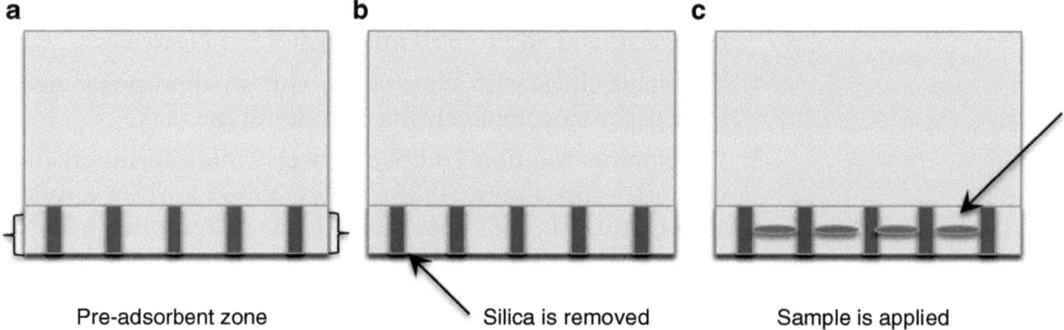

Fig. 1 TLC plate preparation for loading several samples onto the same plate. (**a**) The pre-adsorbent zone is divided into several zones using a fine tip pencil. (**b**) The silica from a width of about 0.5 cm is gently scraped off. (**c**) The sample is applied as a fine streak at about 1.5 cm from the base of the plate

3. Using a pipette with a fine tip, or a Hamilton syringe, carefully remove the lower (organic) phase and transfer to a microcentrifuge tube. The sample can be loaded directly onto the TLC plate or concentrated under a gentle stream of nitrogen at room temperature (*see* **Note 8**).

3.3 Preparation of the TLC Plates (See Fig. 1)

1. At least 2 h before starting, line the TLC tank with filter paper and equilibrate for at least 2 h with chloroform/ethanol/water/tri-ethylamine 30:35:7:35 (v/v/v/v), keeping the tank covered with a lid.
2. To remove impurities from the TLC plates, pre-develop plates with chloroform/methanol 1:1 (v/v) in a clean tank and air-dry in a fume hood.
3. Prepare a 1.8 % boric acid solution in 100 % ethanol. Pour into a glass tray that can easily accommodate the TLC plate. Fill the glass tray to a depth of about 2 cm. Dip the TLC plate in the solution for 2 min, making sure that it is uniformly impregnated with boric acid (*see* **Note 9**). Air-dry the plate for 15 min, then activate it at 100 °C for 15 min. Let it cool.
4. Loading lanes: (a) Using a fine point pencil, divide the pre-adsorbent zone (the lower thick area of the plate) into 2 cm fractions. (b) Scrape a width of ~0.5 cm using a flat spatula or a razor blade. (c) Using a fine tip, spot or streak 20–30 μL of the phospholipid sample onto the pre-adsorbent area at about 1.5 cm from the base. Allow sample to dry (*see* **Note 10**).
5. Gently lower the loaded TLC plate into the pre-equilibrated chromatography tank, making sure the samples are above the surface of the developing solvent (*see* **Note 11**).

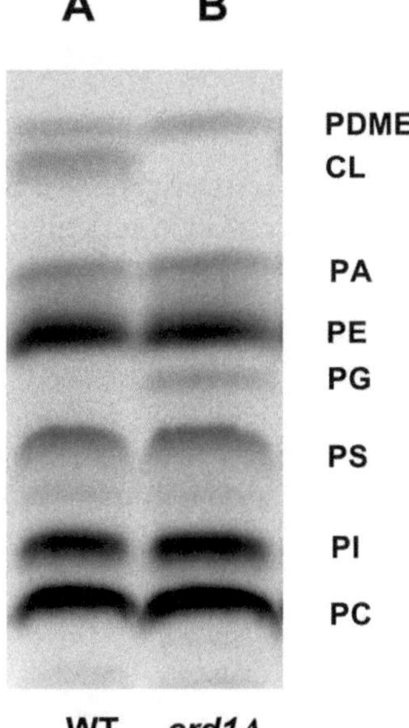

Fig. 2 Separation of phospholipids by 1-D TLC. Cells were grown in YPD, labeled with $^{32}P_i$, washed and digested with zymolyase. Lipids were extracted from spheroplasts with chloroform/methanol 2:1 (v/v), and then analyzed by 1-D TLC. Phospholipids from wild-type cells (**A**), and from *crd1* mutant cells (**B**), which lack CL are depicted. Individual lipid species were identified by the comigration of standards. *PDME* phosphatidyldimethylethanolamine, *CL* cardiolipin, *PA* phosphatidic acid, *PE* phosphatidylethanolamine, *PG* phosphatidylglycerine, *PS* phosphatidylserine, *PI* phosphatidylinositol, *PC* phosphatidylcholine

6. Place the lid on the tank and allow the solvent (the mobile phase) to ascend to about 1 cm from the top of the plate (*see* **Note 12**).
7. Remove the plate from the tank, mark the level of the solvent front by making a small etch on both sides of the plate, and air-dry completely in a fume hood.
8. Cover the TLC plate with plastic wrap, avoiding creases, and place the plate in an imaging cassette, face up. Place the phosphorimaging film on top, lock the cassette, and expose for 15–60 min.
9. Remove the film from the cassette and scan it using a Storm/Typhoon phosphorimager system. Separation of phospholipids from wild-type and *crd1* mutant cells, which lack CL, and accumulate PG, is shown in Fig. 2.
10. Quantify intensity of the bands using available software (e.g., ImageQuant).
11. To determine the *rate of synthesis* of phospholipids, a slightly modified procedure is followed (*see* **Note 13**).

4 Notes

1. Zymolyase is available in two preparations, 20-T and 100-T, with lytic activities of 20,000 units/g and 100,000 units/g, respectively. For this protocol, the less expensive 20-T preparation is sufficient.
2. Just before use, weigh appropriate amount of zymolyase and dissolve in prewarmed zymolyase buffer using gentle pipetting up and down. Do not vortex, as the enzyme may lose activity.
3. The half-life of ^{32}P is relatively short (14.29 days). Therefore, purchase the radioactive $^{32}P_i$ as close as possible to the time of the experiment. Measure and record the initial activity (A_0) as soon as you receive it. Use the following formula to calculate the amount of radioactivity at the time the experiment is begun:

$$A = A_0 \times (0.5)^{t/(1/2)t}$$

where A = current activity, A_0 = initial activity, t = number of days since order was received, ½t = half-life of ^{32}P, which is 14.29 days.
4. Boric acid improves the resolution of the separated phospholipids by acting as an adsorbent modifier [7]. The use of 1.8 % boric acid enhances the separation of PG from PE [8].
5. To ensure consistency, inoculate 6 mL of medium with the overnight culture, measure A_{550} and then aliquot into three tubes.
6. This usually takes about 8 h, but the time could vary depending on the strain used. It is advisable to do a growth curve for each strain before starting labeling experiments, in order to determine the time required for growth of 5–6 generations.
7. As the cells progress from the log phase to the stationary phase, the cell wall becomes more difficult to digest, and the amount of zymolyase required will vary accordingly. For example: use 1.5 mg/mL for cells in early log phase, 2.0 mg/mL for mid-log phase, and 2.5 mg/mL for stationary phase.
8. The samples can be completely dried under nitrogen, resuspended in chloroform/methanol 2:1 (v/v), and stored at 4 °C in a plexiglass box until needed.
9. Aside from boric acid, other chemical modifiers of silica gel have been reported, including EDTA, ammonium sulfate, silver nitrate, or oxalate [9–11].
10. Applying the sample as a fine streak allows for clear distinction of bands that may migrate in close proximity.
11. Before placing the loaded TLC plate in the tank, make sure there is enough developing solvent in the tank. Most of what

is added earlier to equilibrate the tank is absorbed by the filter paper during the equilibration process. It is important that the sample spots remain above the surface of the developing solvent. Otherwise, the samples could leach into the solvent instead of ascending up the TLC plate.

12. This takes 1.5–2 h. Keep checking the level of the solvent to ensure that samples do not reach the top.
13. The protocol described above is used for steady state labeling, which allows the detection of phospholipids present in the cells. However, to determine the *rate of synthesis* of phospholipids, pulse-labeling should be used. The procedure is the same as above except for the following: (a) the number of cells should be higher thus, the culture volume should be 10 mL instead of 2 mL. (b) The amount of $^{32}P_i$ should be higher, use 50 μCi/mL culture instead of 10 μCi/mL culture. (c) Cells are labeled for 15 min only, after which they are harvested and phospholipids are extracted. This is done for every time point during which the rate of synthesis is being monitored.

Acknowledgment

The Greenberg lab is supported by NIH grant DK081367, and R.M.D and A.S.J. are the recipients of Wayne State University Biology Department Graduate Enhancement grants.

References

1. Schlame M, Ren M (2006) Barth syndrome, a human disorder of cardiolipin metabolism. FEBS Lett 580:5450–5455
2. Chicco AJ, Sparagna GC (2007) Role of cardiolipin alterations in mitochondrial dysfunction and disease. Am J Physiol Cell Physiol 292:C33–C44
3. Lenk GM, Ferguson CJ, Chow CY, Jin N, Jones JM, Grant AE et al (2011) Pathogenic mechanism for Charcot-Marie-Tooth disease CMT4J. PLoS Genet 7:1–13
4. Nitsch RM, Blusztajn JK, Pittas AG, Slack BE, Growdon JH, Wurtman RJ (1992) Evidence for membrane defect in Alzheimer disease brain. Proc Natl Acad Sci USA 89:1671–1675
5. Innis SM, Davidson GF, Chen A, Dyer R, Melnyk S, James J (2003) Increased plasma homocysteine and S-adenosylhomocysteine and decreased methionine is associated with altered phosphatidylcholine and phosphatidylethanolamine in cystic fibrosis. J Pediatr 143:351–356
6. Izmailov NA, Shraiber MS (1938) The application of analysis by drop-chromatography to pharmacy *Farmazia* 3:1–7 [in Russian]. English translation: Pelick N, Bollinger HR, Mangold HK (1966). In: Giddings JC, Keller RA (eds) Advances in chromatography, vol 3. Marcel Dekker, New York, p 85
7. Fine JB, Sprecher H (1982) Unidimensional thin-layer chromatography of phospholipids on boric acid-impregnated plates. J Lipid Res 23:660–663
8. Vaden DL, Gohil VM, Gu Z, Greenberg ML (2005) Separation of yeast phospholipids using one-dimensional thin-layer chromatography. Anal Biochem 338:162–164
9. Allan D, Cockcroft S (1982) A modified procedure for thin-layer chromatography of phospholipids. J Lipid Res 23:1373–1374
10. Myher JJ, Kuksis A (1995) General strategies in chromatographic analysis of lipids. J Chromatogr B Biomed Appl 671:3–33
11. Singh AK, Jiang Y (1995) Quantitative chromatographic analysis of inositol phospholipids and related compounds. J Chromatogr B Biomed Appl 671:255–280

Chapter 3

Analysis of Membrane Lipid Biogenesis Pathways Using Yeast Genetics

Martina Gsell and Günther Daum

Abstract

The yeast *Saccharomyces cerevisiae* has become a valuable eukaryotic model organism to study biochemical and cellular processes at a molecular basis. A common strategy for such studies is the use of single and multiple mutants constructed by genetic manipulation which are compromised in individual enzymatic steps or certain metabolic pathways. Here, we describe selected examples of yeast research on phospholipid metabolism with emphasis on our own work dealing with investigations of phosphatidylethanolamine synthesis. Such studies start with the selection and construction of appropriate mutants and lead to phenotype analysis, lipid profiling, enzymatic analysis, and in vivo experiments. Comparing results obtained with wild-type and mutant strains allows us to understand the role of gene products and metabolic processes in more detail. Such studies are valuable not only for contributing to our knowledge of the complex network of lipid metabolism, but also of effects of lipids on structure and function of cellular membranes.

Key words Lipids, Phospholipids, Mutants, Enzyme, Yeast

Abbreviations

DMPE	Dimethylethanolamine
PC	Phosphatidylcholine
PE	Phosphatidylethanolamine
PI	Phosphatidylinositol
PS	Phosphatidylserine

1 Introduction

Lipids are a versatile class of biomolecules. Variations in aliphatic chain composition but also in head groups of polar lipids result in the existence of more than 1,000 different lipid species in any eukaryotic cell [1]. This large variation is accomplished by a relatively small set of enzymes because cells use only ~5 % of their genes and gene products to synthesize all these lipids [2].

Doron Rapaport and Johannes M. Herrmann (eds.), *Membrane Biogenesis: Methods and Protocols*,
Methods in Molecular Biology, vol. 1033, DOI 10.1007/978-1-62703-487-6_3,

It is well known that lipids fulfill three major functions. First, they are used as stores of energy mostly in the form of triacylglycerols (TG) which accumulate in lipid droplets under normal conditions [3]. Secondly, the matrix of cellular membranes is formed by polar lipids, especially glycerophospholipids, sphingolipids, and sterols. Lipids are specifically distributed among organelles where they contribute to certain cellular processes such as protein sorting or traffic in the secretory pathway. Finally, lipids can act as cellular messengers. The latter function may be accomplished by lipids such as diacylglycerols, polyphosphoinositides, and sphingolipids.

In this report, we focus on the biosynthesis of glycerophospholipids, which are major structural lipids in eukaryotic membranes, and on use of the yeast, *S. cerevisiae*, as a valuable experimental system to study this process. The tractable genetics of *S. cerevisiae* has allowed identification and characterization of many structural and regulatory genes involved in synthesis and metabolic conversion of phospholipids [4, 5]. Phospholipid synthesis in yeast is governed by a network of reactions which are subject to strict regulation by genetic and biochemical mechanisms. Moreover, distinct spatial organization within the cell plays an important role in the coordinated process of lipid synthesis. Not surprisingly, phospholipid synthesis is also linked to the metabolism of other major lipid classes including fatty acids, triacylglycerols, sterols, and sphingolipids [6]. It has also to be noted that phospholipid synthesis in yeast is affected by growth conditions which influence the expression of enzymes and/or modulate their catalytic activities. As examples, expression of phospholipid biosynthetic genes in yeast is controlled by carbon sources, nutrient availability, growth phase, pH, and temperature. Finally, posttranslational modifications of gene products, especially phosphorylation of key proteins involved in phospholipid synthesis, affect metabolism of phospholipids and the balance between certain lipid precursors and final products of lipid biosynthetic pathways [4, 7–12].

The use of *S. cerevisiae* gene deletion strains has become a standard method for biochemical, cell biological, and molecular biological studies. Standard methods of gene deletion based on PCR techniques and yeast cell transformation are well established [13, 14], and complete collections of yeast deletion mutants covering nonessential genes are available [15]. Here, we describe the use of such mutants to study yeast phospholipid metabolism with emphasis on the biosynthesis of one of the major yeast phospholipids, phosphatidylethanolamine (PE). All methods presented here are easy to use and can be applied with minor modifications to mutants that affect other aspects of phospholipid metabolism.

Figure 1 gives an overview of phospholipid biosynthetic pathways and the contribution of individual enzymes in yeast. The functions of most of these enzymes have been confirmed by analysis of mutants and/or by biochemical studies. The complex

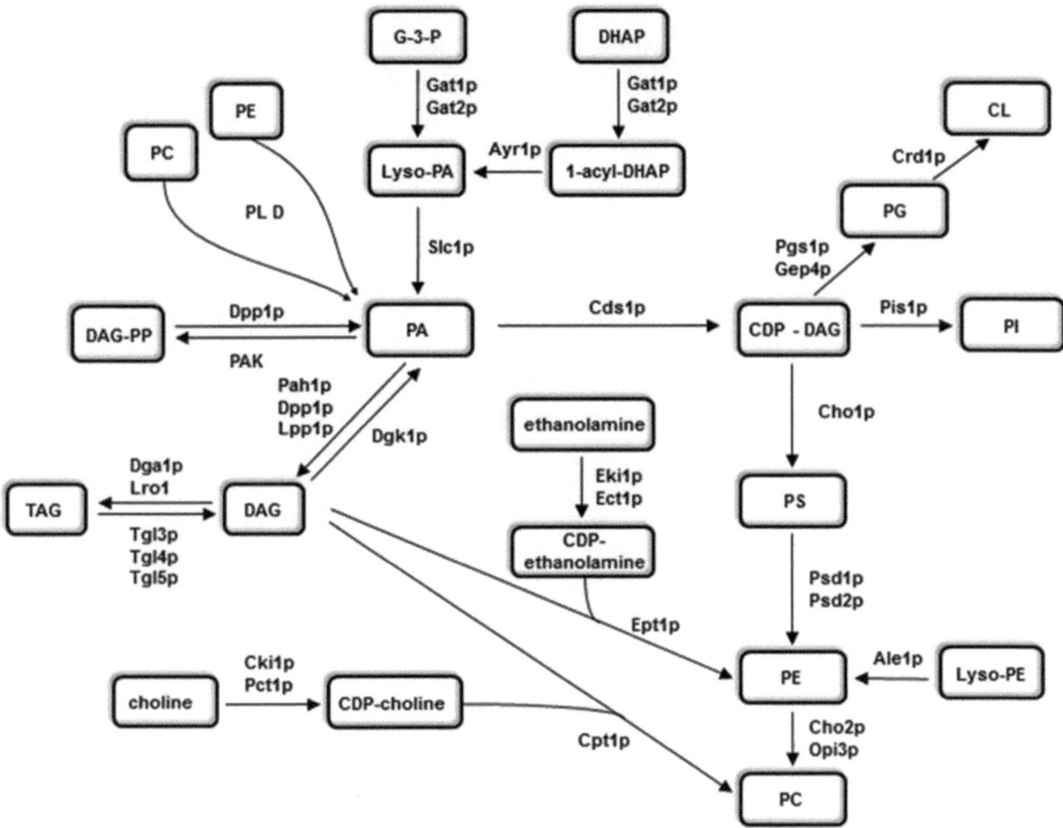

Fig. 1 Phospholipid synthesis pathways in *S. cerevisiae*. The figure shows a simplified overview of phospholipid synthesis pathways with enzymes involved. *CDP-DAG* CDP-diacylglycerol, *CL* cardiolipin, *DHAP* dihydroxyacetone phosphate, *PA* phosphatidic acid, *PC* phosphatidylcholine, *PE* phosphatidylethanolamine, *PG* phosphatidylglycerol, *PI* phosphatidylinositol, *PS* phosphatidylserine, *TAG* triacylglycerol

regulatory mechanisms in phospholipid biogenesis has recently been reviewed in Carman and Han [16]. Major phospholipids of the yeast are phosphatidylcholine (PC), phosphatidylethanolamine (PE), phosphatidylserine (PS), and phosphatidylinositol (PI), which are found at variable amounts in all yeast subcellular membranes. Other lipids such as lyso-phospholipids (LPL), phosphatidic acid (PA), cardiolipin (CL), and phosphatidylglycerol (PG) are only minor components in total cell extracts but may accumulate in certain subcellular compartments or domains [17–22].

Among the phospholipids described above, PE plays specific roles because it is unique regarding its biophysical properties as a non-bilayer (hexagonal phase) forming lipid [23]. Furthermore, a complex network of reactions distributed among different organelles leads to its formation. Four pathways contribute to yeast PE biosynthesis, which are: (1) decarboxylation of PS catalyzed by phosphatidylserine decarboxylase 1 (Psd1p) in the inner mitochondrial membrane [24], (2) decarboxylation of PS by Psd2p in a Golgi/vacuolar compartment [25], (3) incorporation of ethanolamine through the CDP-ethanolamine branch of the Kennedy pathway [26] in the endoplasmic reticulum [27, 28], and (4) synthesis of PE through acylation of lyso-PE catalyzed by the acyl-CoA-dependent acyltransferase Ale1p in the mitochondria-associated membrane (MAM) [29, 30]. These four pathways contribute to PE synthesis with different efficiencies [17], but they compensate each other in case of mutations introduced into one or the other route of PE formation.

Here we describe standard methods commonly used for the analysis of mutants affected in PE metabolism. Description of these techniques will start with growth phenotype analysis of mutant strains and lead to lipid profiling, in vivo pathway analysis, and measurement of enzymatic activities in vitro. To study the four pathways of PE synthesis described above (*see* Fig. 1), usage of different mutants is required. Whereas *Δpsd1*, *Δpsd2,* and *Δale1* single deletions are sufficient to interrupt the respective pathways, a *Δcki1Δdpl1Δeki1* triple mutant is necessary to silence the CDP-ethanolamine branch of the Kennedy pathway of PE synthesis. Reasons for this requirement are the overlapping substrate specificity of the choline and ethanolamine kinases Cki1p and Eki1p on one hand [31, 32] and the fact that ethanolamine phosphate can be provided through sphingolipid degradation by the action of dihydrosphingosine phosphate lyase, Dpl1p [33, 34]. Combination of the different mutations allows turning down more than one of the four possible pathways of PE synthesis. It has to be noted, however, that a minimal level of PE is essential for viability, and all PE biosynthetic pathways cannot be interrupted at the same time.

Phenotype analysis is one of the fundamental tools of genetics. In many cases, a particular phenotype or a set of phenotypic features is indicative of the function of a gene product. Usually, phenotypic analysis of a mutant strain starts with assays of growth characteristics, as shown for mutant strains bearing deletions of genes that function in the different PE biosynthesis pathways (Fig. 2). As mentioned above, a minimal level of PE is essential for yeast cell viability. Mutants lacking one of the PS decarboxylases, Psd1p or Psd2p, respectively, grow like wild type as long as they are cultivated on a fermentable carbon source such as glucose. On a non-fermentable carbon source, e.g., lactate, the requirement for PE increases. Under these conditions, even the single deletion of *PSD1* leads to a growth defect. This result indicates that Psd1p is the major supplier of PE. Enhanced proliferation of mitochondria in cells grown on non-fermentable carbon sources can be correlated with the importance of PE for cell respiration [35]. A *Δpsd1Δpsd2* double mutant is viable but becomes auxotrophic for ethanolamine or choline on glucose media [36], indicating that the other two pathways of PE formation, especially the CDP-ethanolamine pathway, can compensate for this defect. When cells are grown under conditions that demand more PE with lactate as a carbon source, rescue of *Δpsd1Δpsd2* by ethanolamine

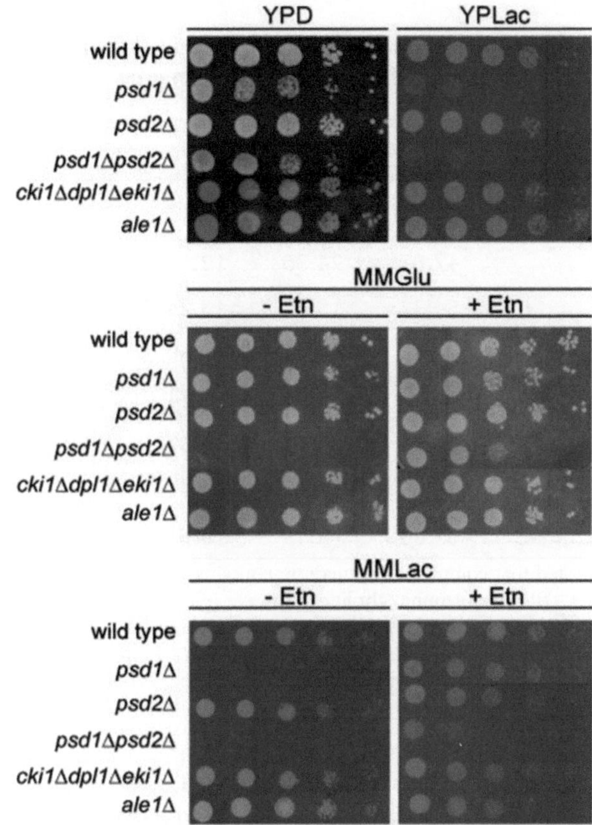

Fig. 2 Growth of yeast strains bearing defects in phosphatidylethanolamine biosynthesis depends on the carbon source. Cell suspensions of strains listed in the figure were spotted on YPD, YPLac, MMGlu, and MMLac with or without 5 mM ethanolamine. Incubation was carried out at 30 °C. *YPD* complex glucose media, *YPLac* complex lactate media, *MMGlu* minimal glucose media, *MMLac* minimal lactate media, *Etn* ethanolamine. Reproduced from [36] with permission of the publisher

supplementation is much less effective [35]. These results also indicate that the efficiency of PE import into mitochondria is limited. Taken together, these simple but well-designed growth phenotype analyses lead to a number of conclusions related to the physiological relevance of PE biosynthetic pathways in yeast [27, 35, 37].

The next step to understand the role of lipid biosynthetic enzymes in cellular metabolism is lipid profiling. For our routine analysis of yeast phospholipids we use two-dimensional thin-layer chromatography and lipid phosphorus estimation as outlined in detail in Subheading 3.2.3. In wild-type homogenate, the major phospholipids are PC, PE and PI and PS (Table 1) [17]. PC accounts for approximately 50 % and PE for approximately 25 % of total cellular phospholipids. In *Δpsd1* and *Δpsd2*, and even more pronounced in *Δpsd1Δpsd2* deletion strains, a reduction of the

Table 1
Lipid composition of cell homogenate from wild-type (FY1679; Euroscarf, Frankfurt, Germany) and yeast strains bearing defects in PE formation grown on YPD

	% of total phospholipids							
Strain	**PC**	**PE**	**PI**	**PS**	**LPL**	**DMPE**	**PA**	**CL**
FY1679	47.57	24.36	12.19	6.89	1.40	5.49	0.87	1.23
psd1Δ	54.32	18.62	13.24	7.65	1.17	3.58	0.66	0.75
psd2Δ	54.65	16.95	13.60	8.93	1.61	2.77	0.43	1.06
psd1Δpsd2Δ	64.28	7.46	11.74	10.69	1.43	1.31	1.63	1.48
cki1Δdpl1Δeki1Δ	48.24	28.91	11.57	4.27	1.20	2.82	1.25	1.74
ale1Δ	49.85	22.53	11.97	6.75	2.50	3.88	1.00	1.52
psd1Δpsd2Δale1Δ	67.80	5.53	10.91	11.51	1.85	0.58	1.11	0.70

It has to be noted that complex YPD medium contains small amounts of ethanolamine and choline which serve a substrate for the CDP-ethanolamine/choline pathway

Reproduced with adaptations from [17] with permission of the publisher

PC phosphatidylcholine, *PE* phosphatidylethanolamine, *PI* phosphatidylinositol, *PS* phosphatidylserine, *LPL* lysophospholipids, *DMPE* dimethylphosphatidylethanolamine, *PA* phosphatidic acid, *CL* cardiolipin

cellular PE content was observed [17–22]. In contrast, the phospholipid patterns of *Δcki1Δdpl1Δeki1* (CDP-ethanolamine pathway mutant) and *Δale1* strains largely resembled wild type. This result confirmed the dominant role of Psd1p and Psd2p in cellular PE formation and suggested that the CDP-ethanolamine pathway and acylation of lyso-PE are only of minor importance in cells harboring functional PS decarboxylases, at least under standard culture conditions (YPD, complex glucose media) [17]. Such data allow estimation of the contributions of different PE biosynthetic routes to total formation of PE in yeast under different physiological conditions.

As mentioned above, growth of yeast cells on non-fermentable carbon sources, e.g., lactate, challenges the cellular requirement of PE mainly through the enhanced proliferation of mitochondria. Under such stringent conditions (minimal medium/lactate with ethanolamine), the cellular PE level of *Δcki1Δdpl1Δeki1* was reduced to 17 % of total phospholipids compared to 15 % in *Δpsd1Δpsd2* and 25 % in wild type. We can conclude from this result that also the CDP-ethanolamine pathway, besides Psd1p, is an important route for cellular PE formation. The decrease in cellular PE was mainly compensated by elevated levels of PC. Interestingly, an increased level of cellular PE was found in *Δale1* [37]. Ale1p had been identified as a lyso-PE acyltransferase [29, 30]. The reason for the increased PE level in the deletion strain may be the broad substrate specificity of Ale1p [38–40] and/or compensation of the *Δale1* deletion by other PE forming routes [37].

Whereas lipid profiling of yeast cells provides information about a steady state situation in the cell, in vivo labeling experiments reflect the dynamic processes of lipid formation and/or turnover. Formation of PE in vivo can be measured by labeling experiments with radioactive serine introduced into the decarboxylation/methylation route of aminoglycerophospholipid biosynthesis; and with radiolabeled ethanolamine incorporated into PE via Kennedy pathway (*see* Fig. 1, and Subheading 3.3) [19]. When yeast cells are cultivated in the presence of [^{3}H]serine, PS is formed as the first component in the lipid biosynthetic sequence, which is then converted by PS decarboxylases Psd1p and/or Psd2p to PE. Further conversion of PE leads to PC through threefold methylation. It has to be noted that enzymes involved in the decarboxylation/methylation route of aminoglycerophospholipid biosynthesis are localized to different subcellular compartments, namely the ER and mitochondria or Golgi, respectively. Consequently, translocation of substrates and intermediates between organelles is required. Not unexpectedly, incorporation of radiolabeled serine into PE strongly depends on the presence or absence of Psd1p and Psd2p. Deletion of each of these enzymes, respectively, reduces incorporation of label into PE. The amount of [^{3}H]serine incorporated into PE in the *Δpsd1* strain is decreased to ~30 % and in *Δpsd2* mutants to ~70 % of wild type [19, 21, 22, 25]. The observation that deletion of *PSD1* has a stronger effect on PE synthesis than deletion of *PSD2* is in line with growth phenotype and lipid analyses described above. In a *Δpsd1Δpsd2* mutant, which completely lacks PS decarboxylase activity, [^{3}H]serine is significantly accumulated in PS, but nevertheless low levels of radioactivity are incorporated into PE. This observation supports the view that radiolabel in the form of [^{3}H]serine incorporated into sphingolipids through the action of serine palmitoyltransferase ends up in ethanolamine phosphate during sphingolipid turnover [25]. Ethanolamine phosphate can directly serve a substrate for enzymes of the Kennedy pathway. Double labeling of whole cells with [^{3}H]serine and [^{14}C]ethanolamine in combination with the use of mutants bearing defects in the different PE biosynthetic pathways allows us to estimate the contribution of each pathway to total cellular PE formation.

Labeling of aminoglycerophospholipids in vivo using intact, metabolizing yeast as described above can be complemented by experiments using permeabilized yeast cells [41]. These cells do not proliferate but still metabolize. This technique allows dissection of individual steps of the biosynthetic routes of PS, PE, and PC formation. The advantage of this system is the possibility to introduce reagents which cannot enter intact cells. Again, the use of different labeled aminoglycerophospholipid precursors and mutants contributes to a better understanding of the complete PE biosynthetic network.

Finally, in vitro measurements of enzyme activities can be used to confirm the efficiency of biosynthetic steps and the effects of mutations. It has to be noted that enzyme activities measured in vitro do not necessarily reflect the situation in vivo due to regulatory effects that may escape detection in the test tube. In the case of PS decarboxylases, however, results obtained in vitro largely match in vivo data [22, 25]. The assay used for these measurements (*see* Subheading 3.4) is a standard procedure described by Kuchler et al. [42] and modified by Birner et al. [18]. PS decarboxylase activity can be measured in total cell extracts as well as in isolated subcellular fractions.

In summary, the use of mutants in yeast lipid research has led to better knowledge about individual steps of lipid formation, and also to a deeper understanding of the links between pathways in a more global way. As an example, Horvath et al. [37] recently described a novel physiological link between triacylglycerol (TG) and PE metabolism in yeast. The bridging enzyme between these two pathways is the phospholipid:diacylglycerol acyltransferase Lro1p, which forms TG from diacylglycerol and a fatty acid derived from a phospholipid, preferentially from PE. Using the set of PE biosynthesis mutants described above, the contribution of the four different PE biosynthetic pathways to TG formation was analyzed. Interestingly, it was found that the CDP-ethanolamine pathway of PE formation contributes most to the cellular TG level, whereas mutations in the other pathways for PE synthesis display only minor effects. Such experiments broaden our view of lipid metabolism and set the stage to understand this complex network of reactions including their regulation in more detail.

2 Materials

2.1 Equipment and Supplies

1. Microsyringe (Hamilton, Bonaduz, Switzerland).
2. 12 mL Pyrex glass vials with Teflon liner caps.
3. Table-top shaker for test tubes (IKA® Vibrax VXR).
4. Glass tubes (20 mL) with ground neck.
5. Silica gel 60 TLC plates (Merck, Darmstadt, Germany).
6. TLC chamber (Springfield Mill, UK) with saturation paper (e.g., Whatman filter paper).
7. Iodine vapor chamber.
8. Incubator (Heraeus).
9. Table-top centrifuge (Hettich Rotina 46 R, Heraeus Fresco17).
10. Merckenschlager homogenizer (Braun-Melsungen) with fitting glass bottles, and glass beads (0.25–0.30 mm diameter; Sartorius).

11. Scintillation counter Packard 1500 Tri-Carb®.
12. Plastic vials for liquid scintillation counting.

2.2 Reagents

1. Medium for yeast cell cultivation: YPD (2 % glucose, 2 % peptone, and 1 % yeast extract).
2. Medium for yeast cell cultivation: YPLac (2.6 % lactate, 2 % peptone, and 1 % yeast extract, pH 5.5 with KOH).
3. Medium for yeast cell cultivation: MMGlu (2 % glucose, 0.67 % yeast nitrogen base, and amino acid mixture).
4. Medium for yeast cell cultivation: MMLac (2.6 % lactate, 0.67 % yeast nitrogen base, and amino acid mixture, pH 5.5 with KOH).
5. Solvents: chloroform and methanol, analytical grade.
6. Washing solution for lipid extraction: 0.034 % $MgCl_2$; 2 N KCl/MeOH (4:1; v/v).
7. Washing solution for lipid extraction: artificial upper phase ($CHCl_3$/MeOH/H_2O; 3:48:47; per vol).
8. 0.26 % Ammonium heptamolybdate tetrahydrate/ANSA (500:22; v/v).
9. ANSA solution: 40.0 g $K_2S_2O_5$, 0.63 g 8-anilio-1-naphthalenesulfonic acid, 1.25 g Na_2SO_3 in 250 mL water.
10. l-[^{3}H]serine (21.99 Ci/mmol, Perkin Elmer, Boston, MA).
11. [^{14}C]ethanolamine (2.9 mCi/mmol, Perkin Elmer, Boston, MA).
12. Scintillation cocktail (Packard Bio-Science, Groningen, The Netherlands) with 5 % H_2O.

3 Methods

3.1 Growth Phenotype Analysis (See Note 1)

1. 5 mL YPD are inoculated with 10 μL of an overnight culture and incubated for ~16 h at 30 °C with shaking.
2. The culture is diluted to an OD_{600} 1 with sterile water and dilutions (1, 1/10, 1/100, 1/1,000, 1/10,000) are prepared in 96-well microtiter plates.
3. 4 μL of the suspensions are spotted on agar plates. Alternatively, a sterile stamp can be used.
4. Agar plates are incubated at 30 °C for a few days depending on the medium and the strain.

3.2 Lipid Analysis

3.2.1 Preparation of Cell Homogenate

1. Homogenate is prepared from a minimum of 50 mL culture of full or selective media. Cells are inoculated from a 48 h preculture to an OD_{600} of 0.1 and grown to the early stationary phase at 30 °C with shaking (*see* **Note 2**).
2. After harvesting at 4,500 × *g* for 5 min, cells are washed with distilled water.

3. Cells are suspended in 10 mM Tris/HCl, pH 7.5, 1 mM EDTA (mL/g CWW) and protease inhibitor phenylmethylsulfonyl fluoride (stock solution 1 M PMSF in DMSO) (2 μL/g cell wet weight)
4. Cells are disintegrated with glass beads in a Merckenschlager homogenizer for 3 min under CO_2 cooling.
5. The cell extract is transferred to a fresh tube and cleared of glass beads, unbroken cells, and cell debris by centrifugation at 2,500 × *g* for 5 min at 4 °C. The supernatant fraction represents the homogenate.
6. Proteins are quantified using the method of Lowry et al. [43] or Bradford [44].

3.2.2 Lipid Extraction

1. Lipids from the homogenate are extracted using the method of Folch et al. [45]. Solvents should be handled with care and work should be performed in a fume hood.
2. Following the procedure of Folch et al. [45] an aliquot of the sample (~3 mg protein) is added to 3 mL of $CHCl_3$:MeOH (2:1; v/v) in a Pyrex glass tube.
3. Lipids are extracted to the polar organic phase by vigorous shaking with a table-top rotary shaker (IKA® Vibrax VXR) at room temperature for 30 min.
4. Proteins and polar substances are removed by consecutive washing steps with 1 mL 0.034 % $MgCl_2$, 1 mL of 2 N KCl/MeOH (4:1; v/v), and 1 mL of an artificial upper phase ($CHCl_3$:MeOH:H_2O; 3:48:47; per vol). These solutions are added to the extracts and incubated with shaking for 10 min (*see* **Note 3**)
5. After each washing step samples are centrifuged for 3 min at 2,500 × *g* in a table-top centrifuge, and the aqueous phase is removed by aspiration (*see* **Note 3**).
6. Washing steps are repeated until no protein intermediate layer is formed any more.
7. Finally, lipids are dried under a stream of nitrogen and stored at −20 °C.

3.2.3 Thin-Layer Chromatography and Phospholipid Quantification

1. Phospholipids are separated by 2D TLC due to different properties of their head groups. Lipids are dissolved in 50 μL $CHCl_3$/MeOH (2:1; v/v) and applied as single spot to a TLC plate (10 × 10 cm) approximately 1–1.5 cm distant from a corner. TLC plates can be loaded using a Hamilton syringe or a sampler pipette.
2. For the first dimension, $CHCl_3$/MeOH/25 % NH_3 (65:25:6; per vol) is used as a solvent, and for the second dimension $CHCl_3$/acetone/MeOH/acetic acid/H_2O (50:20:10:10:5; per vol). Separations usually take 50 min/10 cm distance on TLC plates for each dimension.

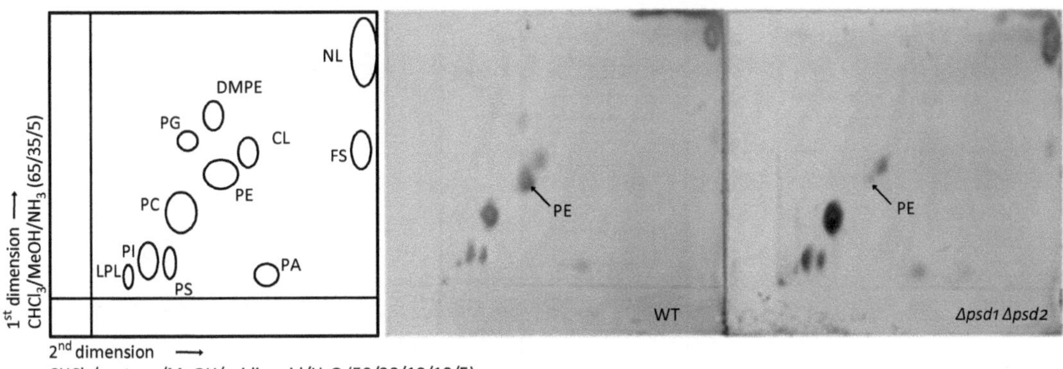

Fig. 3 Separation of yeast phospholipids by two-dimensional TLC. *Left*: Schematic overview of phospholipid separation by two-dimensional TLC. *Middle*: Two-dimensional TLC of phospholipids from wild-type yeast. *Right*: Two-dimensional TLC of phospholipids from a Δ*psd1*Δ*psd2* double mutant. The *arrows* point to the spot of PE. *LPL* lyso-phospholipids, *PI* phosphatidylinositol, *PS* phosphatidylserine, *PA* phosphatidic acid, *PC* phosphatidylcholine, *PE* phosphatidylethanolamine, *FA* fatty acids, *PG* phosphatidylglycerol, *DMPE* dimethylethanolamine, *CL* cardiolipin

3. Phospholipids are visualized by staining with iodine vapor in a saturated chamber for some minutes (Fig. 3). Spots are marked with a pencil. The iodine vapor develops after putting a spoonful of solid crystals of iodine into the chamber. Staining of TLC plates should be done in a fume hood.
4. For destaining of spots, TLC plates are incubated in a heating chamber (~50–60 °C) for a few minutes.
5. Phospholipids can be quantified from TLC plates after removal of the iodine staining. The plate is moistened with deionized water, phospholipid spots are scrapped off and transferred to a phosphate-free glass tube with ground neck (*see* **Note 4**).
6. The lipid phosphorus of the respective spot can be measured by subjecting the sample to hydrolysis as described by Broekhuyse [46]. In brief, 0.2 mL of conc. H_2SO_4/72 % $HClO_4$ (9:1; v/v) are added to each sample. Hydrolysis is performed at 180 °C in a heating block for 30 min. Please note that this step has to be performed in a hood due to formation of acidic fumes!
7. Samples are cooled to room temperature, and 4.8 mL freshly prepared 0.26 % ammonium heptamolybdate tetrahydrate/ANSA (500:22; v/v) is added. Tubes are closed with phosphate-free glass caps, and after vigorous vortexing, samples are heated to 100 °C for 30 min in an oven.
8. Finally, samples are cooled to room temperature and centrifuged briefly in a table-top centrifuge at 1,000×*g* to sediment the silica gel. The intensity of the blue color in the supernatant is a measure for lipid phosphorus. Samples are measured spectrophotometrically at a wavelength of 830 nm using a blank spot from the TLC plate without phospholipid as a control. Inorganic phosphate is used as a standard.

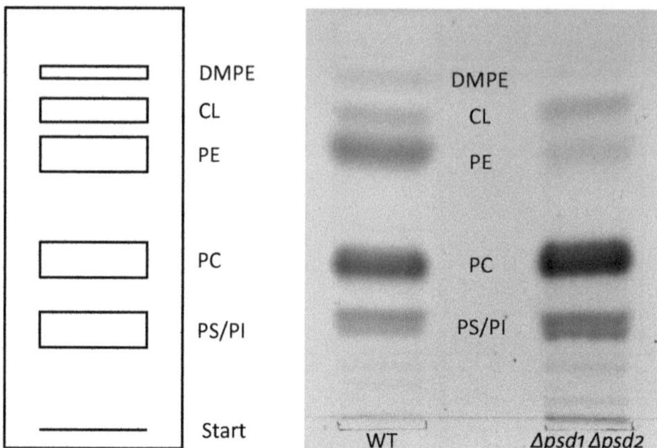

Fig. 4 Separation of phospholipids by one-dimensional TLC. $CHCl_3$/MeOH/25 % NH_3 (65:25:6; per vol) is used as solvent. *Left*: Schematic overview of phospholipid separation by one-dimensional TLC. *Right*: One-dimensional TLC of phospholipids from wild-type and *Δpsd1Δpsd2* double mutant

3.3 In Vivo Labeling of Aminoglycerophospholipids

1. Cells are inoculated from a 48 h preculture to an OD_{600} of 0.1 and grown to stationary phase at 30 °C in YPD with shaking.
2. 10 OD units of cells are divided into three equal portions and put into three sterile Pyrex tubes, harvested by centrifugation and suspended in 500 μL of fresh YPD.
3. Then, cells are incubated for 30 min at 30 °C with shaking.
4. 10 μCi l-[^{3}H]serine or 2 μCi [^{14}C]ethanolamine are added to the cultures (*see* **Note 5**).
5. The cultures are incubated for 10 min, 20 min, and 30 min, respectively, at 30 °C with shaking and then immediately cooled to 4 °C to stop cellular metabolism (*see* **Note 6**).
6. After centrifugation and washing with ice cold water, cells are deep frozen with liquid nitrogen.
7. To extract lipids, cells are vortexed for 60 min together with 1.5 mL of glass beads and 2 mL $CHCl_3$/MeOH (2:1; v/v).
8. After centrifugation, the supernatant is transferred to a fresh Pyrex tube.
9. Glass beads are washed with 2 mL $CHCl_3$/MeOH (2:1; v/v) and centrifuged; and supernatants are combined.
10. Supernatants are washed as described in Subheading 3.2.2.
11. Lipids are dissolved in 50 μL $CHCl_3$/MeOH (2:1; v/v) and spotted on TLC plates.
12. For the analysis of PS, PE, and PC, lipids are separated by one-dimensional TLC using $CHCl_3$/MeOH/25% NH_3 (50:25:6; per vol) as developing solvent (Fig. 4).

13. Phospholipids are visualized by staining with iodine vapor in a saturated chamber for some minutes and marked with a pencil. For destaining, TLC plates are incubated in a heating chamber (~50–60 °C) for a few minutes.
14. Bands of PS, PE, and PC are scraped off the plate and transferred into liquid scintillation vials with LSC SAFETY Cocktail (Packard Bioscience B.V., Meriden, USA) plus 5 vol% water and incubated for at least 1 h.
15. Radioactivity is determined by liquid scintillation counting in a Packard 1500 Tri-Carb® liquid scintillation analyzer.

3.4 PS Decarboxylase Activity Assay In Vitro

3.4.1 Preparation of the Substrate

1. [^{3}H]Phosphatidylserine is used as a substrate for PS decarboxylase measurements. It is synthesized in vitro in a solution consisting of 200 μL 1 M Tris/HCl (pH 8), 200 μL 50 mM NH_2OH (in 2 % Triton X-100), 200 μL 6 mM $MnCl_2$, 10 μL 20 mM CDP-DAG, 200 μL 5 mM l-serine, and 20 μCi l-[^{3}H]serine.
2. To start the reaction, 1 mL of wild-type homogenate containing 10 mg protein (*see* Subheading 3.2.1) is added and the sample is vortexed.
3. Samples are agitated for 60 min in a 30 °C water bath using a magnetic stirrer.
4. The reaction is stopped by adding 20 mL $CHCl_3$/MeOH (2:1; v/v) and vigorous vortexing. Then, the sample is incubated for 10 min at room temperature.
5. After centrifugation for 5 min at 2,500×*g* in a table-top centrifuge, the aqueous phase is removed by aspiration.
6. The lower polar phase is washed five times with 2 N KCl/MeOH (4:1; v/v).
7. Finally, the product is dried under a stream of nitrogen and dissolved in 2 mL $CHCl_3$/MeOH (2:1; v/v).
8. The radioactivity of the sample is determined by liquid scintillation counting in a Packard 1500 Tri-Carb® liquid scintillation analyzer using LSC SAFETY Cocktail (Packard Bioscience B.V., Meriden, USA) plus 5 vol% water.

3.4.2 Enzyme Assay

1. For PS decarboxylase assays, the substrate (displaying radioactivity corresponding to 666 dps; disintegrations per second) is put into a glass tube with ground neck and dried under a stream of nitrogen. 500 μL 0.2 M Tris/HCl (pH 7.2), 20 mM EDTA is added, and the sample is ultrasonicated for 7 min in a water bath.
2. The glass tube containing the substrate is put into a water bath (30 °C) with agitation, and the reaction is started by adding 500 μL of homogenate (2 mg of protein in 10 mM Tris/HCl, pH 7.5, 1 mM EDTA).

3. At different time points (0, 10 and 20 min), 200 μL samples are transferred to a Pyrex tube containing 4 mL $CHCl_3$/MeOH (2:1; v/v) to stop the reaction.
4. Extraction of lipids is performed as described in Subheading 3.2.2.
5. After drying samples under a stream of nitrogen they are dissolved in 50 μL $CHCl_3$/MeOH (2:1; v/v) and spotted onto a TLC plate.
6. PE formed during incubation is isolated by TLC and analyzed as described above (Subheading 3.3).

4 Notes

1. A summary of phenotypes which can be easily screened and is associated with primary or suppressor mutations was provided by Hampsey [47].
2. It is very important that all strains tested are in the same growth phase, because the lipid composition varies strongly at different stages of yeast cultivation.
3. For subsequent analysis of lipids by TLC, it is very important to remove all protein aggregates by discarding the upper aqueous phase as well as the protein interface layer. Alternatively, the lower polar phase can be transferred to a fresh Pyrex tube by using a glass pipette.
4. Note that all tubes and pipettes have to be free of phosphate contaminations. Such contamination would lead to incorrect measurements. For this reason, phosphate-free detergent and ddH_2O has to be used for all steps.
5. As a variation of this assay 0.1 μCi [methyl-^{14}C]choline chloride or 0.5 μCi [1-^{14}C]acetic acid can be used to label lipids. Exogenously added choline incorporates into PC via CDP-choline pathway, and acetate is used as a substrate for fatty acid synthesis.
6. Time points can be varied depending on the problem to be addressed. Early time points mainly represent the synthesis rate of a phospholipid. To obtain steady state data cells are incubated with labeled precursors for several hours. As an example, 20 mL YPD are inoculated to OD_{600} 0.1 from a 48 h preculture, and 10 μCil-[^{3}H]serine, 2 μCi [^{14}C]ethanolamine, 0.1 μCi [methyl-^{14}C]choline chloride or 0.5 μCi [1-^{14}C]acetic acid, respectively, per mL culture are added. Then, cells can be cultivated to the late logarithmic or even to the stationary growth phase. 10 OD units are harvested in a Pyrex tube and lipids are extracted as described above.

Acknowledgments

This work was supported by the Austrian Science Fund FWF (project 21429 and DK Molecular Enzymology W901-B05 to GD). The authors are grateful to Edina Harsay for critical reading of this manuscript.

References

1. Sud M, Fahy E, Cotter D et al (2007) LMSD: LIPID MAPS structure database. Nucleic Acids Res 35:D527–D532
2. Meer GV, Voelker DR, Feigensonvan GW (2008) Membrane lipids: where they are and how they behave. Nat Rev Mol Cell Biol 9:112–124
3. Athenstaedt K, Daum G (2005) Tgl4p and Tgl5p, two triacylglycerol lipases of the yeast *Saccharomyces cerevisiae* are localized to lipid particles. J Biol Chem 280:37301–37309
4. Gaspar ML, Aregullin MA, Jesch SA et al (2007) The emergence of yeast lipidomics. Biochim Biophys Acta 1771:241–254
5. Carman GM, Zeimetz GM (1996) Regulation of phospholipid biosynthesis in the yeast *Saccharomyces cerevisiae*. J Biol Chem 271: 13293–13296
6. Rajakumari S, Grillitsch K, Daum G (2008) Synthesis and turnover of non-polar lipids in yeast. Prog Lipid Res 47:157–171
7. Carman GM, Henry SA (1999) Phospholipid biosynthesis in the yeast *Saccharomyces cerevisiae* and interrelationship with other metabolic processes. Prog Lipid Res 38:361–399
8. Chen M, Hancock LC, Lopes JM (2007) Transcriptional regulation of yeast phospholipid biosynthetic genes. Biochim Biophys Acta 1771:310–321
9. Carman GM, Han GS (2007) Regulation of phospholipid synthesis in *Saccharomyces cerevisiae* by zinc depletion. Biochim Biophys Acta 1771:322–330
10. Patton-Vogt J (2007) Transport and metabolism of glycerophosphodiesters produced through phospholipid deacylation. Biochim Biophys Acta 1771:337–342
11. Santos-Rosa H, Leung J, Grimsey N et al (2005) The yeast lipin Smp2 couples phospholipid biosynthesis to nuclear membrane growth. EMBO J 24:1931–1941
12. Li G, Chen S, Thompson MN, Greenberg ML (2007) New insights into the regulation of cardiolipin biosynthesis in yeast: implications for Barth syndrome. Biochim Biophys Acta 1771:432–441
13. Hinnen A, Hicks JB, Fink GR (1978) Transformation of yeast. Proc Natl Acad Sci USA 75:1929–1933
14. Struhl K, Stinchcomb DT, Scherer S et al (1979) High-frequency transformation of yeast: autonomous replication of hybrid DNA molecules. Proc Natl Acad Sci USA 76:1035–1039
15. Winzeler EA, Shoemaker DD, Astromoff A et al (1999) Functional characterization of the *S. cerevisiae* genome by gene deletion and parallel analysis. Science 285:901–906
16. Carman GM, Han GS (2011) Regulation of phospholipid synthesis in the yeast *Saccharomyces cerevisiae*. Annu Rev Biochem 80:859–883
17. Schuiki I, Schnabl M, Czabany T et al (2010) Phosphatidylethanolamine synthesized by four different pathways is supplied to the plasma membrane of the yeast *Saccharomyces cerevisiae*. Biochim Biophys Acta 1801: 480–486
18. Birner R, Nebauer R, Schneiter R et al (2003) Synthetic lethal interaction of the mitochondrial phosphatidylethanolamine biosynthetic machinery with the prohibitin complex of *Saccharomyces cerevisiae*. Mol Biol Cell 14:370–383
19. Bürgermeister M, Birner-Grünberger R, Nebauer R et al (2004) Contribution of different pathways to the supply of phosphatidylethanolamine and phosphatidylcholine to mitochondrial membranes of the yeast *Saccharomyces cerevisiae*. Biochim Biophys Acta 1686:161–168
20. Gohil VM, Thompson MN, Greenberg ML (2005) Synthetic lethal interaction of the mitochondrial phosphatidylethanolamine and cardiolipin biosynthetic pathways in *Saccharomyces cerevisiae*. J Biol Chem 280: 35410–35416
21. Nebauer R, Schuiki I, Kulterer B et al (2007) The phosphatidylethanolamine level of yeast mitochondria is affected by the mitochondrial components Oxa1p and Yme1p. FEBS J 274:6180–6190

22. Storey MK, Clay KL, Kutateladze T et al (2001) Phosphatidylethanolamine has an essential role in *Saccharomyces cerevisiae* that is independent of its ability to form hexagonal phase structures. J Biol Chem 276:48539–48548
23. Hui SW, Stewart TP, Yeagle PL et al (1981) Bilayer to non-bilayer transition in mixtures of phosphatidylethanolamine and phosphatidylcholine: implications for membrane properties. Arch Biochem Biophys 207:227–240
24. Zinser E, Sperka-Gottlieb CDM, Fasch EV et al (1991) Phospholipid synthesis and lipid composition of subcellular membranes in the unicellular eukaryote *Saccharomyces cerevisiae*. J Bacteriol 173:2026–2034
25. Trotter PJ, Voelker DR (1995) Identification of a non-mitochondrial phosphatidylserine decarboxylase activity (*PSD2*) in the yeast *Saccharomyces cerevisiae*. J Biol Chem 270:6062–6070
26. Kennedy EP, Weiss SB (1956) The function of cytidine coenzymes in the biosynthesis of phospholipides. J Biol Chem 222:193–214
27. Birner R, Daum G (2003) Biogenesis and cellular dynamics of aminoglycerophospholipids. Int Rev Cytol 225:273–323
28. Daum G, Lees ND, Bard M et al (1998) Biochemistry, cell biology and molecular biology of lipids of *Saccharomyces cerevisiae*. Yeast 14:1471–1510
29. Riekhof WR, Voelker DR (2006) Uptake and utilization of lyso-phosphatidylethanolamine by *Saccharomyces cerevisiae*. J Biol Chem 281:36588–36596
30. Riekhof WR, Wu J, Jones JL et al (2007) Identification and characterization of the major lysophosphatidylethanolamine acyltransferase in *Saccharomyces cerevisiae*. J Biol Chem 282: 28344–28352
31. Kim K, Kim K-H, Storey MK et al (1999) Isolation and characterization of the *Saccharomyces cerevisiae EKI1* gene encoding ethanolamine kinase. J Biol Chem 274: 14857–14866
32. Hosaka K, Kodaki T, Yamashita S (1989) Cloning and characterization of the yeast *CKI* gene encoding choline kinase and its expression in *Escherichia coli*. J Biol Chem 264: 2053–2059
33. Saba JD, Nara F, Bielawska A et al (1997) The *BST1* gene of *Saccharomyces cerevisiae* is the sphingosine-1-phosphate lyase. J Biol Chem 272:26087–26090
34. Gottlieb D, Heideman W, Saba JD (1999) The *DPL1* gene is involved in mediating the response to nutrient deprivation in *Saccharomyces cerevisiae*. Mol Cell Biol Res Commun 1:66–71
35. Trotter PJ, Pedretti J, Yates R et al (1995) Phosphatidylserine decarboxylase 2 of *Saccharomyces cerevisiae*. Cloning and mapping of the gene, heterologous expression, and creation of the null allele. J Biol Chem 270: 6071–6080
36. Horvath SE, Wagner A, Steyrer E et al (2011) Metabolic link between phosphatidylethanolamine and triacylglycerol metabolism in the yeast *Saccharomyces cerevisiae*. Biochim Biophys Acta 1811:1030–1037
37. Birner R, Bürgermeister M, Schneiter R et al (2001) Roles of phosphatidylethanolamine and of its several biosynthetic pathways in *Saccharomyces cerevisiae*. Mol Biol Cell 12: 997–1007
38. Stahl U, Stalberg K, Stymne S et al (2008) A family of eukaryotic lysophospholipid acyltransferases with broad specificity. FEBS Lett 582:305–309
39. Jain S, Stanford N, Bhagwat N et al (2007) Identification of a novel lysophospholipid acyltransferase in *Saccharomyces cerevisiae*. J Biol Chem 282:30562–30569
40. Tamaki H, Shimada A, Ito Y et al (2007) *LPT1* Encodes a membrane-bound *O*-acyltransferase involved in the acylation of lysophospholipids in the yeast *Saccharomyces cerevisiae*. J Biol Chem 282:34288–34298
41. Achleitner G, Zweytick D, Trotter P et al (1995) Synthesis and intracellular transport of aminoglycerophospholipids in permeabilized cells of the yeast, *Saccharomyces cerevisiae*. J Biol Chem 270:29836–29842
42. Kuchler K, Daum G, Paltauf F (1986) Subcellular and submitochondrial localization of phospholipid-synthesizing enzymes in *Saccharomyces cerevisiae*. J Bacteriol 165: 901–910
43. Lowry OH, Rosebrough NJ, Farr AL et al (1951) Protein measurement with the folin phenol reagent. J Biol Chem 193:265–275
44. Bradford MM (1976) A rapid and sensitive method for the quantitation of microgram quantities of protein utilizing the principle of protein-dye binding. Anal Biochem 72: 248–254
45. Folch J, Lees M, Sloane Stanley GH (1957) A simple method for the isolation and purification of total lipids from animal tissues. J Biol Chem 226:497–509
46. Broekhuyse RM (1968) Phospholipids in tissues of the eye. I. Isolation, characterization and quantitative analysis by two-dimensional thin-layer chromatography of diacyl and vinyl-ether phospholipids. Biochim Biophys Acta 152:307–315
47. Hampsey M (1997) A review of phenotypes in *Saccharomyces cerevisiae*. Yeast 13:1099–1133

Part II

Analysis of Structure and Topology of Membrane Proteins

Chapter 4

Using 2D Crystals to Analyze the Structure of Membrane Proteins

Ian Collinson, Janet Vonck, and Dilem Hizlan

Abstract

Electron crystallography is a powerful technique for studying the structure and function of membrane proteins, not only in the ground state, but also in active conformations. When combined with high-resolution structures obtained by X-ray crystallography, electron crystallography can provide insights into the mechanism of the protein. In this chapter we discuss obtaining a three-dimensional map of membrane proteins by electron crystallography and how to combine these maps with atomic resolution models in order to study the function of membrane proteins. We argue that this approach is particularly powerful as it combines the high resolution attainable by X-ray crystallography with the visualization of the subject in the near-native environment of the membrane, by electron cryo-microscopy. This point has been illustrated by the analysis of the protein translocation complex SecYEG.

Key words Electron crystallography, Cryo-electron microscopy, 2D crystals, SecYEG, Membrane dynamics, Conformational changes, Active state, Native structure, Three-dimensional maps

1 Introduction

The past 20 years have seen an explosion in the number of available membrane protein structures. It can no longer be said that our knowledge of membrane proteins lags behind their soluble counterparts. The progress made in our understanding of key biological processes such as ATP synthesis, respiration, photosynthesis, transport, and signaling is staggering. But even if critical information is available on the architecture of a particular membrane protein, there often remain outstanding questions on its molecular mechanism. The problem stems from the fact that the atomic snap-shots produced by X-ray diffraction analysis do not report on dynamic processes. Therefore, the elucidation of the dynamic molecular mechanism underlying membrane localized biological systems provides us with a new challenge. ATP synthase, the respiratory complexes, ion channels, and transporters all exhibit major energy-dependent conformational changes that have yet to be fully described.

Doron Rapaport and Johannes M. Herrmann (eds.), *Membrane Biogenesis: Methods and Protocols*, Methods in Molecular Biology, vol. 1033, DOI 10.1007/978-1-62703-487-6_4, © Springer Science+Business Media, LLC 2013

A wide range of biochemical and biophysical methods are available for the analysis of these dynamic reactions, such as chemical cross-linking, fluorescence, electron spin resonance (ESR). The results based on the dynamics and kinetics of the system provide invaluable information to be reconciled with the structure. Ideally, a series of high-resolution structures of defined intermediates of the reaction cycle of a given membrane bound machinery would be additionally required to understand the dynamic molecular mechanism. However, this is not usually feasible. The main problem stems from the fact that high-resolution structures of membrane proteins usually require the production of three-dimensional (3D) crystals for X-ray diffraction. The growth of such crystals often requires high concentrations of detergents and other destabilizing conditions (high salt, high or low pH, divalent metal ions, heavy metals, etc.). Therefore, when it is at all possible the structures usually only resolve the ground state or inactive forms of the membrane protein. In the absence of the stabilizing environment of the membrane, it is exceptionally difficult to obtain structural information of the active state, or illuminating intermediates of the active state, due to the nonnative environment of a 3D-crystal.

The answer lies in the analysis of samples in the native environment of the membrane, often exploited in the biochemical and functional analysis of a given membrane system. Biophysical approaches require the reincorporation of the purified material back into the membrane for subsequent analysis by, for example, solid state NMR, ESR, fluorescence (and Förster resonance energy transfer), which all potentially provide valuable information of the active process. Cryo-electron microscopy (cryo-EM) provides the means to directly visualize membrane proteins in their native environment and thus the possibility to image the protein in an activated state. Electron microscopy has been used now for many years in the analysis of membrane proteins. Indeed, it provided the very first evidence that they could be composed of transmembrane (TM) α-helices. Henderson and Unwin's structure of bacteriorhodopsin was seminal, showing the structure of a 7-helix bundle in the membrane [1] to (eventually) atomic resolution [2]. 3D structures of photocycle intermediates were also produced by electron crystallography [3, 4]. Atomic resolution structures were subsequently produced for several aquaporins and for light harvesting complex II from plants [5–9]. For many other membrane proteins, it has been possible to calculate structures to an intermediate resolution of 6–10 Å, sufficient to distinguish elements of secondary structure, such as TM α-helices, but insufficient to resolve amino-acid side chains, required for a molecular understanding of the system. This is in direct contrast with the high-resolution structure of nonnative 3D crystals.

Thankfully, structures determined by the X-ray crystallography and electron microscopy usually look the same. Therefore, the combination of the two provides us with the means to determine the basic structure of a membrane protein by the former, and then go on to look at interesting intermediates or substrate complexes that could only be produced and stabilized in the native conditions of the bilayer. This chapter describes a series of protocols as an exemplar for the analysis of dynamic membrane protein machineries by combining the complementary powers of electron microscopy and X-ray crystallography.

1.1 The Dynamic Translocon

Proteins destined for secretion, membrane protein insertion, or for organellar import carry targeting information for recognition by the corresponding translocation machinery. The so-called general secretory or sec pathway is responsible for the bulk of secretion and membrane protein insertion in all cells. We understand most about this pathway due to knowledge of its structure. The X-ray structure revealed the architecture of the inactive state [10], which provides several interesting possibilities for the mechanism of protein translocation. Clearly, the transport of a polypeptide across the membrane requires major conformational changes in the initiation, translocation, and termination phases of the reaction, which are not easily addressed by high-resolution structures determined in detergent solution. Such large conformational changes can potentially be mapped by medium resolution information attainable by cryo-EM. This in turn can be used to re-model the high-resolution X-ray structure, providing quasi-atomic detail of an activated conformation.

The bacterial (*E. coli*) protein channel complex SecYEG was first visualized by 2D-crystallography to 8 Å resolution, revealing all 15 TM segments [11]. The results were, however, insufficient to assign all of the helices or visualize amino-acid side chains. An X-ray structure of the archaeal homologues revealed the same structure at atomic resolution [10]. Recently, we have been able to produce 2D-crystals of the *E. coli* complex bound and activated by a mimic of the substrate pre-protein [12]. This complex could only be made in the membrane and could not have been produced or resolved in detergent solution. Therefore, the only means to examine its structure was by cryo-EM of 2D-crystals. The X-ray structure has been fitted and remodeled into the medium resolution maps of the complex determined with and without the pre-protein. The resultant maps visualize the bound substrate and a conformational change believed to be important in the activation process of the transport complex.

2 Materials

1. Chelating sepharose fast flow was acquired from GE Life Sciences, and used after charging with Ni^{2+}.
2. Superdex 200 High-Resolution (gel filtration) and Q-sepharose High-Performance (anion exchange) media was also obtained from GE Life Sciences.

3. Dodecyl-β-D-maltoside detergent (1 % w/v) (Glycon) was used to extract the membranes and $C_{12}E_9$ (Sigma) was used during the purification (0.1 % w/v during Ni-chelation chromatography and 0.05 % w/v during gel filtration and ion exchange chromatography).
4. *E. coli*-derived lipids phosphatidyl-ethanolamine (PE) were used in the crystallization, also dissolved in $C_{12}E_9$ (0.8 % w/v).
5. Dialysis membranes with a 12–14 k molecular weight cut-off (mwco) were produced by Spectra/Por.
6. Synthetic pre-protein mimics were chemically synthesized. The lamB signal sequence plus the first β-strand of the mature protein: MMITLRKLPLAVAVAAGVMSAQAYA-VDFHG-YARSGIGWTG (SS-β_1).
7. Dialysis buffer (20 mM Tris–Cl pH 8, 130 mM NaCl, 1 mM NaN_3, 1 mM EDTA).
8. For negative staining, 1 % w/v uranyl acetate dissolved in water.
9. For preparing cryo samples by back-injection method, as sugar solution, 4 % w/v trehalose is dissolved in water.
10. JEOL 3000 SFF electron microscope is used for data collection.
11. Kodak SO-163 photographic film.
12. Programs used: 2dx program [13, 14], MRC suite of programs [2, 15–17], CCP4 programs [18], COOT [19], and PYMOL (The PyMOL Molecular Graphics System, Version 1.3.x, Schrödinger, LLC).

3 Methods

3.1 Growth of 2D Crystals of SecYEG in the Presence and Absence of a Pre-protein Mimic

The SecYEG complex was purified by subsequent steps of nickel chelating, gel filtration, and ion exchange chromatography [20]. The final product in $C_{12}E_9$ detergent (0.1 % w/v) solution was mixed with *E. coli*-derived lipids phosphatidyl-ethanolamine (PE) also dissolved in $C_{12}E_9$ (0.8 % w/v). The ratio of lipid to protein is critical for the formation of 2D crystals and usually needs to be screened in any given experiment. Typical ratios of lipid:protein (w:w) of 0.05–0.25 may be expected to yield crystal patches within the membrane. The crystals form following the removal of detergent by extensive dialysis using membranes with a 12–14 k mwco [20].

This procedure had to be adapted for the co-crystallization of the SecYEG complex with the pre-protein. The pre-protein mimics were made synthetically and constituted the 25 amino acids of the signal sequence (SS) portion of LamB, an outer membrane protein of *E. coli*, or the signal sequence plus 15 amino acids of the N-terminus of the mature protein (SSS). A tenfold molar excess of the peptide was added from an aqueous solution to the mixture of

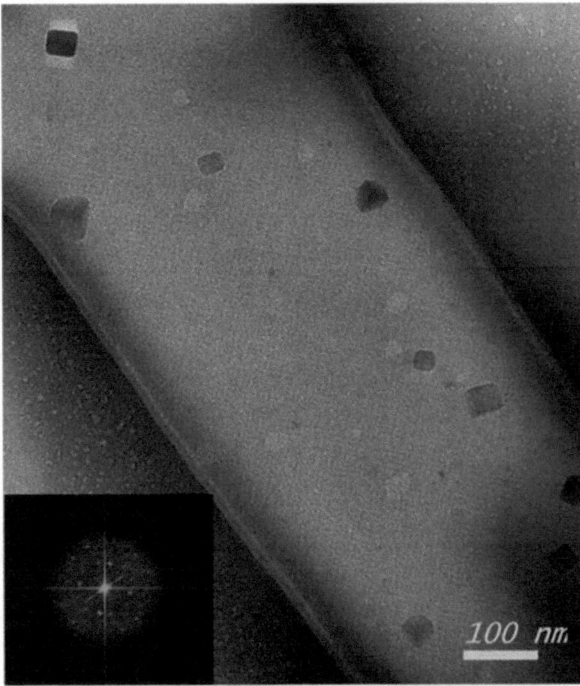

Fig. 1 Image of negatively stained SecYEG-SSS 2D crystal. *Inset*: Diffraction pattern. The SecYEG-SSS crystals are double membrane tubular crystals, as indicated by the *thick edges*

SecYEG (3.4 μM) and lipid. To prevent the loss of the peptide (mw ~ 4.5 k) during the dialysis of the detergent, the peptide was also included in the dialysis buffer (5 μM). The dialysis proceeded for several days at 23 °C, with multiple changes of the outside buffer, followed by storage at 4 °C.

3.2 Quality Assessment of 2D Crystals

Although cryo samples are used for high-resolution data collection, the growth and quality of crystals are checked by negative staining (*see* **Note 1**). With some crystal types, it is possible to observe an ordered structure on charge-coupled device (CCD) images of negatively stained crystals (Fig. 1). However, the best way to investigate the quality of the crystals is to assess the diffraction pattern, either by calculating the Fourier transform (FFT) of a CCD image or by taking images of crystalline areas on photographic film and checking the diffraction pattern with an optical diffractometer. Good quality crystals show evenly distributed sharp spots in all directions, and the higher resolution information is present, the more spots will be visible further out in the diffraction pattern. The resolution limit of negative staining is 15–20 Å, and if spots are observed to this limit, the 2D crystals will usually be ordered to much higher resolution. At this stage, one should prepare a cryo-specimen and validate the presence and the quality of the spots by cryo-EM.

3.3 Preparation of Cryo Samples

Cryo samples of 2D crystals can be prepared in two ways. The first method is by vitrification in the same way as preparing cryo samples for single particle analysis: the sample solution is incubated on a glow-discharged carbon-coated grid for about a minute and washed if necessary. After blotting the excess solution, without letting the sample dry, the sample is quickly frozen in liquid ethane by a plunge-freezing device, then kept frozen in liquid nitrogen. The more common method uses a sugar as cryo-protectant, and the back-injection method is followed in most sample preparations in electron crystallography [21, 22]. In order to prepare the sample, a small piece of carbon film is floated from the mica on a sugar solution, e.g., trehalose or glucose, and picked up on an EM grid using tweezers. After the removal of excess solution, a small drop of the specimen is placed on the grid side of the sandwich, mixed with the remaining sugar and incubated for about a minute. After the removal of excess solution by putting the sandwich, grid side down, on a piece of filter paper, the sample is frozen in liquid nitrogen. This method ensures optimal flatness of the crystals on the support film. Lack of flatness will cause the overlap in the diffraction pattern of data at slightly different tilt angles and introduce blurring of the diffraction spots in the direction perpendicular to the tilt axis [22].

3.4 Electron Microscopy and Scanning

Ideally, data should be collected on a dedicated cryo-EM with a field emission gun and an in-built cryo stage that allows the sample to be imaged for long periods of time without contamination or drift. We use a JEOL 3000 SFF (*see* **Note 2**). We routinely collect data by a spot scanning procedure to minimize beam-induced movement, especially for tilted specimens [23]. The optimal magnification (M) to collect data is determined by the expected resolution and the scan step (d) of the available scanner: the pixel size d/M should be no smaller than 3× the expected resolution. Usually for high-resolution 3D maps, images are taken at magnifications of 60,000–70,000× (*see* **Note 3**). With electron microscopes with fixed-angle specimen cartridges like the JEOL 3000 SFF, specimen tilts of 0°, 10°, 20°, 30°, 45°, and even 60° can be used. With microscopes with a dynamic tilt stage, tilt angles at different intervals can be chosen. As will be discussed in "Subheading 3.9 Merging Tilted Data Sets," with thicker specimens, collecting data at smaller tilt-angle intervals can be helpful with merging.

We collect data on photographic film (Kodak SO-163) (*see* **Note 4**). As explained before, the crystal quality is evaluated by optical diffraction of the negatives. Images with strong sharp spots in all directions should be selected; having spots only in one direction could indicate specimen movement or lack of flatness of the sample. Before scanning the crystalline region, experimenters should decide on the size of the scan area. For a given area, crystals with a small unit cell will give stronger spots than crystals with a large unit cell, and consequently fewer unit cells contributing to the diffraction pattern [16]. The scanned area should include as much as possible of the crystal, and care should be taken not to include too much of the noncrystalline or distorted regions that will reduce the quality of the spots. Before scanning, it is advisable to identify the center and size of the crystalline area by optical diffraction and mark it on the negative in order to place the center of the crystal in the center of the scanned area. The MRC programs require square regions, and 4,000 × 4,000 or 6,000 × 6,000 pixels size is commonly used for scanning (*see* **Note 5**).

3.5 Processing of Individual Images

For each image, the amplitudes and phases of all Fourier components, and the tilt geometry of the crystal that defines the orientation the crystal in three dimensions, have to be determined. The data has to be corrected for lattice distortions (unbending) and for the contrast transfer function (CTF) of the EM. The MRC suite of programs has been developed over many years to process images [2, 15–17]; lately electron crystallographers are using the 2dx software package that is based on these MRC programs, but has a user-friendly graphical interface and can provide different levels of automation [13, 14] (*see* **Note 6**). The general steps of processing an image are as follows:

- Calculation of the Fourier transform (FFT) of the image
- Determination of defocus and tilt geometry calculation based on the defocus gradient
- Indexing of the diffraction pattern and determination of the lattice dimensions
- Tilt geometry calculation according to the change in lattice dimensions
- Preparation of the spot list for the reference area
- Identification and correction of the lattice distortions (iteratively processed in Unbending I and Unbending II steps)
- Extracting phases and amplitudes
- Correction of amplitudes and phases for the CTF
- Calculation of the projection map

Due to lack of space here, we will discuss each of these steps briefly; for more detailed information the reader can refer to the 2dx manuals.

By averaging 2 × 2 pixels of the image, a down-sampled copy of the image and its diffraction pattern (FFT) is calculated (Fig. 2a). These files will be used for many processes in order to save disk space. In the following step, the defocus and astigmatism is calculated. In order to control whether the defocus values are calculated correctly or not, visualizing a periodogram can give a good view of the Thon rings and the placement of the CTF zeros according to the calculated defocus values (Fig. 2b, c).

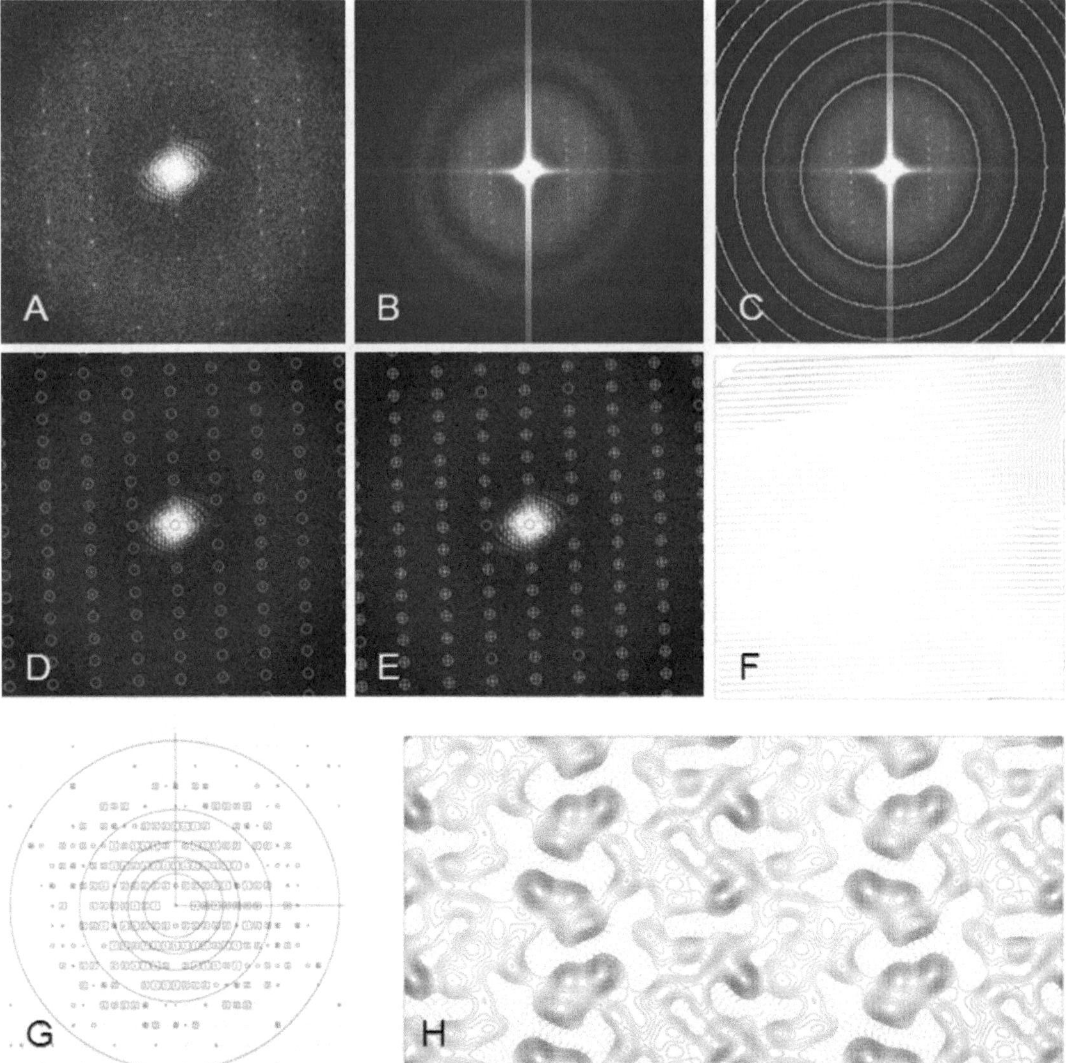

Fig. 2 Processing of an image. (**a**) Diffraction pattern of an image. (**b**) Calculated periodogram showing Thon rings. (**c**) CTF zeros superimposed on Thon rings in order to check the correctness of defocus values. (**d**) *Boxes* indicate the lattice positions. (**e**) *Plus* signs indicate the spots used for calculating the reference. (**f**) Distortion plot that will be used for unbending lattice distortions. (**g**) IQ plot of one data set, *numbers* in boxes indicate the quality of each spot according to their signal-to-noise ratio. (**h**) Calculated projection map

Next, the unit cell dimensions are calculated. In 2dx, the unit cell dimensions can be determined automatically with two different algorithms. If there is information available about the lattice dimensions, the Findlattice, if not the Getlattice, algorithm can be used. In either case one should check whether the defined lattice is the correct one and the spots are indeed in the center of the boxes that show the lattice (Fig. 2d). After the first calculation, the lattice can be refined manually so that sharp and bright spots sit in the center of circles and/or boxes. The values will be more accurate when high-resolution spots are used for the calculation. The "Evaluate Lattice" step in 2dx gives the calculated lattice dimensions from the determined lattice, the theoretical magnification as well as the lattice error, which is calculated as the normalized root-mean-square deviation RMSDN indicating the distance of the peak locations from the chosen lattice [24].

With tilted images of up to 30° the tilt geometry is calculated from the defocus gradient across the image in the previous step. For tilts higher than 30°, the tilt geometry can be calculated accurately from the lattice dimensions.

The next steps are to prepare for the unbending procedures. Blurred spots, especially at high resolution, can be the result of lattice distortions. In order to correct for distortions a reference image is created using sharp, good quality spots of IQ 1–4 (*see* **Note 7**) that are generally present at rather low resolution. A list of such spots is prepared automatically by "Get Spotlist for Unbend I" and can be visualized by the 2dx program (Fig. 2e). Spots can be selected or deselected manually and the list can be saved for the upcoming unbending procedure.

Around the spots, a tight mask of 1 pixel radius is applied, and an inverse FFT creates a tightly filtered image with a perfect lattice. A small reference area is then boxed out as the best diffracting area of this filtered image and the diffraction pattern of this reference is calculated.

If the negative is scanned so that the best crystalline area is in the center of the scanned image, then the boxed reference area should be also in the center of the image. The reference area can have a size of 1/10th to 1/20th of the size of the image [16]. At the same time, the FFT is masked with a loose mask so that the filtered image will also contain the information about the lattice distortions. The radius of this loose mask can be chosen at a range of ~10–20 pixels [16]. Then the cross-correlation map between the reference and the loosely masked transform is calculated by multiplying the loosely masked transform with the complex conjugated Fourier transform of the reference and back-transforming to real space. Cross-correlation of this map with the auto-correlation map of the reference gives correlation peaks that show the location of the unit cells. The height of the peaks shows the amount of the correlation while the vectors show the deviations of the lattices from their positions if they were in a perfect lattice and will be used to unbend lattice distortions (Fig. 2f). Noncrystalline areas in the image will be clearly visible in the correlation maps. ManualMasking-CCmap can be used to define the coordinates around the crystalline area, so that in "Mask Crystal from Polygon" the noncrystalline area can be masked out to reduce noise. After obtaining the masked image, the unbending process should be started over.

Table 1
Statistics of the projection map. Statistics are evaluated in four resolution zones, with the number of spots in each resolution zone (Ndat) and the respective phase residuals (AVRGFOM)

No.	Resolution zone (Å)	Ndat	AVRGFOM (°)
1	∞–14.7	28	8.7
2	14.7–10.4	27	14.7
3	10.4–8.5	24	21.2
4	8.5–7.4	9	35.8

The overall error for 88 reflections was: 16.7°

After the diffraction pattern of the corrected image is calculated, it can be observed that previously diffuse spots get sharper after the unbending procedure. It is recommended that the procedure is repeated two or three times, using the corrected image obtained after each round as reference, until no further improvements occur.

After extracting the phases and amplitudes, the phases need to be corrected for the effects of the CTF. Following the correction, an APH file is obtained that contains the IQ value, phase, and amplitude for each spot, defined with (*h*,*k*) coordinates. For each data set, the spots are also plotted with either resolution rings or CTF zeros superimposed (Fig. 2g).

3.6 Calculation of the Projection Map

Before merging data sets for calculation of a projection map, the plane symmetry group of the crystal is determined. Symmetry can only be identified on non-tilted images, since as the sample gets tilted its projection loses its symmetry. Only the non-tilted data sets that show the highest possible symmetry should be used for merging.

The best data set is used as reference to bring the data sets to a common phase origin before merging. As reference, a data set with evenly distributed, good quality spots to high resolution and the smallest phase error for the determined symmetry group should be used. For phase origin refinement high signal-to-noise ratio spots of IQ 1–4 to a resolution of 8–10 Å can be used. In the first run, step and box-size (for example 3° and 121, respectively) should be chosen so that the full 360° is covered for phase origin search. The calculated phase origin can be inserted as input and two to three rounds with smaller step-sizes (for example 1° or 0.5°) covering a smaller area for fine-tuning of the phase origin should be done (*see* **Note 8**).

After refining the phase origin, the data sets are merged and the spots are averaged. To determine the resolution cut-off the merging statistics can be used. In the example given in Table 1, the phase statistics are examined in four different resolution ranges. When deciding on the resolution cut-off, it is important that in each resolution band there is a significant number of spots and the phase errors are acceptable (random phases are 45° when there is twofold symmetry or 90° when there is not). Also in order to correct for the resolution-dependent fall-off of the image-derived amplitudes, a B-factor can be applied.

After the determination of the resolution cut-off the projection map can be calculated (Fig. 2h). Also a plot showing the quality of each spot will be produced; as before each spot will be represented by boxes and the numbers inside will indicate the phase errors for that particular spot (1 <8°, 2 <14°, 3 <20°, 4 <30°, 5 <40°, 6 <50°, 7 <70°, 8 <90°) (2dx manual).

In addition for crystals with twofold symmetry, phases will be rounded to 0° or 180° and a Figure of Merit plot will be calculated that evaluates the deviation of the actual phase of each spot from the symmetry constrained phase.

3.7 Obtaining Tilted Images

Although the sample preparation techniques are mainly the same, one needs to pay special attention to the following in order to make the image processing and merging more convenient in the future steps.

Extra care should be given to prepare a flat grid that is free of bends so that the tilt axis and angle will be the same throughout the grid area. The carbon film should be free of wrinkles and the crystals flat, which can be most easily achieved by using the back-injection method as described above. If an EM without a eucentric tilt holder is used, the defocus will change perpendicular to the tilt axis, so it is convenient to collect images at grid squares that are neighboring each other along the tilt axis, before moving to the next row of grid squares in the direction perpendicular to the tilt axis.

3.8 Processing Tilted Images

For processing of tilted images the same procedure as for untilted images is followed, but in addition the tilt geometry has to be determined, which will be discussed here.

Once the negatives are checked by optical diffraction, it is advisable to mark the direction of the tilt axis and the high and low defocus sides of the negatives. Since the tilt geometry needs to be carefully identified, it is important to validate whether the tilt geometry identified on negatives and the ones calculated by the program are in agreement.

In common practice, images are collected at tilt angles of 10°, 20°, 30°, 45° and if possible 60°. The user can decide to collect data sets at tilt angles between these values, which can be needed, especially with thick specimens.

For tilted images of less than 30°, the tilt geometry is calculated from the defocus gradient along the image, while for higher tilt angles the change in lattice dimensions is used for this purpose.

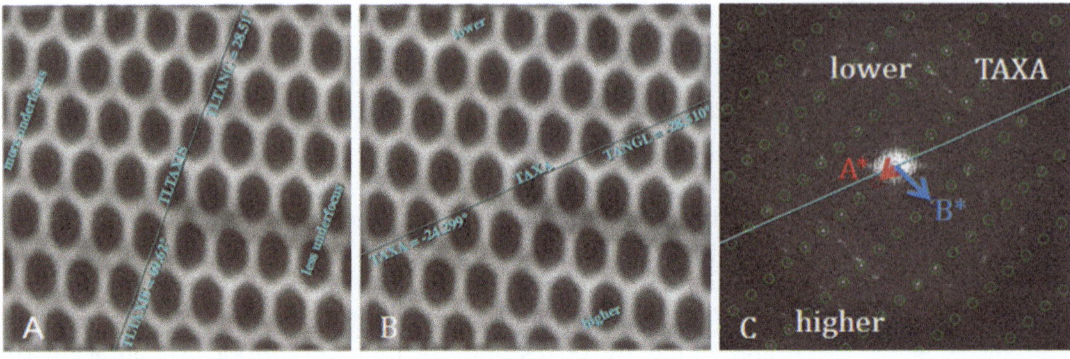

Fig. 3 Tilt geometry of tilted images. Spot-scan images are scanned 4,000 × 4,000 pixels according to the location of the crystalline area. (**a**) TLTAXIS and TLTANG refer to the tilt geometry of the sample in the microscope. Along the tilt axis the defocus is constant, while the sample area higher in the microscope has less defocus, while the area lower in the microscope has more defocus. (**b**) TAXA and TANGL values for the crystal shown on the image. (**c**) The crystal geometry depends on the orientation of the crystal axes on the image, TAXA is shown in respect to the *a** (red) and *b** (blue) vectors of the lattice

2dx calculates tilt geometries with both methods and uses the results of either method depending on the tilt angle. However, the tilt geometries calculated with either method should not be significantly different.

The tilt geometry is defined with the following parameters:

TLTAXIS: the direction of the tilt axis on the image. This can be manually identified by observing the negative by optical diffraction as the axis where the defocus remains the same. Perpendicular to the tilt axis one can observe the defocus gradient.

TLTANGL: The angle between the *X* axis of the image and the tilt axis.

TLTAXA: on the image, with the projection of 3D vectors, the vector from the tilt axis to A*, the direction of the first crystal axis.

TAXA: the angle between the tilt axis and the A* vector on the crystal; measured from the tilt axis in the direction where A* to B* is positive.

TANGL: the tilt angle for the crystal, the amount is equal to TLTANGL, however the sign is determined by the orientation of the crystal and the indexing (*see* **Note 9**) (2dx manual and scripts).

In Fig. 3, the tilt geometry of an image of SecYEG-SSS sample taken at 30° is shown. The tilt axis and angle of the image is shown in Fig. 3a with the low and high defocused areas of the image. As discussed above, TAXA and TANGL values refer to the crystal (Fig. 3b), the overlap of TAXA and diffraction pattern can be seen in Fig. 3c. Determination of the sign of TANGL is discussed in **Note 9** using the same image's tilt geometry as an example.

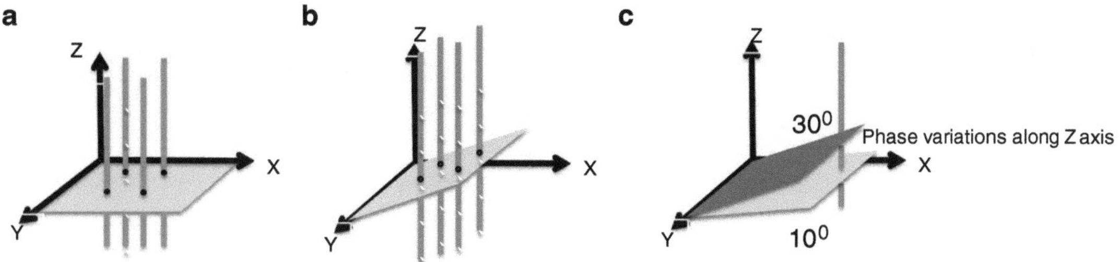

Fig. 4 The 3D FFT of a 2D crystal is represented as continuous lattice lines along the *Z* axis. (**a**) The diffraction pattern of an untilted image is a central section in the *X*–*Y* plane and consists of spots in the *X*–*Y* plane. (**b**) The diffraction pattern of a tilted image is a central section cutting the lattice lines at different *Z* values. In order to build a 3D map, tilted data sets at different angles and orientations need to be collected to fill the Fourier space. (**c**) Two data sets, one tilted 10°, the other 30° along *Y*. The two FFTs have a common line of data along the line of intersection, in this case the *Y* axis. When it is difficult to merge two tilted data sets due to rapid phase variations along the *Z* axis, it is advisable to collect data at tilts in between, in this case for example at 20°, in order to have more common data points

3.9 Merging Tilted Data Sets

The information in a 3D Fourier transform of a 2D crystal is present as continuous lattice lines in the *Z* dimension which represent themselves as spots on the FFT of an image, which is a central section of 3D Fourier space. As more tilted data sets are included, information along the *Z* axis is acquired to form the 3D map (Fig. 4a, b).

Once we have calculated the projection map, we can use this map as a reference for phase origin refinement of the tilted data. Starting with the non-tilted data sets, data sets at lower tilt values are searched for common phase origin and the tilt geometry of each data set is refined and finally merged to a 3D data set (*see* **Note 10**). Once a merged 3D data set is calculated, as done for the calculation of projection maps, it is used as the new reference for fine-tuning of the phase origin and tilt geometries of the image data sets. Then the next higher tilt data sets are included into the process of alignment, and this cycle continues until all the tilted data sets are aligned to a common phase origin and the tilt geometries are refined. For thick specimens, the phase variations along the *Z* axis are quite dramatic compared to thin specimens (*see* **Note 11**). For that reason it is recommended to slowly increase the tilt angles of the data sets. As the tilt angles increase, it might be difficult to merge data sets when there is not enough common data. If this occurs, collecting a few data at slightly lower tilts will help in merging the higher tilted data sets (*see* **Note 12**) (Fig. 4c). Data sets with phase residuals of up to 35–40° can be included. Due to lack of enough spots in certain resolution zones or to CTF errors, phase residuals can be high for high-resolution spots. For this reason, apart from the overall phase residual, the phase residuals in each resolution zone should be checked for each data set and if necessary different resolution

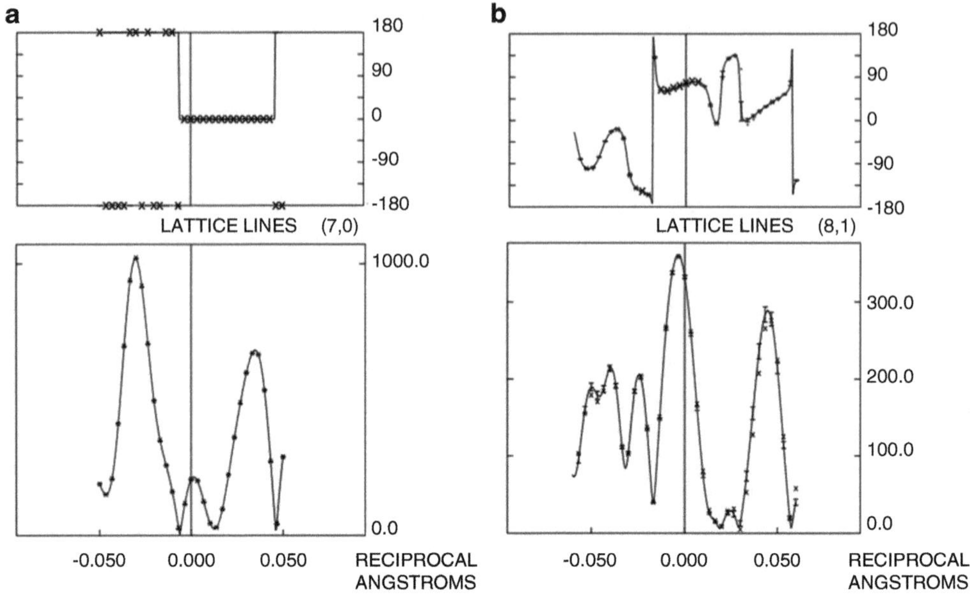

Fig. 5 Examples of lattice lines. Phases (*top*) and amplitudes (*bottom*) of lattice lines (7,0) in (**a**) and (8,1) in (**b**). The imposed twofold symmetry restricts the phases to 0° or 180° for (7,0). In thinner samples than the SecYEG crystal shown here, the changes in phases along the lattice lines are less dramatic

cut-offs can be used for each image. Next to phase origin and tilt geometry refinement, beam tilt refinement needs to be performed for data better than about 5 Å resolution [16].

After merging, the data needs to be weighted before preparing for lattice line fitting and the lattice lines are prepared by MRC programs latlinprescal and latlinek, respectively (Fig. 5). Unwanted spots can be removed by the prepmklcf program depending on their figure of merit (*see* **Note 13**). Spots that are defined as (*h*,*k*,*z*) until this point need to be converted so that output file is in the standard crystallographic format and has H, K, L, A, PHI, FOM values which can be used by the CCP4 programs [18] (*see* **Note 14**). The output is ready for calculation of the map, which can be visualized by programs like COOT [19] or PYMOL (The PyMOL Molecular Graphics System, Version 1.3.x, Schrödinger, LLC).

Maps calculated by electron crystallography have a lower resolution perpendicular to the crystal plane, due to the missing cone in Fourier space caused by a lack of high tilt data sets (in most cases data at more than 45° tilt are missing). The resolution on the *Z* axis can be calculated from the point spread function [25].

3.10 Combination of 3D Maps with X-Ray Structures

The typical resolution of 3D maps calculated by electron crystallography is routinely ~7–10 Å [11, 26, 27], which is much lower than that obtained by X-ray crystallography. However since 2D crystals are usually made in the presence of lipids, which reconstitute the native structure of the bilayer, docking in of X-ray

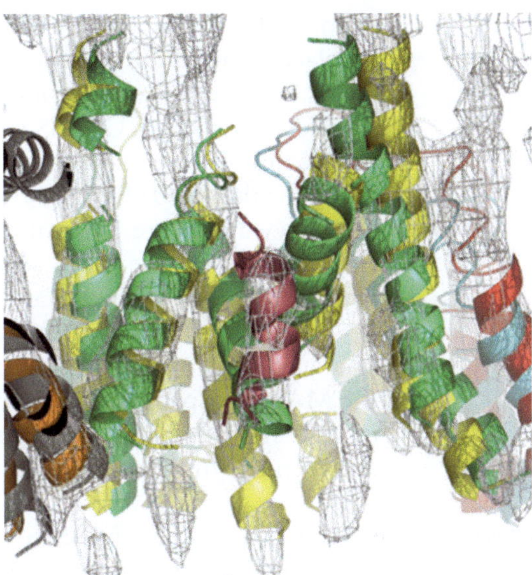

Fig. 6 Detailed side view of map density of the SecYEG complex bound to the pre-protein mimic, looking into the lateral gate from the outside. Fitted *E. coli* homology model of the unbound (SecY, E, and G are shown in *green*, *dark gray*, *cyan*, respectively) and bound complexes (SecY, E, and G are shown in *yellow*, *orange*, *red*, respectively) are superimposed on each other for comparison. Pre-protein mimic is shown in *dark pink*. The map is contoured at 1.5 SD and shown in *gray*

structures can provide information about the natural state of the protein, e.g., the dimeric nature of proteins observed with NhaA and SecYEG [11, 26].

If the crystal structure of the protein in question is available, this can be docked in the 3D map. If there is no crystal structure, a homology model can also be used. For example, in 2004 the X-ray structure of the protein conducting channel of *Methanococcus jannaschii* was solved [10]. Based on that structure a model of *Escherichia coli* SecYEG was built [28] based on the earlier published EM map [11]. For our work with SecYEG-SSS we have also inserted polyalanine helices into the model in order to model the missing additional transmembrane helices present in the *E. coli* channel. COOT was used in order to merge the docked model and helices into one pdb file (Fig. 6) [12].

3.11 Applying Noncrystallographic Symmetry

When there is noncrystallographic symmetry (NCS) present in 2D crystals, it is possible to apply this symmetry to the asymmetric units [11, 27]. For example in the case of SecYEG-SSS 2D crystals, two monomers of a dimer are related to each other by twofold NCS. In order to apply NCS, the operators, which can be calculated by PROFESS from the CCP4 Program Suite (or FINDNCS from CCP4), have to be determined precisely. Firstly, with the help of the COOT program, heavy atoms are placed on the same residue of different asymmetric units for example a heavy atom can be placed on the 50th residue of each asymmetric unit. FINDNCS needs at least three sites for each NCS asymmetric unit; with our calculations with PROFESS we have used eight sites in each asymmetric unit. These atoms can be written to a pdb file that is used as input for the PROFESS program. It is important that almost all of the heavy atoms are being used for the NCS operator calculation; these results will be given at the top of the list and will give the rotation (in Euler or polar angles) and translation vectors.

The results can be confirmed for example by applying the NCS operators to dimers or to 3D maps of 2D-crystals with the help of the MAPROT program of CCP4. An application of this is explained in the next section for calculating the correlation coefficient between the monomers of a dimer.

3.12 Calculation of Correlation Coefficient of Different Asymmetric Units

In the case of SecYEG-SSS crystals, 2D crystals contained two asymmetric units: two monomers of a dimer. In order to calculate the correlation coefficient between the monomers, firstly the two monomers needed to be isolated and superimposed on one another. The programs used for these calculations are available in the CCP4 Program Suite.

An X-ray structure or model is placed into a defined cell by PDBSET in P1, followed by creating a mask around the structure by NCSMASK. The map from which the monomer will be cut out is also placed into the same cell as the structure or model by MAPMASK. The created mask is used to cut one monomer from the map by MAPROT using the mode "From" one monomer at a time. First without rotation/translation one monomer is cut out into a map file, then using the twofold symmetry axis the second monomer is cut out, rotated and translated so that the two monomers are superimposed. The cut out densities of each monomer are placed into separate map files with the same cell dimensions, and by this step the densities are ready to be compared by the OVERLAPMAP program. OVERLAPMAP in the "corr section" mode correlates the maps section by section; when the process is written to a log file the user can obtain the correlation coefficient of each section and a total correlation coefficient of the two densities.

3.13 Concluding Remarks

The eventual visualization and fitting of the atomic models to the medium resolution structures of the SecYEG complex determined with and without the pre-protein mimic have allowed us to interpret the early stages of the active translocation cycle in quasi-molecular detail. Moreover, they have been instrumental in the formulation of a new hypothesis for protein transport, whereby the signal sequence unlocks the channel prior to the channel opening, intercalation of the mature part of the pre-protein, and translocation (Fig. 6).

4 Notes

1. We usually use uranyl acetate (1 % w/v) for negative staining.
2. Currently available microscopes include the FEI Tecnai G2 Polara, the FEI Titan Krios, and the JEOL JEM 3200 FSC.
3. The commonly used Zeiss scanner has a step size of 7 μm, so that a magnification of 70,000× allows a resolution up to 3 Å.
4. If film is not available, a CCD camera or direct electron detector can be used. However, for 2D crystal images the imaging area is critical to gain a high enough signal from sufficient unit cells, and for high-resolution work at least 4,000 × 4,000 pixels are needed. Extreme care should be taken to center the crystalline area on the detector. With these cameras, the image is immediately available, and the quality can be assessed by calculating an FFT, which provides a large advantage. Also the time-consuming and expensive steps of developing and digitizing the film are not necessary.
5. It is important to scan the negatives always in the same orientation, in order to avoid problems with determination of tilt geometry and handedness.
6. The 2dx software and documentation can be downloaded from http://www.2dx.unibas.ch/.
7. The IQ values are a measure for the quality of the spot in terms of their signal-to-noise ratio: IQ = 7(A/B), where A is the background-corrected amplitude and B the background near the spot, so a spot of IQ 1 is at least seven times above the background while an IQ 7 spot has a signal-to-noise ratio of 1 and IQ 8 is below the background [2].
8. If the phase origin can't be found, it is possible that the crystal was indexed wrong. The option "reverse *h* and *k*" can be tried, especially if the axes are very similar or for highly tilted crystals. With certain low symmetries, like p1 or $p12_1$, it can be necessary to use the "Flip around A axis" or "Sign exchange" (SGNXCH) option to determine the phase origin.
9. The sign of TANGL is calculated by the formula: "A is above" × "SignTLTAXA" × "Hand." The tilt geometry given in the example in Fig. 3: TLTAXIS : 69.62, TLTANG : 28.51, TLTAXA : −27.19, TAXA : −24.3. In the given example; HAND = 1 (Handedness of the lattice A* to B*, right handed = 1, left handed = −1); A is above = 1 (A is on the higher side, so central line of the FFT on the *X*–*Y* plane cuts the *Z* axis at negative values for the spot (1,1)); SignTLTAXA = −1; As a result the sign of TANGL is = 1 × (−1) × 1 = (−).
10. As higher tilt data sets are included, there can be changes in the tilt geometry values of lower tilt data sets. As all data will be

refined against each other, variations are to be expected. However, sometimes certain data sets show drastic changes. It is advisable to keep a record of tilt geometry values at each round in order to identify these data sets and exclude them if necessary.

11. Another important point about merging tilted data sets from a thick specimen is to use a rather small Z^* window, to allow for the more rapid phase changes along the lattice lines.
12. With SecYEG-SSS crystals, merging tilted data became difficult due to the thickness of the specimen. Each data set was carefully checked for the correctness of tilt geometry and defocus values. However, even slight errors in tilt geometry caused difficulty in merging.

    At the same time each data set was checked for whether the SGNXCH option was to be used or not. For that reason especially for higher tilted data sets, we have searched plus/minus 5–10° of the defined tilt geometry for the best fit of the data set to merged data.
13. Lattice line plots should be checked for dramatic phase and amplitude changes. When the FOM values are also poor, the user can reject these unreliable spots.
14. In order to convert to CCP4 format, the continuous lattice line data on z has to be converted to discrete spots by including a unit cell dimension l. This is usually chosen to be about 2× the expected crystal thickness; it is important that it is not smaller than the real thickness.

Acknowledgments

We would like to thank Prof. Dr. Werner Kühlbrandt for his support and Dr. Özkan Yildiz for his help with computer programs.

References

1. Henderson R, Unwin PNT (1975) Three-dimensional model of purple membrane obtained by electron microscopy. Nature 257: 28–32
2. Henderson R, Baldwin JM, Ceska TA, Zemlin F, Beckmann E, Downing KH (1990) A model for the structure of bacteriorhodopsin based on high resolution electron cryomicroscopy. J Mol Biol 213:899–929
3. Subramaniam S, Henderson R (2000) Molecular mechanism of vectorial proton translocation by bacteriorhodopsin. Nature 406:653–657
4. Vonck J (2000) Structure of the bacteriorhodopsin mutant F219L N-intermediate revealed by electron crystallography. EMBO J 19: 2152–2160
5. Kühlbrandt W, Wang DN, Fujiyoshi Y (1994) Atomic model of plant light-harvesting complex by electron crystallography. Nature 367:614–621
6. Cheng A, van Hoek AN, Yeager M, Verkman AS, Mitra AK (1997) Three-dimensional organization of a human water channel. Nature 387:627–630
7. Li H, Lee S, Jap BK (1997) Molecular design of aquaporin-1 water channel as revealed by electron crystallography. Nat Struct Biol 4:263–265
8. Walz T, Hirai T, Murata K, Heymann JB, Mitsuoka K, Fujiyoshi Y, Smith BL, Agre P, Engel A (1997) The three-dimensional structure of aquaporin-1. Nature 387:624–627

9. Gonen T, Cheng Y, Sliz P, Hiroaki Y, Fujiyoshi Y, Harrison SC, Walz T (2005) Lipid-protein interactions in double-layered two-dimensional AQP0 crystals. Nature 438:633–638
10. van den Berg B, Clemons WM Jr, Collinson I, Modis Y, Hartmann E, Harrison SC, Rapoport TA (2004) X-ray structure of a protein-conducting channel. Nature 427:36–44
11. Breyton C, Haase W, Rapoport TA, Kühlbrandt W, Collinson I (2002) Three-dimensional structure of the bacterial protein-translocation complex SecYEG. Nature 418:662–664
12. Hizlan D, Robson A, Whitehouse S, Gold VA, Vonck J, Mills D, Kühlbrandt W, Collinson I (2012) Structure of the SecY complex unlocked by a preprotein mimic. Cell Rep 1: 21–28
13. Gipson B, Zeng X, Stahlberg H (2007) 2dx_merge: Data management and merging for 2D crystal images. J Struct Biol 160:375–384
14. Gipson B, Zeng X, Zhang Z, Stahlberg H (2007) 2dx—User-friendly image processing for 2D crystals. J Struct Biol 157:64–72
15. Amos LA, Henderson R, Unwin PN (1982) Three-dimensional structure determination by electron microscopy of two-dimensional crystals. Prog Biophys Mol Biol 39:183–231
16. Henderson R, Baldwin JM, Downing K, Lepault J, Zemlin F (1986) Structure of purple membrane from *Halobacterium halobium*: recording, measurement and evaluation of electron micrographs at 3.5Å resolution. Ultramicroscopy 19:147–178
17. Crowther RA, Henderson R, Smith JM (1996) MRC image processing programs. J Struct Biol 116:9–16
18. Collaborative Computational Project, N (1994) The CCP4 suite: programs for protein crystallography. Acta Crystallogr D Biol Crystallogr 50:760–763
19. Emsley P, Cowtan K (2004) Coot: model-building tools for molecular graphics. Acta Crystallogr D Biol Crystallogr 60:2126–2132
20. Collinson I, Breyton C, Duong F, Tziatzios C, Schubert D, Or E, Rapoport T, Kühlbrandt W (2001) Projection structure and oligomeric properties of a bacterial core protein translocase. EMBO J 20:2462–2471
21. Glaeser RM, Downing KH (1990) The "specimen flatness" problem in high-resolution electron crystallography of biological macromolecules. In: Peachy LD, Williams DB (eds) XIIth international congress for electron microscopy. San Francisco Press, Inc., Seattle, pp 98–99
22. Vonck J (2000) Parameters affecting specimen flatness of two-dimensional crystals for electron crystallography. Ultramicroscopy 85:123–129
23. Downing KH (1991) Spot-scan imaging in transmission electron microscopy. Science 251:53–59
24. Zeng X, Gipson B, Zheng ZY, Renault L, Stahlberg H (2007) Automatic lattice determination for two-dimensional crystal images. J Struct Biol 160:353–361
25. Unger VM, Schertler GFX (1995) Low resolution structure of bovine rhodopsin determined by electron cryo-microscopy. Biophys J 68:1776–1786
26. Williams KA (2000) Three-dimensional structure of the ion-coupled transport protein NhaA. Nature 403:112–115
27. Vonck J, Krug von Nidda T, Meier T, Matthey U, Mills DJ, Kühlbrandt W, Dimroth P (2002) Molecular architecture of the undecameric rotor of a bacterial Na^+-ATP synthase. J Mol Biol 321:307–316
28. Bostina M, Mohsin B, Kühlbrandt W, Collinson I (2005) Atomic model of the *E. coli* membrane-bound protein translocation complex SecYEG. J Mol Biol 352:1035–1043

Chapter 5

Crystallization of Membrane Proteins

Florian G. Müller and C. Roy D. Lancaster

Abstract

The crystallization of membrane proteins is an essential technique for the determination of atomic models of three-dimensional structures by X-ray crystallography. The compositions of solutions of purified membrane proteins are altered, so as to transiently induce supersaturation, a requirement for crystal nucleation and growth. The establishment of the precise optimal crystallization conditions has to be performed individually by a combination of systematic approaches and trial-and-error. These procedures have become more efficient due to the introduction of laboratory automation. Here we describe the crystallization of the dihaem-containing quinol:fumarate reductase (QFR) membrane protein complex and illustrate key factors important in the screening process.

Key words Atomic model, Detergents, Laboratory automation, Membrane protein crystals, Small-amphiphile concept, Supersaturation, Three-dimensional structure, X-ray crystallography

1 Introduction

A prerequisite for an atomic-level understanding of the mechanism of action of membrane proteins is the availability of accurately determined three-dimensional structures. The most productive technique, by far, in the determination of atomic models of membrane protein structure is X-ray crystallography. The method involves the production of the membrane protein of interest, either homologously in its natural source or heterologously in an optimal expression system, followed by its purification and crystallization. Although all three steps are far from trivial, the latter is particularly demanding, considering that the first membrane protein structure, that of the photosynthetic reaction center [1], was determined in 1985, more than a quarter of a century after that of the first soluble protein, myoglobin [2].

Protein crystallization [3] in general requires supersaturation, meaning that the protein concentration in a droplet is transiently higher than the protein's solubility [4]. The formation of crystallization nuclei requires high degrees of supersaturation,

Doron Rapaport and Johannes M. Herrmann (eds.), *Membrane Biogenesis: Methods and Protocols*, Methods in Molecular Biology, vol. 1033, DOI 10.1007/978-1-62703-487-6_5, © Springer Science+Business Media, LLC 2013

followed by crystal growth, requiring lower supersaturation degrees. Supersaturation can be achieved by reducing the solubility of the protein through varying various physicochemical parameters such as the pH value, the ionic strength, or the dielectric properties of the medium. This is achieved by gradually varying the buffer composition and/or the concentration of salts (e.g., ammonium sulfate, sodium chloride), high molecular weight organic polymers (e.g., polyethylenglycols (PEGs), or organic molecules (e.g., 2-Methyl-2,4-pentanediol (MPD), ethanol) [4]. The components required for crystallization have to be established empirically. Based on the analysis of previous crystallization conditions, a number of crystallization screens have been developed which are commercially available and can be used for initial screens.

The most popular technique of achieving supersaturation, far more popular than batch methodology or microdialysis, is vapor diffusion [4]. In an air-tight compartment (Figs. 1a, b), a small volume of protein solution, generally with a low concentration of precipitating agent (e.g., salt), is equilibrated via the vapor phase with a large volume of reservoir solution, generally with a high concentration of precipitating agent. Vapor diffusion from the protein droplet results not only in lower protein solubility, resulting from the increasing salt concentration in the droplet, but also yields a higher protein concentration, thus intensifying supersaturation. The protein can either be hanging from a cover slip ("hanging drop," Fig. 1a) or be located in a depression well ("sitting drop," Fig. 1b) above the reservoir solution. Prior to performing crystallization screens, it is advisable to check that the protein sample is suitably concentrated for the purpose using a pre-crystallization test (PCT, Subheading 3.1) [5].

The crystallization of integral membrane proteins [6–8] is inherently more difficult, given that the treatment of the membrane integral domain introduces additional dimensions to the crystallization problem, generally involving the use of detergents [9]. In spite of more elaborate classification schemes [10], an early classification of membrane protein crystals [11], still popular to date, distinguishes "type-I" crystals, consisting of stacked two-dimensional crystals (Fig. 1c) and "type-II" membrane protein crystals (Fig. 1d), i.e., true three-dimensional crystals of membrane protein detergent complexes.

Successful crystallization strategies, such as the lipidic cubic phase methodology [12], generally resulting in so-called type-I membrane protein crystals, and non-covalent [13] as well as covalent chaperone strategies [14], are beyond the scope of this short chapter and have been reviewed elsewhere. The single most successful concept in the crystallization of membrane proteins has been the introduction of the "small-amphiphile concept," introduced by Hartmut Michel in the early 1980s [11, 15]. Small amphiphiles (cf. Subheading 2.2) such as heptane-1,2,3-triol or benzamidine reduce the number of detergent molecules bound by

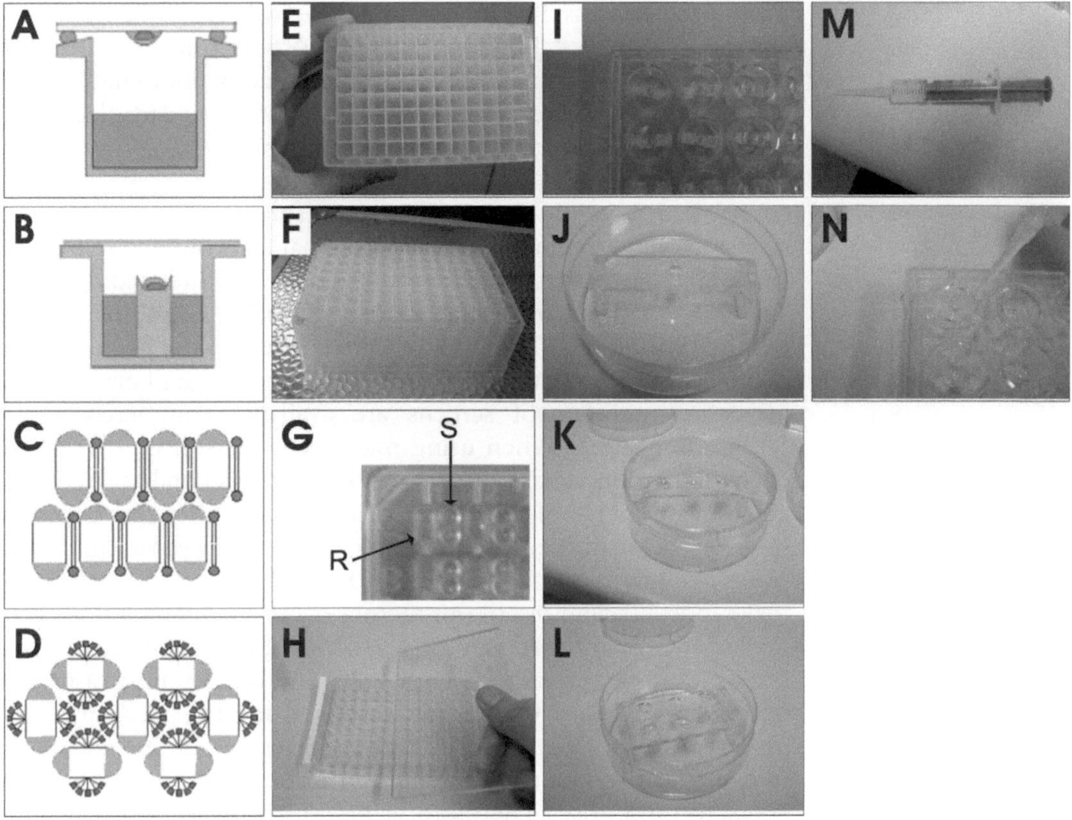

Fig. 1 Materials and methods. (**a**, **b**) Schematic representations of (**a**) a hanging drop on a cover slip in an air-tightly closed crystallization chamber. (**b**) A sitting drop on a bridge in a crystallization chamber closed air-tightly with a cover slip. (**c**, **d**) type I (**c**) and type-II (**d**) membrane protein crystals (redrawn from reference [11]. (**e**, **f**) 96 deep-well block (**e**) with lid (**f**) to hold the commercially available crystallization screens from 15 mL Falcon containers and for use with the Phoenix robot. (**g**) 96-well sitting drop crystallization plate (MRC2) with two sitting drop chambers ("S") per reservoir chamber ("R") for replication of crystallization conditions. (**h**) Sealing of 96-well plate sitting drop (MRC 2) with the foil by using the tool. (**i**–**l**) Various bridges glued into different crystallization chambers. (**i**) Micro-bridge with one cavity in one crytallization chamber of a 24-well plate, (**j**) Bridge with one cavity in a single-reservoir plate, (**k**) Bridge with three cavities in a single-reservoir plate, (**l**) Two bridges with three cavities in a single-reservoir plate. (**n**, **m**) Greasing of 24-well plate. (**m**) Silicone grease inside a 10 mL syringe with truncated yellow pipette tip, (**n**) greasing of the edges of a reservoir chamber (container) of the 24-well plate with syringe

the membrane protein [16] and can therefore be crucial for enabling polar protein-protein interactions for the stabilization of a type-II membrane protein crystal lattice. Heptane-1,2,3-triol has been shown to be essential in combination with detergents such as *N*,*N*-dimethyldodecylamine-*N*-oxide (LDAO), *N*,*N*-dimethylundecylamine-*N*-oxide (UDAO), or nonyl-β-D-glucoside (NG) for the crystallization of photosynthetic reaction centers [15], bacterial light-harvesting complexes [17], and rhodopsin [18], respectively, but it is without positive effect when combined with mild detergents such as alkyl maltosides. Benzamidine has been used successfully in combination with octyl-β-D-glucoside (OG) for the crystallization of bacteriorhodopsin [19, 20] and bacterial light-harvesting complexes [21] and in combination with a detergent mixture of dodecyl-β-D-maltoside and decyl-β-D-maltoside for the crystallization of quinol:fumarate reductase (QFR) [22].

If there is no prior knowledge available on the crystallization of the target protein or of a closely related protein, the 96-well crystallization format on a 200–400 nL scale (cf. Subheadings 3.3.1 and 3.3.2) is used for sparse matrix screening [3, 4] of a broad range of conditions. These have been selected on the basis of previous successes in producing crystals of a broad range of proteins. An increasing number of screens are available commercially, c.f. Subheading 2.3.1. When using multiple screens, it is advisable to check for redundancies using the PICKscreens database [23]. To take full advantage of these screens, a minimal buffer concentration (e.g., 20 mM) in the protein sample buffer is desirable. In case of protein solubility problems, 50–100 mM NaCl can be added. To make most efficient use of our membrane protein samples, we generally perform initial screens on the nL scale using robotic dispensing of solutions, cf. Subheadings 2.3.2 and 3.3. In most cases, it is necessary to translate the initial hits from nL-sized drops to μL-sized drops. Due to the different surface-to-volume ratios, differences in the kinetics of vapor diffusion, supersaturation, and crystallization arise. Therefore, a broad grid screen in a 24-well format (cf. Subheadings 3.3.3 and 3.3.4), involving incremental variation of the pH value (in ±1.0 pH unit steps), the PEG (in ±2 % steps), and salt (in ±100 mM steps) concentrations, is often necessary, followed by a finer, more focused screen in ±0.5 % PEG concentration, ±10 mM salt concentration, and ±0.5 pH value steps in order to optimize the conditions [24]. Frequently, it is necessary to scale up the crystallization droplet further and move to the 1- to 6-well format (20–50 μL, cf. Subheading 3.3.5). Due to the low surface tension of the detergent-containing drops, we implement the latter format exclusively as a sitting drop setup. Various compounds such as glycerol (10–20 %), ethanol (5–10 %), and DMSO (5–10 %) can also be used as additives, sometimes being essential for crystallization or useful for optimization. If they are volatile, they must also be included in the reservoir solution. Commercial volatile screens are also available. Sometimes oxidants (Sub-heading 2.2; or antioxidants) are essential, thus possibly stabilizing a single protein conformation required for crystal packing. For the stabilization of such a single protein conformation, the addition of a substrate analogue (cf. Subheading 2.2) or an inhibitor can be helpful.

Once crystals have been obtained, one can distinguish (colorless) protein from nonprotein crystals without access to an inhouse X-ray facility. Protein crystals contain large solvent channels that can be penetrated by dyes [25, 26] like methylene blue [27], resulting in blue crystals (cf. Subheading 3.4.2). Such solvent channels are absent in small molecule and inorganic crystals, which therefore do not take up the blue color.

In terms of examples, this chapter focuses on dihaem-containing QFR as a representative of the electron-transferring membrane proteins from respiration. Progress in other fields of membrane protein structure determination, such as G-protein-coupled receptors [28] as well as channels and transporters [29], has been reviewed recently.

2 Materials

2.1 Pre-crystallization Test

1. Reagents [30].
   - A1—0.1 M Tris–HCl pH 8.5, 2.0 M Ammonium sulfate.
   - B1—0.1 M Tris–HCl pH 8.5, 1.0 M Ammonium sulfate.
   - A2—0.1 M Tris–HCl pH 8.5, 0.2 M Magnesium chloride hexahydrate, 30 % w/v Polyethylene glycol 4,000.
   - B2—0.1 M Tris–HCl pH 8.5, 0.2 M Magnesium chloride hexahydrate, 15 % w/v Polyethylene glycol 4,000.
2. Plain glass cover slips.
3. Microscope (e.g., SZX 9/Olympus).

2.2 Preparation for Crystallization of Dihaem-Containing QFR Protein Additives and Equipment

1. Small amphiphile:
   - Benzamidine hydrochloride hydrate.
   - Heptane-1, 2, 3-triol (>98 %).
2. Oxidant:
   - Prepare a 100 mM stock solution of potassium hexacyanoferrate (III) ($K_3[Fe(CN)_6]$) with pure water.
3. Substrate analogue quinone for dihaem-containing QFR:
   - Prepare a 200 mM stock solution of Vitamin K_2 (Menaquinone-4,) with dimethyl formamide.
4. Benchtop centrifuge (in a cold room).

2.3 Crystallization

2.3.1 Crystallization Accessories

1. Crystallization Kits.
   - Mb class I suite (Qiagen/Cat.No.130711).
   - MembFac (Hampton Research/HR2-114).
   - Structure Screen 2 (Molecular Dimensions Limited/MD1-02).
   - MemStart & MemSys HT96 (Molecular Dimensions Limited/MD1-33) [8].
   - MemStart & MemGold (Molecular Dimensions Limited/MD1-41) [31].

  - MemStart & MemGold2 (Molecular Dimensions Limited/MD1-63) [32].
  - JBScreen Cryo 1–4 (Jena Bioscience/CC-103, CC-104, CC-105, CC-106)).
2. 96 deep-well block (96-well Assay Block, 2 mL/Costar).
3. Corning® 96 Square Well Storage Block Mat, Nonsterile (Product #3083) (96-well lid for deep well block).
4. 8-Channel pipette (0–200 μL).
5. Crystallization plate storage incubator (e.g., Rigaku Gallery700) with automated imaging system (e.g., Rigaku Minstrel HT) or climate cabinet (18 or 4 °C) and stereo microscope with polarization filter (e.g., SZX 9/Olympus) and camera (e.g., DX 40–1020 CL/Kappa).

2.3.2 Crystallization by Vapor Diffusion

96-Well Hanging Drop

1. Automated liquid handling system, e.g., Phoenix RE (Art Robbins, Rigaku Edition).
2. Additional 96-deep-well block (DWB) (cf. Subheading 2.3.1) for preparation of the mirror block, a mirror image of the source block, required for the drops on the plate seal.
3. 96-well plate (96-well microtest plates, clear polystyrene, e.g., Sarstedt, No/Ref 82.1581.001).
4. 96-well hanging-drop plate seals (e.g., ViewDrop II™ (25/pack) TTP LabTech Ltd, 4150–05600).
5. Hanging drop closing application (plate-seal holder for hanging drop preparations, e.g., TTP LabTech Ltd, 3019–05101 (= Lucite calibration plate for Phoenix for alignment of the plate seal) and guide base for positioning membrane on plate, e.g., TTP LabTech Ltd, 3019–05012).
6. PCR tube for holding protein sample.

96-Well Sitting Drop

1. Automated liquid handling system, e.g., Phoenix RE (Art Robbins, Rigaku Edition).
2. 96-well plate sitting drop (two drops per well respectively per reservoir, e.g., MRC crytallization plate (MRC 2), Molecular Dimensions, MD 11-00-100).
3. Plate seals (e.g., Greiner Bio-One/676070 or Molecular Dimensions Limited/MD6-01S).
4. Tool for sealing the 96-well plate (*see* **Note 9** and Fig. 1h).
5. PCR tube for holding protein sample.

24-Well Sitting/Hanging Drop

1. 24-well pre-greased plate (e.g., Greiner/GR662050; Intel Economy) or 24-well Crystalgen SuperClear™ Plates, pregreased (e.g., Jena Bioscience,CPL-132) or 24-well Crystalgen SuperClear™ Plate (e.g., Jena Bioscience,CPL-130) with point 8 and 9 (*see* **Note 12**).

2. Micro-bridges (e.g., Hampton Research, HR3-310, HR3-340: round bottom, capacity 35–40 μL).
3. Ethyl acetate.
4. Cover slips (20 × 20 mm).
5. Dichlorodimethylsilane.
6. Trichloroethane.
7. Ethanol (100 %).
8. 5 or 10 mL syringe.
9. Silicone grease (e.g., GE Bayer Silicones/Babysilone—Paste 35 g).
10. Truncated yellow pipette tip.

1- to 6-Well Sitting Drop

1. Crystallization chambers with lid (e.g., Boll, Munich).
2. Crystallization stages with one or three cavities (e.g., W. Jöckel Company, Weiterstadt, Germany).
3. Ethyl acetate.
4. Adhesive tape.

2.4 Distinguishing Protein and Nonprotein Crystals with Methylene Blue Dye

Dye solution, e.g., 3,7-bis(Dimethylamino)phenazathionium chloride (= methylene blue) solution, (e.g., Izit Crystal Dye (Hampton Research, HR4-710)).

3 Methods

3.1 Pre-crystallization Test [30]

1. Pipet 0.5–1.0 mL PCT Reagent (*see* Subheading 2.1) A1 into the reservoir A1 of the 24-well pre-greased plate. Pipet 0.5–1.0 mL of PCT reagent A2 into the reservoir A2 of the 24-well pre-greased plate.
2. Pipet 0.5–1 μL of protein solution onto the center of a glass cover slip.
3. Pipet a volume equal to that used in **step 2**, of PCT Reagent A1 from reservoir A1 into the sample drop on the siliconized cover slide. Do not mix the drop.
4. Invert the cover slip with the drop over reservoir A1 and seal.
5. Repeat **steps 2–4** for the reagent and the reservoir A2.
6. Wait 30 min.
7. After 30 min, view the two drops using a light microscope with magnification between 20 and 100×. If you have the right protein concentration, you should have one clear drop and another with a light granular precipitate (*see* **Note 1**).

3.2 Preparation for the Crystallization of Dihaem-Containing QFR (Modified from [22])

1. Per 100 μL of protein solution (16–24 mg/mL) (*see* **Note 2**), weigh in 2.4–4.8 mg benzamidine (for dihaem-containing QFR) (alternatively 6 mg heptane-1,2,3-triol).
2. Dissolve the benzamidine in 16–24 mg/mL (*see* **Note 2**) pure membrane protein solution (*see* **Note 3**). For example dihaem-containing QFR (20 mM HEPES pH 7.3, 1 mM DTT, 1 mM EDTA, 20 mM malonate (substrate analogue for first binding site), 0.1 % decyl-β-D-maltoside (DM) and 0.01 % dodecyl-β-D-maltoside (LM)).
3. Add oxidizing agent 100 mM potassium hexacyanoferrate (III) ($K_3[Fe(CN)_6]$) stock solution to a final concentration of 2 mM.
4. Add 200 mM substrate analogue stock solution (Vitamin K_2 for dihaem-containing QFR) to a final concentration of 1 mM.
5. Store it for 30 min at room temperature.
6. Centrifuge it at 4 °C for 10 min at maximum speed in the Eppendorf centrifuge (*see* **Note 3**).

3.3 Crystallization by Vapor Diffusion

3.3.1 96-Well Hanging Drop (400–800 nL)

1. Prepare the mirror block by transferring 200 μL per well of the source 96 deep-well block (DWB) with commercially available crystallization screen (*see* Fig. 1e, f) into the empty additional DWB using the 8-channel pipette, creating a mirror image by reversing the column order, i.e., source block column 1 becomes mirror block column 12, source block column 12 becomes mirror block column 1, etc.
2. Remove the membrane from the 96-well hanging-drop plate seal.
3. Place a few drops of water on Lucite Calibration Plate to ensure temporary adhesion of seal. Place plate seal, adhesive side up (for crystallization drops), on top of Lucite Calibration Plate and align edges of Plate Seal with Lucite Calibration Plate.
4. Load the source block, the 96-well plate (*see* **Note 4**), a PCR tube containing 50 μL (or 90 μL) of protein sample (*see* Subheading 3.2 for preparing protein sample and **Note 8**), the hanging drop plate seal aligned on the lucite calibration plate, and the mirror block onto the specified position of the automated liquid handling system (Phoenix robot). (The specified positions are defined in the pipette program of Phoenix software). Ensure that the face of the hanging drop plate seal is oriented upwards, which after pipetting will face the reservoir chamber of the 96-well plate.
5. Run the Phoenix software protocol "hanging drop with TTP Labtech seal" for 400 nL (or 800 nL) hanging drops. This will pipet 100 μL of the respective reservoir solutions from the source block into the respective reservoir chambers of the 96-well plate, and 200 nL (or 400 nL) of reservoir solution

(from the mirror block) and 200 nL (or 400 nL) of protein solution (from the PCR tube) onto each of the 96 hanging drop positions on the plate seal.

6. Carefully remove hanging drop reservoir plate and plate seal/calibration plate assembly from Phoenix deck.
7. Carefully invert the plate seal/calibration plate assembly and close the plate with the seal using the hanging drop closing application, so that the hanging drops face the respective reservoir chambers.
8. Mark all information on the long exterior wall side of the plate (date, protein, additives, all concentrations, and used screen).
9. Label the plate with a bar code (*see* **Note 5**) and insert into an automated incubator set to 18 or 4 °C. In conjunction with this incubator, the automated imaging system can be scheduled to take pictures of all droplets at regular intervals (*see* **Note 6** and Fig. 2a–c). If the automated incubator is not available, use a 18 or 4 °C climate cabinet (*see* **Note 7**) in combination with a microscope and a camera.

3.3.2 96-Well Sitting Drop (200–400 nL)

1. Put the 96 deep-well block (DWB) with a commercially available crystallization screen (*see* Fig. 1e, f), the 96-well plate (MRC2, *see* **Note 4** and Fig. 1g, h) and a PCR tube containing 30 μL (or 50 μL) protein solution (*see* Subheading 3.2 for preparing protein sample and *see* **Note 8**) at the specified position of the Phoenix robot. (The specified positions are defined in the pipette program of Phoenix software).
2. Use the pipetting protocol for 200 nL (or 400 nL) sitting drop from the Phoenix Software. This will pipet 50 μL reservoir solution from the DWB into the 96 reservoir chambers of the 96-well crystallization plate, and it pipet twice 100 nL (or 200 nL) of reservoir solution (from the 96 DWB) and twice 100 nL (or 200 nL) protein solution (from the PCR tube) in the two sitting drop chambers.
3. Remove the membrane from foil (plate seal).
4. After pipetting, close the 96-well plate air-tight with the plate seal (*see* **Note 9** and Fig. 1h).
5. Mark all necessary information on the long exterior wall side of plate (date, protein, additives, all concentrations, and used screen).
6. Put the plate with a bar code (*see* **Note 5**) in an automated incubator with 18 or 4 °C. In conjunction with this incubator, the automated imaging system can be scheduled to take pictures of all droplets at regular intervals (*see* **Note 6** and Fig. 2a–c). If the automated incubator is not available, use a 18 or 4 °C climate cabinet (*see* **Note 7**) in combination with a microscope and a camera.

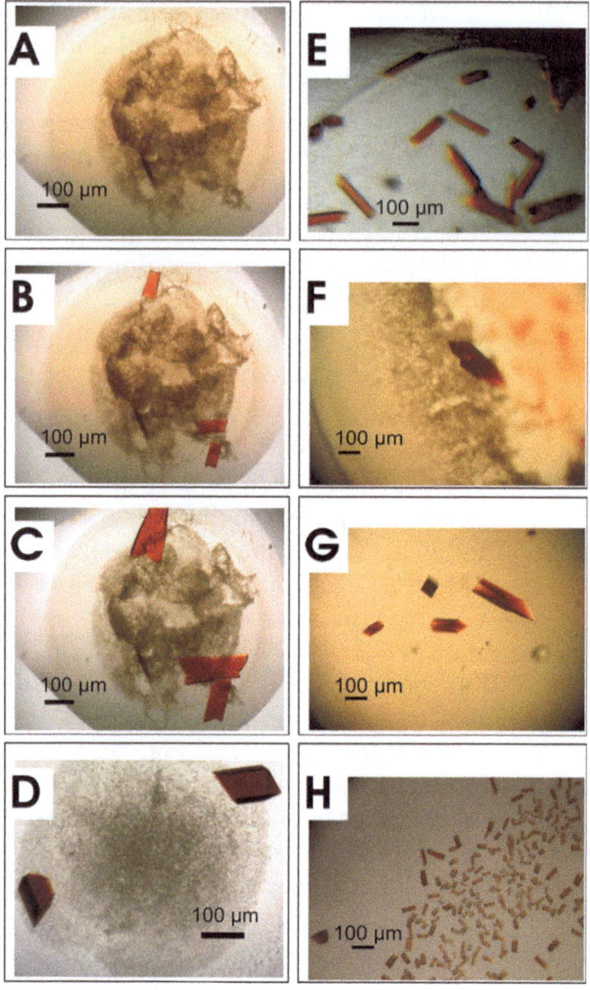

Fig. 2 Results. Automatically taken pictures after 13 (**a**), 20 (**b**), and 27 days (**c**) from a single 96-well 800 nL hanging drop (with 100 μL reservoir solution in reservoir chamber): 400 nL dihaem-QFR protein solution (20 mg/mL in 20 mM HEPES pH 7.3, 20 mM malonate, 1 mM EDTA, 0.1 % DM, 0.01 % LM with additives: 4.8 % benzamidine, 2 mM $K_3Fe(CN)_6$) added to 400 nL Nr. 41 (D1) reservoir solution from Mb class I suite (Qiagen): 0.05 M NaCl, 0.05 M Li_2SO_4, 0.05 M Tris–HCl pH 8.5, 30 % PEG 400. (**d**) Crystallization hit with 96-well 800 nL hanging drop (with 100 μL reservoir solution in reservoir chamber): 400 nL dihaem-QFR protein solution (20 mg/mL in 20 mM HEPES pH 7.3, 20 mM malonate, 1 mM EDTA, 0.1 % DM, 0.01 % LM with the additives: 4.8 % benzamidine, 2 mM $K_3Fe(CN)_6$) added to 400 nL reservoir solution Nr.66 (F6) from Mb class I suite (Qiagen): 0.5 M NaCl, 0.05 M Tris–HCl pH 8.5, 10 %(w/v) PEG 4000. (**e**) Reproduction of crystallization hit (**d**) with 24-well 2 μL hanging drop (1 mL reservoir solution in reservoir chamber). 1 μL dihaem-QFR protein solution (16 mg/mL in 20 mM HEPES pH 7.3, 20 mM malonate, 1 mM EDTA, 0.1 % DM, 0.01 % LM with the additives: 4.8 % benzamidine, 2 mM $K_3Fe(CN)_6$) mixed with 1 μL reservoir solution Nr.66 (F6) from Mb class I suite (Qiagen): 0.5 M NaCl, 0.05 M Tris–HCl pH 8.5, 10 % (w/v) PEG 4000. (**f–h**) Single crystal with 15 % PEG 4000 (**f**) a few big crystals with 16.5 % PEG 4000 (**g**) or plenty of small crystals with 18 % PEG 4000 (**h**) with three different 6-well 40 μL sitting drop crystallization setups. For each drop: 20 μL membrane protein (dihaem-QFR) solution (20 mg/mL (**f**) or 18 mg/mL (**g, h**) in 20 mM HEPES pH 7.3, 20 mM malonate, 1 mM EDTA, 0.1 % DM, 0.01 % LM with additives (4.8 % benzamidine, 2 mM $K_3Fe(CN)_6$) mixed with 20 μL reservoir solution Nr. 66 (F6) from Mb class I suite (Qiagen) but with a higher PEG concentration: 0.5 M NaCl, 0.05 M Tris–HCl pH 8.5, 15 % (**f**) or 16.5 % (**g**) or 18 % (**h**) PEG 4000 (w/v) (with 5, 6.5, and 8 percentage points, respectively, more PEG 4000 in the 8 mL reservoir solution compared to the reservoir conditions for (**d**) and (**e**))

3.3.3 24-Well Hanging Drop (2–5 μL)

1. Insert cover slips in 2 % dichlorodimethylsilane dissolved in trichloroethane for silanizing. Wash silanized cover slips in 100 % ethanol (*see* **Note 10**).
2. If not using 24-well pre-greased plates: Grease the edges of the 24-well plate reservoir chambers with silicone grease (*see* **Note 12** and Fig. 1m, n).
3. Prepare the reservoir solutions by using bi-distilled water and filter it (0.2 μm cut-off). For the reservoir solutions use the crystallization hits of the screens from the 96-well plate (*see* **Note 13**).
4. Pipet 1 mL of each of the self-made reservoir solution into the reservoir chamber of the 24-well plate (*see* **Note 14**)
5. Mix in the center of a silanized single cover slip 1 μL (or 2 μL) of the reservoir solution with 1–1.5 μL (or 2–3 μL) of the protein solution (*see* Subheading 3.2 for preparing protein sample) (*see* **Notes 4** and **10**).
6. Invert the cover slip, put it on the reservoir chamber, of which the reservoir solution for mixing with protein was taken from, and seal it. The mixed drop is hanging on the bottom and oriented into the sealed chamber (Fig. 1a).
7. Mark all necessary information on the long exterior side of plate (date, protein, additives, all concentrations, and used screen).
8. Put the plate with a bar code (*see* **Note 5**) in an automated incubator with 18 or 4 °C. In conjunction with this incubator, the automated imaging system can be scheduled to take pictures of all droplets at regular intervals (*see* **Note 6** and Fig. 2a–c). If the automated setup is not available, use an 18 or 4 °C climate cabinet (*see* **Note 7**) in combination with a microscope and a camera.

3.3.4 24-Well Sitting Drop (2–12 μL)

1. Glue the micro-bridges with ethyl acetate into the reservoir chambers of the 24-well plates (*see* **Note 11** and Fig. 1i–l).
2. If not using 24-well pre-greased plates: Grease the edges of reservoir chambers of the 24-well plate with silicone grease (*see* **Note 12** and Fig. 1m, n).
3. Prepare the reservoir solutions by using bi-distilled water and filter it (0.2 μm cut-off). For the reservoir solutions, use the crystallization hits of the screens from the 96-well plate (*see* **Note 13**).
4. Pipet 1 mL of each of the self-made reservoir solutions into the reservoir chamber of the 24-well plate (*see* **Note 14**).
5. Mix between 1 and 6 μL of the reservoir solution with equal volumes of the protein solution in the cavity of the micro-bridge (*see* Subheading 3.2 for preparing protein sample).

6. Seal the crystallization chambers with the sitting drops with cover slips (Fig. 1b).
7. Mark all necessary information on the long exterior side of plate (date, protein, additives, all concentrations, and used screen).
8. Put the plate with a bar code (*see* **Note 5**) in an automated incubator with 18 or 4 °C. In conjunction with this incubator, the automated imaging system can be scheduled to take pictures of all droplets at regular intervals (*see* **Note 6** and Fig. 2a–c). If the automated incubator is not available, use a 18 or 4 °C climate cabinet (*see* **Note 7**) in combination with a microscope and a camera.

3.3.5 1- to 6-Well Sitting Drop (40–50 μL)

1. Glue one or two bridges (with one or three cavities) with ethyl acetate into the crystallization chamber (*see* **Note 11** and Fig. 1i–l).
2. Prepare the reservoir solution using bi-distilled water and filter it (0.2 μm cut-off). For the reservoir solutions use the crystallization hits of the screen from the 96- or 24-well plate (*see* **Note 13**).
3. Transfer 8 mL of reservoir solution into the crystallization chamber.
4. Mix 20 μL of reservoir solution with 20–30 μL of prepared protein solution (*see* Subheading 3.2 for preparing protein sample) in each of the three or six wells (*see* **Note 4**).
5. Close the crystallization chamber with the lid and seal it airtight with adhesive tape.
6. Put the plate in the climate cabinet at 18 or 4 °C (*see* **Note 7**).
7. Check for crystals using a light microscope with magnification between 20 and 100× after 1, 2, and 4 weeks (*see* **Note 15**). In addition, you could check after 2 and 3 months.

3.4 Distinguishing Protein and Nonprotein Crystals

3.4.1 Crystallization of Reservoir Solution

1. Pipet the volume of reservoir solution which you used before for one of the five methods in Subheadings 3.3.1–3.3.5.
2. Pipet an equal volume of protein buffer with all additives (without protein).
3. Close the crystallization plate as before for one of the five methods in Subheadings 3.3.1–3.3.5.
4. Check for salt crystals after the same time you observed crystals with the hanging or sitting protein crystallization drop.

3.4.2 Methylene Blue Crystal Dye [27]

1. Pipet 1 μL of dye solution into a 10 μL drop or 0.5 μL into a 4 μL drop. The dye does not need to be added to the reservoir. For smaller drops (<4 μL), use a CryoLoop to pick up a small amount of dye within the loop and dip the CryoLoop into the drop to dispense the dye. This can be repeated to achieve the desired shade of blue color in the drop (*see* **Note 16**).

Existing protein, peptide, and nucleic acid crystals in the drop will absorb the dye within 1–24 h, taking on a blue color.

2. As a screening diagnostic, the dye can be mixed (1:9) with the crystallization reagent in the reservoir and added to the drop at the time of initial screen set up, before the appearance of a crystal.

4 Notes

1. If you detect heavy amorphous precipitate at the two protein drops, dilute the protein sample one to one. If you detect two clear drops, concentrate the sample to half of the original volume and repeat the test [30].
2. The most frequently used protein concentration for crystallization is approx. 40–80 μM. This is 8–16 mg/mL for a membrane protein complex with a molecular weight of 200 kDa, which is slightly higher than the recommendation for soluble proteins (10 mg/mL). Try higher concentrations if you do not get any crystals and no heavy precipitate.
3. A very pure homogeneous protein sample is more suitable for the crystallization. The homogeneity of the protein sample could be tested with gel filtration and with a light scattering detector in combination with a refraction index detector. Be economical with the purified protein, because reproduction of crystals with protein from a second purification is sometimes more difficult.
4. It is better to reproduce the crystallization conditions or to crystallize at least two times under the same conditions. There is a high chance to find crystals only in one of two of the same crystallization drops. For replication, you could prepare two 96-well hanging drop plates under the same conditions. For replication of the 24-well hanging drops you could prepare two crystallization drops (mix two times protein with two times of the same reservoir solution) on the cover slide. If you use the 96-well sitting drops MRC2 (Fig. 1g, h) plates, there are two sitting drops per reservoir chamber for replication. If no crystallization-robot is available for screening with 96-well plates, the commercially crystallization screen can be used in a 24 well plate setup (*see* Subheadings 3.3.3 and 3.3.4.).
5. Generate a barcode for the 96- or the 24-well plate with the software program of the incubator and store the plate in the automated incubator.
6. The following schedule could be used for the automatically taken pictures: 0, 1, 2, 3, 5, 7, 9, 20, 30, 60, and 120 days. The picture after day 0 (directly after finalizing the preparation of the crystallization plate) could be important to distinguish

between possible impurities (e.g., granular precipitate) and small crystals, which are growing after hours or 1 day or more days. Crystals could also grow after 2, 4, 8, or more weeks. That is the reason why images are taken after 1 month (Fig. 2c), 2, or more months. Frequently, you can detect crystal growth on a series of subsequently recorded pictures and also observe when the crystal ceases to grow further (Fig. 2a–c).

7. If you use a climate cabinet, the crystallization plates could be inspected under a microscope for impurities in the crystallization drops before introducing them to the climate cabinet. Take pictures of impurities and/or note the kind of impurity and the corresponding well number to be aware of false positive crystal growth. The plates could be screened under the microscope for crystals after 2 days and then in intervals of 1, 2, or more weeks. Using a stereo-microscope with polarisation filters (before and behind the object), transparent crystals can be visualised more easily.
8. For the crystallization drops, use always 10 μL protein solution more than the robot will pipette. This will help, that the robot takes the specified protein volume and to avoid air bubbles. If you use only the volume of the needed protein, the robot will most likely not take everything. The protein tip of the robot might get air inside and therefore will not pipette the correct amount of drops. Furthermore, the robot might pipette air bubbles.
9. Put the seal on the 96-well plate softly. Close the plate with gentle pressure from the center to outside with the appropriate tool (*see* Fig. 1h). Close the 96-well plate again with "hard" pressure to remove "waves" from the foil, which might have developed in the previous step. Attention: When removing the sticky seal from the 96-well plate to replace it with a new one, you could destroy the drops.
10. If you use detergent, the silanized cover slips are useful for crystallization of membrane proteins. By using the silanized cover slips, the hanging drop of the membrane protein will not be flattened by attraction to the cover slip surface, but formed to a round drop.
11. Ethyl acetate can be used to glue the crystallization bridges. Use gloves to fill the ethyl acetate under a fume hood in a glass dish. Dip the bottom edges of the crystallization bridge into the ethyl acetate for a short time. Glue the crystallization bridge directly into the reservoir chamber (*see* Fig. 1i–l).
12. Put silicone grease inside a 10 mL syringe. After that, attach a truncated yellow pipette tip to the syringe. Use the syringe to grease the edges of the reservoir chamber (container) of the 24-well plate (*see* Fig. 1m, n).

Table 1
Pipetting table: The first to the third rows list the final concentrations in the reservoir solution of Tris buffer, NaCl, and PEG 4000, respectively

		1	2	3	4	5	6
1	Buffer Tris–HCl, pH 8.5 (M)	0.05	0.05	0.05	0.05	0.05	0.05
2	NaCl (M)	0.5	0.5	0.5	0.5	0.5	0.5
3	PEG 4000 (%)	6 %	8 %	10 %	12 %	14 %	16 %
4							
5	Tris (μL)	250	250	250	250	250	250
6	NaCl (μL)	125	125	125	125	125	125
7	PEG 4000 (μL)	150	200	250	300	350	400
8	H_2O (μL)	475	425	375	325	275	225
9	Final volume (μL)	1,000	1,000	1,000	1,000	1,000	1,000

The fourth row is left blank intentionally for separation purposes. The fifth to eighth rows display the required volumes of the respective stock solution (0.2 M Tris–HCl, pH 8.5, 4 M NaCl, 40 % PEG 4000) and of ultrapure water. The ninth row lists the total final volume per reservoir

13. You could screen or rather change the polyethylene glycole (PEG) concentration from the crystallization hit of the 96-well screen (Fig. 2d). Therefore, you could choose a row with six conditions on a 24-well plate: Increment the PEG concentration in 2 % (or 2.5 %) increments, starting with 4 (or 5) percentage points less PEG than in the crystallization hit from 96-well screen and ending at 6 (or 7.5) percentage points more (from the commercially available screen) (*see* **Note 14**). Lower PEG concentrations tend to result in a few big crystals and higher PEG concentrations tend to result in more, but smaller crystals (Fig. 2f–h). For replication, you could use two protein-reservoir drops per coverslip. For 40 or 50 μL crystallization drops (*see* Subheading 3.3.5), you need one crystallization chamber for each PEG concentration. In this setup, crystals tend to grow at significantly higher PEG concentrations (Fig. 2f–h) as compared to the corresponding crystallization hit from 96-well screen (Fig. 2d, e). Furthermore, you could screen the pH of the crystallization hit (in 0.5 pH steps for 1–2 two numbers higher and lower as the pH of crystallization hit).
14. Alternatively, you could prepare stock solutions for individual components of the conditions of crystallization hits from the 96-well screen. Add the stock solutions with the correct volume for the correct concentration (for 1 mL total volume per reservoir chamber) separately into the 24-well reservoir chambers as indicated in Table 1. Fill up each reservoir chamber to 1 mL volume with ultrapure water and mix it. For example, a

crystallization hit from the Mb class I suite Screen (Qiagen) of dihaem-containing QFR (Fig. 2d, e): Nr. 66 (F6): 0.5 M Sodium chloride, 0.05 M Tris–HCl pH 8.5, 10 %(w/v) PEG 4000.

15. 40 μL sitting drops can produce large single crystals (Fig. 2f). In a 40 μL drop, a few (Fig. 2g) multiple larger or plenty of smaller crystals (Fig. 2h) could grow, which all are suitable for X-ray structure analysis.
16. If too much dye is added to the drop, diluting the drop too much, the crystal(s) may dissolve due to the decreased relative supersaturation (lower sample and reagent concentration). One may attempt to grow again the crystal(s) by simply sealing the experiment and allowing the drop to re-equilibrate with the reservoir [27].

Acknowledgments

We thank Dr. Yvonne Carius for discussions and Saarland University, the Faculty of Medicine (HOMFOR), the state of Saarland (LFFP 11/02), and the Deutsche Forschungsgemeinschaft (DFG, grants INST 256/275-1 FUGG, GK 845-Lancaster, and GK1326-Lancaster) for supporting our research.

References

1. Deisenhofer J, Epp O, Miki K, Huber R, Michel H (1985) Structure of the protein subunits in the photosynthetic reaction center of *Rhodopseudomonas viridis* at 3Å resolution. Nature 318:618–624
2. Kendrew JC, Bodo G, Dintzis HM, Parrish RG, Wyckoff H, Phillips DC (1958) 3-Dimensional model of the myoglobin molecule obtained by X-ray analysis. Nature 181:662–666
3. Bergfors TM (2009) Protein crystallization, 2nd edn. International University Line, La Jolla
4. McPherson A (1990) Current approaches to macromolecular crystallization. Eur J Biochem 189:1–23
5. Watson AA, O'Callaghan CA (2005) Crystallization and X-ray diffraction analysis of human CLEC-2. Acta Crystallogr Sect F Struct Biol Cryst Commun 61:1094–1096
6. Michel H (ed) (1990) Crystallization of membrane proteins. CRC, Boca Raton
7. Hunte C, von Jagow G, Schägger H (2003) Membrane protein purification and crystallization. A practical guide, 2nd edn. Academic, San Diego
8. Iwata S (ed) (2003) Methods and results in the crystallization of membrane proteins. International University Line, La Jolla
9. Privé GG (2007) Detergents for the stabilization and crystallization of membrane proteins. Methods 41:388–397
10. Schulz GE (2011) A new classification of membrane protein crystals. J Mol Biol 407: 640–646
11. Michel H (1983) Crystallization of membrane proteins. Trends Biochem Sci 8:56–59
12. Caffrey M (2003) Membrane protein crystallization. J Struct Biol 142:108–132
13. Hunte C, Michel H (2002) Crystallisation of membrane proteins mediated by antibody fragments. Curr Opin Struct Biol 12:503–508
14. Lieberman RL, Culver JA, Entzminger KC, Pai JC, Maynard JA (2011) Crystallization chaperone strategies for membrane proteins. Methods 55:293–302
15. Michel H (1982) 3-Dimensional crystals of a membrane protein complex—the photosynthetic reaction center from *Rhodopseudomonas viridis*. J Mol Biol 158:567–572

16. Gast P, Hemelrijk P, Hoff AJ (1994) Determination of the number of detergent molecules associated with the reaction center protein isolated from the photosynthetic bacterium Rhodopseudomonas viridis—effects of the amphiphilic molecule 1,2,3-heptanetriol. FEBS Lett 337:39–42
17. Koepke J, Hu XC, Muenke C, Schulten K, Michel H (1996) The crystal structure of the light-harvesting complex II (B800-850) from *Rhodospirillum molischianum*. Structure 4: 581–597
18. Palczewski K, Kumasaka T, Hori T, Behnke CA, Motoshima H, Fox BA, Le Trong I, Teller DC, Okada T, Stenkamp RE, Yamamoto M, Miyano M (2000) Crystal structure of rhodopsin: a G protein-coupled receptor. Science 289:739–745
19. Schertler GFX, Bartunik HD, Michel H, Oesterhelt D (1993) Orthorhombic crystal form of bacteriorhodopsin nucleated on benzamidine diffracting to 3.6Å resolution. J Mol Biol 234:156–164
20. Essen LO, Siegert R, Lehmann WD, Oesterhelt D (1998) Lipid patches in membrane protein oligomers: crystal structure of the bacteriorhodopsin-lipid complex. Proc Natl Acad Sci USA 95:11673–11678
21. McDermott G, Prince SM, Freer AA, Hawthornthwaitelawless AM, Papiz MZ, Cogdell RJ, Isaacs NW (1995) Crystal structure of an integral membrane light-harvesting complex from photosynthetic bacteria. Nature 374:517–521
22. Lancaster CRD, Kröger A, Auer M, Michel H (1999) Structure of fumarate reductase from *Wolinella succinogenes* at 2.2Å resolution. Nature 402:377–385
23. Hedderich T, Marcia M, Köpke J, Michel H (2011) PICKScreens, a new database for the comparison of crystallization screens for biological macromolecules. Cryst Growth Des 11:488–491
24. Newby ZER, O'Connell JD, Gruswitz F, Hays FA, Harries WEC, Harwood IM, Ho JD, Lee JK, Savage DF, Miercke LJW, Stroud RM (2009) A general protocol for the crystallization of membrane proteins for X-ray structural investigation. Nat Protoc 4:619–637
25. Sumner JB, Dounce AL (1937) Crystalline catalase. Science 85:366–367
26. Wilkosz PA, Chandrasekhar K, Rosenberg JM (1995) Preliminary characterization of EcoRI*-DNA co-crystals—incomplete factorial design of oligonucleotide sequences. Acta Crystallogr D Biol Crystallogr 51: 938–945
27. IZIT User guide (http://hamptonresearch.com/documents/product/hr003025_4-710_izit_user_guide.pdf)
28. Lebon G, Warne T, Tate CG (2012) Agonist-bound structures of G protein-coupled receptors. Curr Opin Struct Biol 22:482–490
29. Sciara G, Mancia F (2012) Highlights from recently determined structures of membrane proteins: a focus on channels and transporters. Curr Opin Struct Biol 22:476–481
30. PCT User Guide (http://hamptonresearch.com/documents/product/hr000257_2-140_user_guide.pdf)
31. Newstead S, Ferrandon S, Iwata S (2008) Rationalizing alpha-helical membrane protein crystallization. Protein Sci 17:466–472
32. Parker JL, Newstead S (2012) Current trends in alpha-helical membrane protein crystallization: an update. Protein Sci 21:1358–1365

Chapter 6

Molecular Dynamics Simulations of Membrane Proteins

Kristyna Pluhackova, Tsjerk A. Wassenaar, and Rainer A. Böckmann

Abstract

Molecular dynamics simulations are a powerful tool for complementing experimental studies, providing insights in biological processes at the molecular and atomistic level, at timescales from picoseconds to microseconds. Simulations are useful for testing hypotheses and can provide explanations for experimental observations as well as suggestions for further experiments. This does require that the simulation setup allows assessment of the question addressed. For example, it is evident that for simulation of a protein in its functional state the protein model and the environment have to mimic the biological situation as close as possible. In this chapter, a general strategy is presented for setting up and running simulations of membrane proteins of known structure in biological membranes of diverse composition and size.

Key words Membrane proteins, Lipid bilayers, Molecular dynamics simulation, All-atom force field, Coarse-grained force field, MARTINI, Backmapping

1 Introduction

In molecular dynamics (MD) simulations, particles are represented by soft spheres that interact through bonded and nonbonded interactions. This classical approximation allows using Newton's equations of motion to determine the time evolution of a system, from which inferences can be made about the mechanics and kinetics of processes observed. The forces between particles are calculated for every configuration, based on a set of functions and associated semiempirical parameters, together forming a force field. Force fields typically contain parameters for bonds, angles, and torsion angles, as well as for nonbonded Coulombic and van der Waals interactions, but may contain additional terms, for example, explicit parameters for hydrogen bonds.

Atomistic MD simulations can currently be performed for system sizes of up to a million atoms and simulation times in the microsecond range. Recent developments in specialized hardware

Doron Rapaport and Johannes M. Herrmann (eds.), *Membrane Biogenesis: Methods and Protocols*, Methods in Molecular Biology, vol. 1033, DOI 10.1007/978-1-62703-487-6_6,

have even enabled performing simulations of membrane proteins for hundreds of microseconds [1]. Yet such resources are not commonly available and the computational resources required for investigating relevant biological processes may often be considered excessive. The length and time scales accessible may be extended significantly by so-called coarse-grained models at the cost, however, of accuracy and resolution. Both atomistic and coarse-grained models are now widely used in the field of membrane simulations (for a review *see* [2]). The type of simulation to be chosen depends very much on the particular problem and the following questions should be considered: What is the timescale of the processes to be studied? How large should the membrane environment be chosen? Is sufficient sampling in the simulation expected?

In all-atom simulations, the standard technique to study membrane proteins in a lipid bilayer is based on the insertion of the protein of interest into a pre-equilibrated bilayer of given composition and size, moving the lipids out of the way (*see*, e.g., [3]). A different strategy in use is based on building a bilayer around the protein, either by placing lipid by lipid around the protein [4] or by spontaneous aggregation of lipids to form a micelle or a bilayer around the membrane protein [5, 6]. The latter methods require comparably long simulation times, i.e., of up to hundreds of nanoseconds for the simulation of the combined system, requiring several days of computational time on a high-performance compute cluster. An additional problem arises when the membrane to be inserted has a mixed composition. For single-component membranes, a merged system will be close to equilibrium, but in multicomponent membranes, specific interactions between the protein and the different lipids may cause the merged system to be far from equilibrium, requiring up to microseconds for resorting of the lipids.

Coarse-grained simulations, in contrast, are very fast but lack the atomistic details. We here introduce a combined technique that uses a coarse-grained (CG) representation (MARTINI model [7, 8]) to set up a protein-membrane system of arbitrary composition in a very efficient way. The MARTINI model maps, on average, four heavy atoms to a coarse-grained atom. The smaller number of atoms in the MARTINI model, the increased time step for CG models in the numerical solution of the equations of motion, and the decreased friction in CG models with respect to an atomistic description allow for a speedup of simulations by 2–3 orders of magnitude, making it possible to equilibrate systems of considerable size and complexity. The equilibration of the protein-membrane system in the CG model is followed by automatic backmapping of the system to a corresponding all-atom representation, thereby enabling atomistic molecular dynamics simulations (for an example *see* Fig. 1).

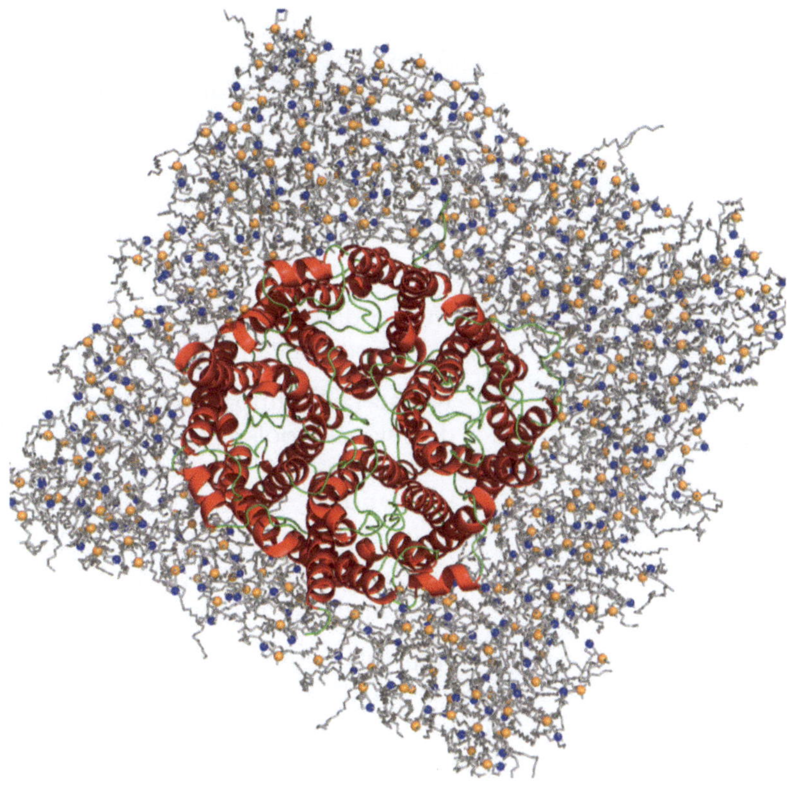

Fig. 1 Top view of tetrameric aquaporin 1 in a POPC bilayer consisting of 326 lipid molecules after 100 ns of simulation. Water molecules and ions are omitted for clarity. Aquaporins are integral membrane proteins that regulate the flow of water through membranes in the direction of a concentration gradient

2 Materials

The procedures laid down in this chapter require the availability of several software components, all of which are freely available for academic purposes, although some may require registration. The commands are provided assuming a Unix-like environment (Linux/BSD) and user-level knowledge of such systems. The following programs are used:

- GROMACS [9] (http://www.gromacs.org/Downloads): A suite of programs for molecular dynamics simulations.
- Modeller [10] (http://salilab.org/modeller/download_installation.html): A program for homology modeling. Note that it requires registration, but it is free for academic use.

- DSSP [11] (ftp://ftp.cmbi.ru.nl/pub/software/dssp/): A program for determining secondary structure of proteins. The program is available as a prebuilt binary for Windows as well as for Linux.
- Martinize.py [12] (http://md.chem.rug.nl/cgmartini/index.php/downloads/tools/204-martinize): A program for generating coarse-grained structures and topologies from atomistic structure files of proteins and nucleic acids.

Two additional programs are required, which are made available to support the procedure described here:

- Insane [13] (http://md.chem.rug.nl/cgmartini/index.php/downloads/tools/239-insane): A program for building membranes of arbitrary composition for coarse-grained simulations.
- Backward [14] (http://md.chem.rug.nl/cgmartini/index.php/downloads/tools/240-backward): A program for mapping an atomistic representation to a coarse-grained system.

A further requirement is the polarizable MARTINI force field, version 2.1 [7, 8], which can be obtained from http://md.chem.rug.nl/cgmartini/index.php/component/content/article/1-latest-news/224-m22.

Additional topology and structure files, as well as simulation parameters and scripts, are available at http://www.biotechnik.nat.uni-erlangen.de/research/boeckmann/downloads/mps/

For visualization of molecular dynamics trajectories and for displaying protein or membrane structures, it is advised to install vmd (Visual Molecular Dynamics) [15] program: http://www.ks.uiuc.edu/Research/vmd/.

3 Methods

The steps described below are written to be used on a Linux-based workstation. On other platforms, the syntax may be different. Following every step, it is important to check for errors and warnings reported by the software used. It is vital to assert that the results from every step are correct to avoid wasting time and computational resources on propagating errors.

3.1 Preparing the Structure of the Membrane Protein

1. Start with downloading a structure of the membrane protein of your choice from the Protein DataBank (http://www.rcsb.org) or from the Orientations of Proteins in Membranes database (http://opm.phar.umich.edu/) (*see* **Note 1**).

2. Preparation
Before building a system from your structure, make sure to answer the following questions:
   - Is the structure complete or are residues or parts thereof missing?
   - Are there ligands, cofactors, (metal) ions, or posttranslational modifications, such as palmitoylation, which are required for the process of interest?
   - What is the oligomeric state of the protein? Is it active as a monomer? Does it require other proteins for stability and/or function?
   - Which amino acids belong to the extracellular part, which belong to the intracellular part, and which constitute the transmembrane region of the protein?
   - What is the composition of the membrane required for the process to be investigated? Are models for the components required available in the force field of choice?
   - What is the time scale required for the process of interest?
3. Correcting for missing heavy atoms from side chains.
If the protein is missing atoms other than hydrogens, these have to be built in using modeling software. Hydrogen atoms will be built automatically based on the heavy atom positions during generation of a topology in **step 5**, and are of no concern.
The program Modeller comes with a tool for filling in missing heavy atoms automatically. It is assumed that the protein structure is contained in a file in PDB format, called *protein.pdb*. The syntax for fixing the atoms then is
mod9.11 model-missingatoms.py
The program writes out a file *protein_allatom.pdb* containing only heavy atoms.
In some cases it may be convenient to manually repair a structure, for example, using GUI-based software, such as SwissPDB Viewer [16] or PyMOL [17] (*see* **Note 2**). In PyMOL, there is a mutagenesis wizard that can be used for rebuilding side chains, by mutating the broken residue to the same type, allowing choosing between different rotamers.
4. Adding missing residues
It happens regularly that certain portions of a protein cannot be resolved in experiments due to increased motility. For the termini this may not matter much, but missing loops in the protein sequence need to be handled with care. Modeller has an additional module to rebuild missing loops based on the sequence of the protein. Structures obtained from the PDB have the sequence of the protein listed, which can be extracted using

```
mod9.11 get_sequence.py
```

Note that this requires the structure file to be named *protein.pdb*. The sequence is written to a file named *protein.seq*. This file should be copied twice to a file with the name *alignment.ali*, in which the 'structure' line in the second copy should be replaced by 'sequence', and the missing residues should be inserted. Below is an example of such a file for the Melectin peptide with missing residues at positions 5–7 (Ile-Leu-Lys):

```
>P1;protein
structureX:protein:   1 : :+16  : :::-1.00:-1.00
GFLSKVLPKVMAHMK*
>P1;protein_fill
sequence::::::::
GFLSILKKVLPKVMAHMK*
```

The next step, building the missing residues, requires editing the script *http://www.biotechnik.nat.uni-erlangen.de/research/boeckmann/downloads/MPS/model-missingres.py*. For the example given above, the residue range should be set to residues 5–7 (line 13):

```
return selection(self.residue_range('5','7'))
```

In addition, the number of models to be generated has to be specified (lines 20–21):

```
a.starting_model= 1    # index of the first model
a.ending_model= 5    # index of the last model
```

After editing the script, Modeller can be run to fix the structure. The input structure file should be named *protein.pdb* and the file *alignment.ali* needs to be available:

```
mod9.11 model-missingres.py
```

As specified in the script 5 different structure files will be generated, named *protein_fill.BL$modellercode.pdb*. From these files, the best candidate should be chosen, based on visual inspection in, e.g., PyMOL. The selected structure should be renamed to *protein_corrected.pdb*.

5. Topology and protonation states
At this point, a correct starting structure should be available, from which the simulation system can be built. The first step in a simulation is bringing the structure in line with the force field chosen and constructing a topological description of the protein. In GROMACS, the tool for this operation is called *pdb2gmx*. This program also adds hydrogen atoms to the protein structure based on the positions of the heavy atoms and can be instructed to ignore hydrogens that are present in the file (*-ignh*). The program is run in interactive mode, which

asks the user to specify protonation states for titratable residues (*-inter*) and termini (*-ter*):

```
pdb2gmx -f protein_corrected.pdb -ter -ignh -inter
```

Standard protonation at neutral pH may be chosen by removing the *-inter* option from the above command, but mind that this will neglect p*K*a changes due to the environment (*see* **Note 3**).

The program will ask for the protein force field and water type for the simulation. Mixing different, possibly incompatible force fields and lipid or solvent models should always be avoided, as it may lead to artifacts, e.g., in the protein–lipid interaction. For membrane protein simulations, currently the CHARMM36 [18] and the Amber force field [19, 20] offer parameters for both lipids and proteins. For a united atom force field, in which nonpolar hydrogens in methyl or methylene groups are not explicitly simulated, the GROMOS96 force field (preferentially 54A7) [21, 22] is widely being used.

The command given above should yield an output structure named *conf.gro* and a corresponding topology in a file named *topol.top*.

6. Orientation of the protein in space.
   In the case of membrane proteins, the orientation of the protein with respect to the membrane requires some attention. The protein should be positioned with the transmembrane region aligned with the membrane plane. Several methods are available for positioning the protein. One of such methods is used as a basis for the Orientations of Proteins in Membranes database, where the oriented structures of membrane proteins available in the Protein DataBank are made available. However, it is often sufficient to orient the protein along its principal axes, although the validity of the result requires to be checked. Aligning the protein with GROMACS can be done using:

```
editconf -f conf.gro -princ -o protein_princ.pdb
```

If the resulting orientation is not correct, it is also possible to add the option *–rotate* to the command line, which takes three arguments, denoting rotation around *x*, *y*, and *z*, respectively.

At this point, the protein should be prepared, including the topology and the proper orientation with respect to the membrane. The next step is setting up and adding the membrane.

3.2 Membrane-Protein System Setup

The combined protein–lipid bilayer system is set up using a so-called coarse-grained (CG) representation. This simplifies and significantly accelerates the generation and equilibration of a lipid bilayer of arbitrary size and composition. After equilibration of the coarse-grained protein-membrane-solvent system, the system is converted back to atomistic for subsequent high-resolution MD simulations.

This section can be skipped if the goal is to insert the membrane protein into a pre-existing atomistic lipid bilayer (*see* **Note 4**).

1. Coarse-graining of the system and adding a lipid bilayer, water, and ions.

   First the protein needs to be converted to coarse-grained (CG) representation. This is done using the program *martinize.py* that was introduced as a one-step solution for generation of a coarse-grained structure and topology from a protein structure. The program uses the DSSP program for determining the secondary structure, the path to which needs to be provided to the program. For explanation of the options, the user is referred to the programs internal help (*./martinize.py –h*).

   ```
   ./martinize.py -f protein.pdb -nt -v -o topol.
   top -x protein_CG.pdb -p ALL -ff martini21p -dssp
   dssppath/dsspcmbi
   ```

   Following this step, the structure is energy-minimized in vacuum to resolve stretched bonds that may be introduced by the conversion. This is done in two steps, using gromacs. In the first step the structure, topology, and run parameters are combined into a run input file, using *grompp*. The second step runs the simulation using *mdrun*.

   ```
   grompp -f martini_min.mdp -p topol.top -c pro-
   tein_CG.pdb -o min_protein.tpr
   mdrun -v -deffnm min_protein
   ```

   Next a coarse-grained lipid bilayer is added around the protein, as well as water and ions. This step uses the program Insane (INSert (in) membrANE), which is a versatile tool for building coarse-grained membranes and solvent.

   ```
   ./insane.py -f min_protein.gro -o withbilayer.
   gro -pbc rectangular -x 10 -y 10 -z 10 -l POPC
   -sol PW -salt 0.15
   ```

   In the example above, a cubic box of 10 nm side length is chosen, a POPC lipid bilayer is added as well as water and ions (NaCl) at a physiological concentration of 0.15 M (*see* **Note 5**). *See* **Note 6** for setting asymmetric bilayers and defining the relative abundance of lipids in each leaflet.

   The resulting protein/bilayer assembly should be checked in a molecular viewer. For visualization of bonds between coarse-grained beads one can use vmd together with the script *cg_bonds.tcl*. To load this script in VMD, type in the console of VMD:

   ```
   source cg_bonds.tcl
   cg_bonds -tpr topol.tpr -gmx gromacspath/gmx-
   dump -cutoff 12
   ```

2. Bilayer and water equilibration.

   The added bilayer and water resemble a crystalline arrangement (Fig. 2), which is quickly dispersed in a simulation run with

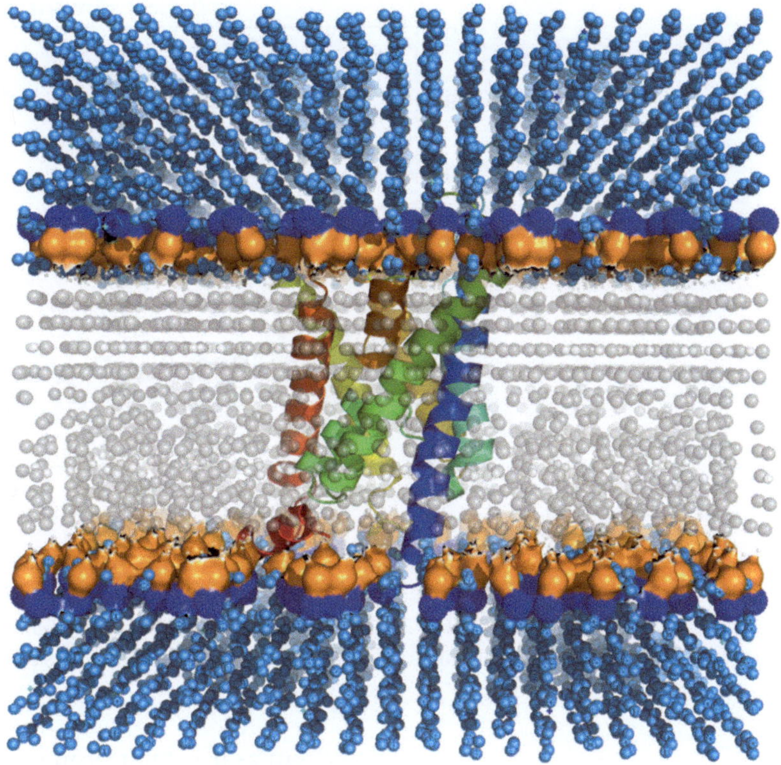

Fig. 2 Side view of monomeric aquaporin 1 (in cartoon representation) with generated CG surroundings (247 POPC molecules (lipid tails displayed as *grey balls*, headgroups as *orange*/*blue* surfaces) in the bilayer, water (*blue spheres*), and ions (omitted for clarity)). The initial crystal-like structure will vanish within a few tens of picoseconds of position restraint simulation

position restraints on the protein. In this type of simulation, the CG protein atoms are fixed in space using harmonic restraints. The insertion depth of the protein in the bilayer can still be adjusted, though, by motion of the surrounding lipids. Before running the simulation, it has to be asserted that the topology file of the coarse-grained system contains definitions for all components, including solvent and lipids used.

The system can then be energy-minimized, using a sequence of commands similar to that given before:

```
grompp -f martini_min.mdp -p topol.top -c with-bilayer.gro -o min_all.tpr
mdrun -v -deffnm min_all
```

Subsequently, the position restraint simulation is run, again in a similar manner:

```
grompp -f posre.mdp -p topol.top -c min_all.gro -s CG_posre.tpr
mdrun -v -deffnm CG_posre
```

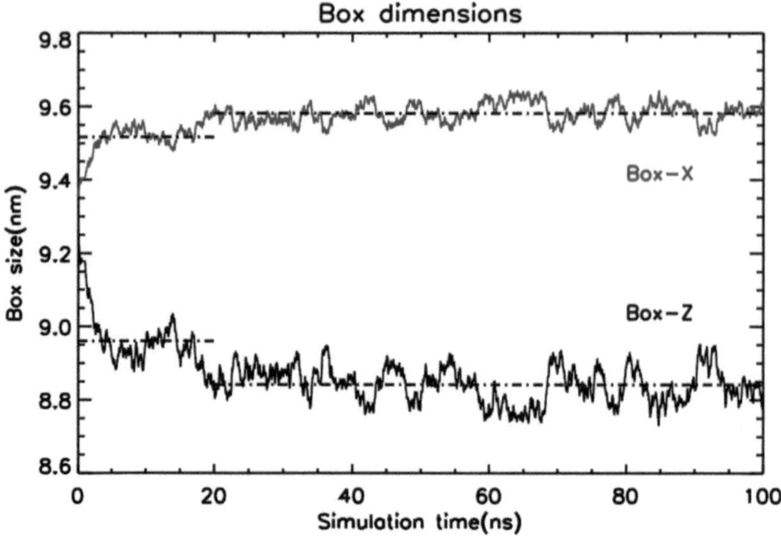

Fig. 3 The time evolution of the box dimensions for a monomeric aquaporin 1 embedded in a POPC bilayer, as shown in Fig. 2. Note that the *Y* dimension is set equal to *X*. The box size reaches the equilibrium value in approximately 20 ns. The total 100 ns of CG simulation required less than 4 h on 24 compute cores

The system as it is built, using *martinize* and *insane*, may still be strained. To relax the system, a series of short runs have to be performed in the isothermal-isobaric ensemble (NpT, at constant temperature and pressure) with position restraints. In these runs the integration time step is gradually increased from 1 fs to 2 fs, 5 fs, 10 fs, and finally to 20 fs. The relaxation can be checked from the convergence of the simulation box size using the *g_energy* tool of GROMACS (compare Fig. 3).

The equilibration of a membrane may take between tens to hundreds of nanoseconds, or several microseconds for complex, multicomponent membranes.

3. Backmapping to all-atom representation.
In order to study the protein mechanics and protein–lipid interactions at atomic resolution, the system has to be converted to an all-atom (e.g., Amber or CHARMM force field) or to a united atom representation (GROMOS), using a procedure called reverse transformation or backmapping. It may be useful to delay this step and perform a first stage of unrestrained coarse-grained simulation, to obtain a rough view of the energy landscape. Structures can then be selected from the resulting trajectory and converted back to atomistic. In this way, larger conformational changes may be identified which should, however, be interpreted carefully due to the limited accuracy of current CG force fields for proteins as compared to atomistic force fields.

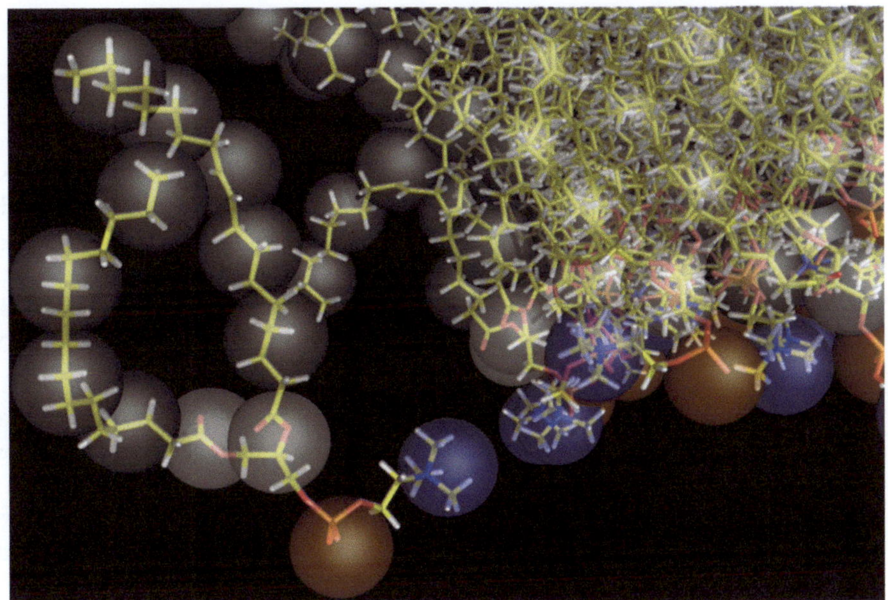

Fig. 4 Backmapped all-atom representation of a MARTINI POPC membrane

A detailed description of the procedure used here for backmapping will be described elsewhere [14]. At the core of this method is the program *backward*, which constructs an atomistic starting structure from the coarse-grained positions. Unique about this procedure is that it requires only the coarse-grained structure and the atomistic topology for the conversion. The method is tailored to work with a native version of GROMACS, although the protocol can be easily implemented for other MD packages, without requiring changes in the code.

Here, the MARTINI structure *CG_posre.gro* is converted to CHARMM36 all-atom representation and written to *aa_charmm.gro*:

```
./initram.sh -f CG_posre.gro -o aa_charmm.gro
-to charmm36 -p topol.top
```

Figure 4 shows part of a CG POPC bilayer together with the atomistic structure resulting from the backmapping.

3.3 Membrane-Protein Simulation

1. Relaxation

When a system is built by direct embedding in an atomistic membrane or by backmapping from a coarse-grained structure, the system can be strained due to the treatment. This strain needs to be dissipated before running the production simulation, which is achieved by a series of simulations in which the system is gradually relaxed to the production simulation conditions. This means that at this point a clear notion of the simulation parameters is required. In particular, the following points need attention, which are controlled in the simulation by parameters in the simulation parameter (.mdp) file (*see* **Note 7**):

- Treatment of nonbonded interactions. Several schemes are available for determining nonbonded interactions. Which scheme should be used depends on the force field chosen. Here, using CHARMM36, the particle mesh Ewald (PME) [23] method should be used for Coulombic interactions and a shifted Lennard-Jones potential is used for van der Waals interactions. This is preset in the parameter files provided.
- Temperature and pressure control. Due to numerical methods used, it is necessary to control the temperature and pressure by coupling to an external bath. In most cases, physiological conditions should be used, meaning a pressure of 1 atm and a temperature of 298 K for laboratory conditions or 310 K for in vivo conditions. Several methods are available for control of temperature and pressure. The current opinion favors the use of the Parrinello–Rahman algorithm for control of pressure and either the Nosé–Hoover or the Bussi algorithm for control of temperature. As these may not be stable if the system is far out of equilibrium, it is commonly advised to use the Berendsen algorithm for equilibration towards the target values, in particular for the pressure. The different thermodynamic properties of the membrane with the protein and the solvent may cause energy to flow from one part to another, which has to be avoided by separately coupling protein and the lipids to one temperature bath, and water and ions to a second bath.

The sequence for relaxation typically comprises the following steps:

- MD at constant temperature and volume (no pressure coupling) with position restraints (*see* **Note 8**) on the protein to allow equilibration of the environment around the protein, without disturbing the protein's internal state:

```
grompp -f nvt.mdp -c aa_charmm.gro -p topol.top
-o nvt.tpr
mdrun -deffnm nvt
```

- MD at constant temperature and pressure with position restraints on the protein to allow further relaxation of the environment:

```
grompp -f npt.mdp -c nvt.gro -p topol.top
-o npt.tpr
mdrun -deffnm npt
```

- Short unrestrained MD under production conditions

```
grompp -f pre.mdp -c npt.gro -p topol.top
-o pre.tpr
mdrun -deffnm pre
```

2. Production simulation
With the system being relaxed, it is time to set up the production run. This requires specifying the length of the simulation by setting the time step to use and the number of steps to simulate. The total time required depends on the typical time of the process studied. Other points of attention are the frequency to write out structures and energies, whether and how often to write out velocities, and which groups to use for splitting the total energy into intra-group and inter-group interaction energies. The latter can be useful for investigating protein–lipid, protein–protein, or protein–ligand interactions. These settings are listed in the simulation parameter file *md.mdp*. The production simulation is run using the same grompp/mdrun combination as before:

```
grompp -f md.mdp -c pre.gro -p topol.top -o md.
tpr
mdrun -deffnm md
```

If after the production simulation it appears that the system has not converged sufficiently, then the run should be extended.

3.4 Simulation Analysis

The analysis of biomolecular simulations largely depends on the particular system and the problems to be addressed with the help of MD simulations. The following will thus only provide a basic framework for the analysis of protein-membrane systems. The GROMACS package provides many tools for various analyses (*see* www.gromacs.org); it also provides a framework for coding of specialized routines in a very efficient way.

1. Simulation quality assurance.
Check the output logfile for error messages (file *md.log*) and visually inspect the simulation trajectory (*traj.xtc*), i.e., the file containing the system coordinates at specified time steps of your simulation (specified in the mdp file) and/or the final structure of your simulation in comparison to the starting structure. Particular attention should be paid to the formation of unexpected pores in the membrane, a separation of the lipid leaflets, ion aggregation on the protein surface, or loss of protein secondary structure. Such events may indicate problems in the chosen force field (combination) or the setup of the mixed protein–lipid system.

    Inspection of the potential energies, temperature, volume, or box sizes (using the *g_energy* tool of GROMACS) provides valuable information about the equilibration of the biomolecular

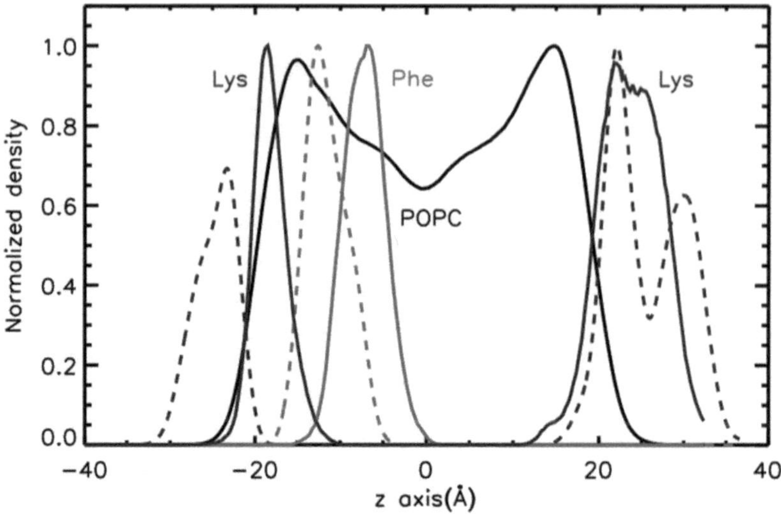

Fig. 5 Example: Density profile of the yeast mitochondrial outer membrane protein FIS1 in a POPC lipid bilayer. The insertion depths of Lys and Phe residues in the lipid bilayer are tracked in time (interrupted line beginning of the simulation, full line position after 300 ns)

system, e.g., no drift in the lateral box size should be observed after equilibration.

2. Protein structure and quality assurance.
As a first step, calculate the root mean square deviation (RMSD) of the protein as a function of simulation time (use the GROMACS tool *g_rms*). The RMSD is a measure for the deviation of the protein structure from the crystal structure. The fluctuations may be analyzed using the GROMACS tool *g_rmsf*. It provides hints about less or highly flexible regions. The secondary structure and its time evolution is analyzed by the GROMACS tool *do_dssp* for each residue.

Check the distance between the images of your protein in the periodic simulation setup (GROMACS tool *g_mindist* with the flag *-pi*). The distance should at least be larger than twice the largest cutoff distance for nonbonded interactions (*see* **Note 9**).

The tilt angle of the whole protein or of its individual helices differs usually for different membrane thicknesses and lipid types. Its analysis (using, e.g., the GROMACS tool *g_bundle*) is instructive to, e.g., learn about the preferred lipid environment of the membrane protein.

3. Lipid bilayer around the protein.
The insertion depth of the protein in the membrane, i.e., the distance of the centers of mass of the protein and the bilayer, can be analyzed with the tool *g_dist*. For locating the insertion depths of individual residues in more detail the tool *g_density* is perfectly suited (compare the example given in Fig. 5).

Calculation of the diffusion coefficients of lipids as a function of the distance from the membrane protein provides insight into the strength of the lipid–protein interaction and about protein-induced long-range ordering effects in the membrane. Use the GROMACS tool *g_msd* (option *-lateral z*) to compute the diffusion coefficients of lipids (choose the center of mass of the lipid headgroups as a reference, and substract the overall lateral motion of the whole leaflet).

GridMAT-MD [24] is a simple program calculating the area per lipid and bilayer thickness for protein-membrane complexes. Check the webpage http://www.bevanlab.biochem.vt.edu/GridMAT-MD/index.html for details and download.

In order to conclude on the fluidity of your membrane the deuterium order parameters can be calculated for each individual chain (*sn*1 and *sn*2) by using *g_order*.

4 Notes

1. Check for ligands used, e.g., for improved crystallization, chemically modified amino acids, or nonstandard amino acid names in the pdb file. Make sure that your final structure file contains only polypeptide chains.
2. PyMOL [17] is a program for molecular visualization and allows for the preparation of artistic figures (www.pymol.org). Educational subscriptions are available at no cost.
3. If you suspect some amino acids adopting a nonstandard protonation, calculate first the p*K*a of the titratable groups. One example of such a program is the freely available mcce (Multi-Conformation Continuum Electrostatics) [25, 26].
4. If you already have a bilayer of interest at hand, the protein is not too large, and the orientation of the protein in the membrane is known, the easiest possibility to insert the protein into the lipid bilayer is the GROMACS tool *g_membed* [3]. In this tool, the protein is first shrunken and inserted into the bilayer, removing overlapping lipids and water molecules. Subsequently, the protein is slowly resized to its original size, moving the lipids out of the way.

   In a first step, orient the protein to the membrane normal and translate it to the membrane center (using *editconf*). Then run *g_membed*:

```
grompp -f membed.mdp -p topol.top -c solvated.pdb
g_membed -f topol.tpr -p topol.top -xyinit 0.1
-xyend 1.0 -nxy 1000 -v -n index.ndx
```

5. The currently supported lipid types inside http://md.chem.rug.nl/cgmartini/index.php/downloads/tools/239-insane are:

   Phospholipids: DPPC, DHPC, DLPC, DMPC, DSPC, POPC, DOPC, DOPC, DAPC, DUPC, DPPE, DHPE, DLPE, DMPE, DSPE, POPE, DOPE, PPCS, DOPG, POPG, DOPS, and POPS

   Glycolipids: DSMG, DSDG, DSSQ GM1, DGDG, MGDG, SQDG, CER, GCER, DPPI, PI, and PI34

   Sterols: CHOL

   However, be aware that not for all of the above mentioned lipids all-atom topologies are currently available.

6. The syntax of *insane.py* to generate a mixed symmetric bilayer is:

   ```
   ./insane.py -f protein.pdb -o withbilayer.gro -pbc rectangular -x 10 -y 10 -z 10 -l POPC:70 -l POPS:10 -l CHOL:20 -sol PW -salt 0.15
   ```

   For preparing an asymmetric bilayer additionally provide the flags *-u* (upper bilayer) und *-l* (lower bilayer):

   ```
   ./insane.py -f protein.pdb -o withbilayer.gro -pbc rectangular -x 10 -y 10 -z 10 -l POPC:70 -l POPS:10 -l CHOL:20 -sol PW -u POPC:80 -u CHOL:20 -salt 0.15
   ```

   It may be convenient to have the sum of relative abundances in each leaflet to sum up to 100.

7. For an overview and explanation of all options for the simulation parameter file check the GROMACS manual or the webpage http://manual.gromacs.org/online/mdp_opt.html.

8. A file with restraints on selected atoms can be generated using the gromacs tool *genrestr*. As an input it requires a structure file as well as an index file with a group of atoms that should be restrained. Index files are most effectively generated using the *make_ndx* tool of GROMACS.

9. For charged systems it is recommended to have a minimum distance between the images of the protein of four times the Debye screening length (0.7–0.8 nm at physiological salt concentration of 0.15 M) to avoid self-interaction.

Acknowledgment

This work was supported by a grant from the Deutsche Forschungsgemeinschaft (BO 2963/2-1) to RAB.

References

1. Jensen MO, Jogini V, Borhani DW, Leffler AE, Dror RO, Shaw DE (2012) Mechanism of voltage gating in potassium channels. Science 336:229–233
2. Tieleman DP (2012) Computer simulation of membrane dynamics. In: Comprehensive biophysics, vol 5. Elsevier
3. Wolf MG, Hoefling M, Aponte-Santamaría C, Grubmüller H, Groenhof G (2010) g_membed: efficient insertion of a membrane protein into an equilibrated lipid bilayer with minimal perturbation. J Comput Chem 31:2169–2174
4. Kandt C, Ash WL, Tieleman DP (2007) Setting up and running molecular dynamics simulations of membrane proteins. Methods 41:475–488
5. Marrink SJ, Lindahl E, Edholm O (2001) Simulation of the spontaneous aggregation of phospholipids into bilayers. J Am Chem Soc 123:8638–8639
6. Böckmann RA, Caflisch A (2005) Formation of detergent micelles around the outer membrane protein OmpX. Biophys J 88: 3191–3204
7. Marrink SJ, Risselada HJ, Yefimov S, Tieleman DP, de Vries AH (2007) The MARTINI forcefield: coarse grained model for biomolecular simulations. J Phys Chem B 111:7812–7824
8. Monticelli L, Kandasamy SK, Periole X, Larson RG, Tieleman DP, Marrink SJ (2008) The MARTINI coarse grained forcefield: extension to proteins. J Chem Theory Comput 4: 819–834
9. Hess B, Kutzner K, van der Spoel D, Lindahl E (2008) GROMACS 4: algorithms for highly efficient, load-balanced, and scalable molecular simulation. J Chem Theory Comput 4:435–447
10. Sali A, Potterton L, Yuan F, van Vlijmen H, Karplus M (1995) Evaluation of comparative protein modelling by MODELLER. Proteins 23:318–326
11. Kabsch W, Sander C (1983) Dictionary of protein secondary structure: pattern recognition of hydrogen-bonded and geometrical features. Biopolymers 22:2577–2637
12. de Jong DH, Gurpreet S, Bennett WFD, Arnarez C, Wassenaar TA, Schäfer LV, Periole X, Tieleman DP, Marrink SJ (2013) Improved parameters for the martini coarse-grained protein force field. J Chem Theory Comput 9: 687–697
13. Wassenaar TA, Sengupta D, Tieleman DP, Marrink SJ (in preparation) INSANE: fast and versatile generation of custom membranes for molecular simulations
14. Wassenaar TA, Pluhackova K, Böckmann RA, Marrink SJ, Tieleman DP (2013) Going backward: A flexible geometric approach to reverse transformation from coarse grained to atomistic models. (in preparation)
15. Humphrey W, Dalke A, Schulten K (1996) VMD—visual molecular dynamics. J Mol Graph 14:33–38
16. Guex N, Peitsch MC (1997) SWISS-MODEL and the Swiss-PdbViewer: an environment for comparative protein modeling. Electrophoresis 18:2714–2723
17. The PyMOL molecular graphics system, Version 1.5.0.4 Schrödinger, LLC
18. Klauda JB et al (2010) Update of the CHARMM all-atom additive force field for lipids: validation on six lipid types. J Phys Chem B 114:7830–7843
19. Hornak V, Abel R, Okur A, Strockbine B, Roitberg A, Simmerling C (2006) Comparison of multiple Amber force fields and development of improved protein backbone parameters. Proteins 65:712–725
20. Jämbec JPM, Lyubartsev AP (2012) An extension and further validation of an all-atomistic force field for biological membranes. J Chem Theory Comput 8:2938–2948
21. Schuler LD, Daura X, van Gunsteren WF (2001) An improved GROMOS96 force field for aliphatic hydrocarbons in the condensed phase. J Comput Chem 22:1205–1218
22. Poger D, Mark AE (2010) On the validation of molecular dynamics simulations of saturated and cis-mono unsaturated phosphatidylcholine lipid bilayers: A comparison with experiment. J Chem Theory Comput 6:325–336
23. Darden T, York D, Pedersen L (1993) Particle mesh Ewald: an N-log(N) method for Ewald sums in large systems. J Chem Phys 98: 10089–10092
24. Allen WJ, Lemkul JA, Bevan DR (2009) GridMAT-MD: a grid-based membrane analysis tool for use with molecular dynamics. J Comput Chem 30:1952–1958
25. Georgescu RE, Alexov EG, Gunner MR (2002) Combining conformational flexibility and continuum electrostatics for calculating pKa's in proteins. Biophys J 83:1731–1748
26. Alexov E, Gunner MR (1997) Incorporating protein conformational flexibility into pH-titration calculations: results on T4 lysozyme. Biophys J 74:2075–2093

Chapter 7

Site-Specific Fluorescent Probe Labeling of Mitochondrial Membrane Proteins

Christine T. Schwall and Nathan N. Alder

Abstract

The complexity of biological membranes presents technical challenges for the analysis of membrane protein biogenesis and function. Here we describe an in vitro fluorescence-based experimental approach for studying the high-resolution structural features of membrane proteins within isolated mitochondria. By this strategy, membrane proteins are cotranslationally labeled with a fluorescent probe at a specific site by the inclusion of aminoacyl tRNA analogs in a cell-free translation system. Labeled proteins are then targeted to the correct subcompartment within active mitochondria by the endogenous import machinery. For each site-specifically labeled protein, a series of rigorous controls must be conducted to ensure the proper membrane integration, topology, and assembly of each labeled sample. The assays described herein serve as the basis for more sophisticated analyses by which multiple fluorescence-based measurements can render detailed information on the topology, microenvironment, and dynamic conformational changes as they occur in real time.

Key words Cell-free transcription and translation, Fluorescence, Site-specific labeling, Mitochondria, Membrane protein, Analytical fluorescence

1 Introduction

Within the mitochondrial proteome (consisting of ~1,500 individual proteins in man, ~900 in the model eukaryote *Saccharomyces cerevisiae* [1, 2]), a substantial number of proteins (roughly 25 %) are associated with or embedded within membranes [3]. Yet our understanding of the biogenesis, structure, and function of mitochondrial membrane proteins has been hindered by the technical obstacles that plague the study of membrane proteins in general: their hydrophobic lipid-interactive surfaces cause them to aggregate in water and they are recalcitrant to standard solution-based biochemical and biophysical approaches. A further complication arises from the complexity of the native inner and outer membranes (IM and OM, respectively). Both membranes are extremely protein-rich (having a protein:phospholipid ratio of about 3:1 for the

Doron Rapaport and Johannes M. Herrmann (eds.), *Membrane Biogenesis: Methods and Protocols*,
Methods in Molecular Biology, vol. 1033, DOI 10.1007/978-1-62703-487-6_7, © Springer Science+Business Media, LLC 2013

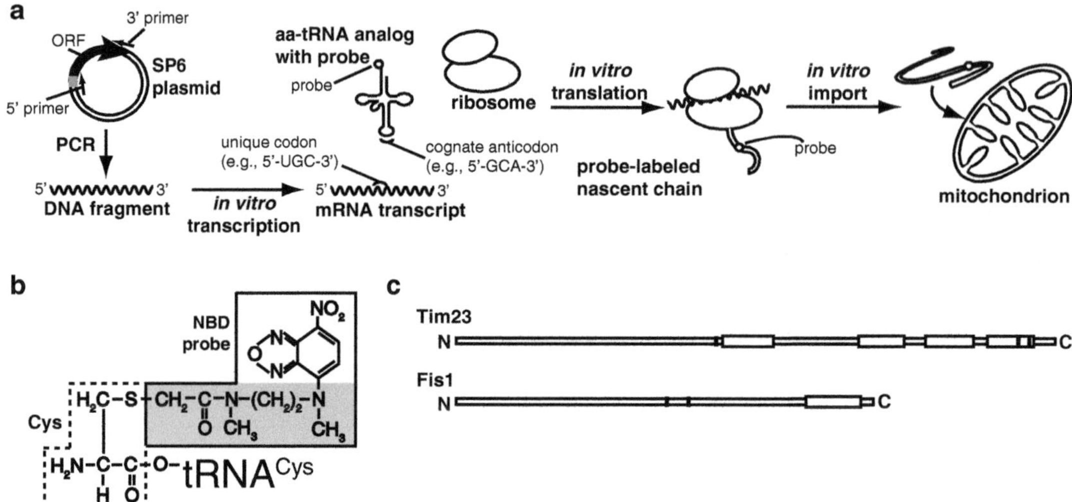

Fig. 1 (**a**) Protocol flowchart. (**b**) Structure of NBD-Cys-tRNA^{Cys} with the cysteine boxed by *dotted lines* and the NBD probe boxed in *solid lines* with the tether highlighted in *grey*. (**c**) Schematic of Tim23 and Fis1 constructs used in this protocol. *Open boxes* represent transmembrane segments; *dark boxes* represent native cysteine sites

IM and 1:1 for the OM) [1]. Hence, this molecular environment makes it difficult to analyze specific membrane proteins—let alone single sites on specific proteins—within the context of their native biomembranes.

We have developed a fluorescence-based experimental approach for the analysis of unique sites on mitochondrial membrane proteins within their biologically relevant lipid bilayer [4]. By this approach (Fig. 1a), the protein of interest is synthesized in an in vitro translation system and nonnatural (fluorescent-labeled) amino acids are incorporated cotranslationally at a specific site using aminoacyl tRNA (aa-tRNA) analogs. These aa-tRNA analogs are typically derived from cysteinyl or lysyl tRNAs (tRNA^{Cys} and tRNA^{Lys}, respectively) that are enzymatically aminoacylated with either cysteine or lysine (yielding Cys-tRNA^{Cys} and Lys-tRNA^{Lys}, respectively). These amino acids are used because their chemical reactivity (a reactive thiol group in the case of cysteine, the primary amine group in the case of lysine) allows them to be covalently bound to reactive fluorescent derivatives (e.g., iodoacetamide esters or maleimides of fluorophores for cysteine, succinimidyl esters for lysine) following their aminoacylation to the cognate tRNA [4–6].

Polypeptides are labeled at single sites by programming the translation reactions with mRNA transcripts that contain a unique in-frame codon that is recognized by the aa-tRNA. For example, the anticodon of our tRNA^{Cys} analog recognizes the 5′-UGC-3′ codon. Using the tRNA^{Cys}-based approach requires that native cysteines be replaced (e.g., by alanine or serine) to remove redundant labeling sites and that a single in-frame 5′-UGC-3′ codon be engineered at the desired probe site by cysteine-scanning mutagenesis. Provided that native cysteines are not critical for structure or function, this is a viable option, particularly given the relative paucity of cysteines in native proteins. Lysines, typically being much more abundant, are more difficult to remove without structural consequence. Hence, suppressor tRNAs based on the $tRNA^{Lys}$ scaffold whose anticodons are engineered to recognize amber stop (5′-UAG-3′) codons have been widely used to incorporate labeled lysine residues [7–9]. In fact, suppressor tRNAs that recognize amber, opal (5′-UGA-3′), or ochre (5′-UAA-3′) codons that are based on both the $tNRA^{Lys}$ and $tRNA^{Cys}$ scaffolds have been successfully used for site-specific labeling [10].

Although many types of fluorescent probes can be employed by our approach, in this chapter we shall focus on the use of one: the environment-sensitive fluorophore 7-nitrobenz-2-oxa-1,3-diazolyl (NBD) (Fig. 1b). NBD fluorescence is dependent upon the polarity and hydrogen bonding capacity of its microenvironment [11, 12]; therefore, it is an excellent reporter for whether a site on a membrane protein faces an aqueous environment or the nonpolar milieu of a lipid bilayer. NBD is also small and uncharged and is therefore less likely to impact the structure or function of the labeled polypeptide than larger fluorophores. Here we describe the site-specific incorporation of NBD into two model proteins with α-helical transmembrane segments: Tim23, the polytopic channel-forming subunit of the TIM23 complex in the inner membrane, and Fis1, a single-spanning component of the fission machinery in the outer membrane (Fig. 1c). In each case, NBD-Cys-$tRNA^{Cys}$ is used to incorporate the probe into unique sites of monocysteine constructs. Following translation, labeled membrane proteins are incorporated into the proper membrane of isolated mitochondria by the endogenous mitochondrial import machinery. After preparative steps, mitochondria containing the labeled protein are subjected to a series of gel-based controls and quantitative fluorescence analysis. This protocol serves as the foundation for more sophisticated spectral analyses designed to investigate the structure and function of proteins in a native membrane environment using a range of independent fluorescence-based approaches [4, 13].

2 Materials

All solutions must be prepared with strict adherence to nuclease-free technique (*see* **Note 1**). Buffers and salts (HEPES, $MgCl_2$, KCl, $Mg(OAc)_2$, KOAc, and KPO_4) are prepared as large (250 ml) stocks and stored at 4 °C. All other reagents are prepared and partitioned into aliquots as indicated below and stored at −80 °C.

2.1 PCR Generation of DNA Fragments and In Vitro Transcription (See Note 2)

1. Plasmid containing the gene of interest inserted behind an SP6 promoter. We use plasmids such as pSP65 or pGEM4Z (Promega) with the open reading frame (ORF) of interest (Tim23 and Fis1, respectively) subcloned into the multicloning site (*see* **Note 3**).
2. Primers for DNA amplification of the ORF. The upstream (5′) oligonucleotide is complementary to the plasmid SP6 promoter and the downstream (3′) oligonucleotide is complementary to the gene sequence downstream of the stop codon (*see* **Note 4**).
3. DNA polymerase and dNTP mix.
4. PCR cleanup kit.
5. PCR thermal cycler and thin-walled PCR tubes (*see* **Note 5**).
6. Refrigerated microcentrifuge.
7. Transcription Buffer (10×): 1 M HEPES-KOH, pH 7.5, 200 mM $MgCl_2$, and 25 mM spermidine.
8. 100 mM DTT.
9. Ribonucleotide (rNTP) mixture: 100 mM each of ATP, GTP, UTP, and CTP in 20 mM Tris–HCl, pH 7.5.
10. 0.1 U/μl Diguanosine Triphosphate [G(5′)ppp(5′)G].
11. 20 U/μl Ribonuclease Inhibitor.
12. 0.05 U/μl pyrophosphatase.
13. SP6 RNA Polymerase (*see* **Note 6**).
14. 3 M NaOAc (pH 5.2).
15. 100 % Ethanol.
16. TE Buffer (10 mM Tris–HCl, pH 7.5, 1 mM EDTA, pH 7.5).
17. Rotary evaporator.

2.2 In Vitro Translation (See Note 7)

1. Translation Buffer (10×): 200 mM HEPES-KOH, pH 7.5, 1 M KOAc, pH 7.5, 25 mM $Mg(OAc)_2$, 2 mM spermidine, 0.08 mM *S*-adenosylmethionine.
2. 200× Protease Inhibitor mixture: 50 μg/ml each of Antipain, Chymostatin, Leupeptin, Pepstatin A, and 5 % (v/v) of Aprotinin.
3. Energy Generating System and amino acids without cysteine (EGS-C): 90 mM HEPES-KOH, pH 7.5, 15 mM ATP, pH 7.5, 15 mM GTP, pH 7.5, 120 mM creatine phosphate, 0.12 U/μl creatine phosphokinase, 0.38 mM each of 19 amino acids (except cysteine).
4. Wheat germ extract (*see* **Note 8**).
5. Aminoacyl tRNA analog NBD-Cys-$tRNA^{Cys}$ (15 pmol/μl) (tRNA Probes Catalog #C-65) (*see* **Note 9**).
6. tRNA Buffer: 2 mM $MgCl_2$, 1 mM KOAc, pH 7.5.

2.3 Mitochondrial Import for Sample Characterization

1. Mitochondria isolated from *Saccharomyces cerevisiae* as described [14] and stored at −80 °C in aliquots of 0.5 mg mitochondrial protein.
2. 2 M sucrose.
3. Bovine serum albumin (fatty acid-free).
4. 1 M potassium phosphate, pH 7.5. Mix 81.0 ml of 1 M K_2HPO_4 and 19.0 ml of 1 M KH_2PO_4.
5. 0.2 M malate.
6. 0.2 M pyruvate.
7. 100 mM ATP, pH 7.5.
8. 2.5 mg/ml Proteinase K.
9. 100 mM phenylmethylsulfonyl fluoride (PMSF) prepared in ethanol.
10. 50 mM carbonyl cyanide 3-chlorophenylhydrazone (CCCP) prepared in ethanol.
11. 3 mg/ml Ribonuclease A.
12. 50 % (v/v) trichloroacetic acid.
13. Fluorescent molecular weight markers.
14. ^{14}C-methylated protein molecular weight markers.
15. Electrophoresis running buffer (25 mM Sigma 7–9 buffer, 0.2 M glycine, 0.1 % (w/v) sodium dodecyl sulfate).
16. SDS-PAGE Sample Buffer (125 mM Tris–Base, 18 % (v/v) glycerol, 3.6 % (w/v) sodium dodecyl sulfate, 0.045 % (w/v) bromophenol blue, 0.1 M DTT).
17. Destain solution (50 % (v/v) methanol, 10 % (v/v) acetic acid).
18. Gel imaging system capable of imaging fluorescent and radiolabeled samples (*see* **Note 10**).
19. Gel drying apparatus with gel dryer and vacuum pump.
20. Imaging screen-K, 20×25 cm phosphor imaging screen and exposure cassette-K (Bio-Rad).

2.4 Mitochondrial Import for Analytical Fluorescence

Please note that the chemicals and buffers utilized in Subheading 2.3, **steps 1–7**, are also needed for Subheading 2.4.

1. Steady state spectrofluorometer (*see* **Note 11**).
2. Quartz 4×4 mm cuvettes (Starna Cells, Inc.).
3. Flea micro stir bars (Bel-Art Products).

3 Methods

3.1 PCR Generation of DNA Fragments and In Vitro Transcription (See Note 12)

1. Assemble the PCR reaction mixture (Table 1) in a PCR tube (*see* **Notes 13** and **14**).
2. Add reaction to PCR thermal cycler programmed as follows: first denature (94 °C, 4 min); 30 cycles of amplification (denature 94 °C, 20 s; anneal 54 °C, 20 s; extension 72 °C, 30 s); final extension (72 °C, 4 min) (*see* **Note 15**).
3. Purify the amplified DNA fragment using a PCR purification kit and elute with 50 µl of RNase-free H_2O (*see* **Note 16**).
4. Assemble the in vitro transcription mixture (Table 2) in a 1.5 ml microfuge tube and incubate the reaction at 37 °C for 1.5 h (*see* **Note 17**).
5. Precipitate mRNA by adding 13.3 µl of 3 M NaOAc (pH 5.2) and 340 µl of 100 % ethanol and incubate on ice for 2 h.
6. Spin samples at maximum speed for 20 min in a microfuge (4 °C) and aspirate supernatant without disturbing the pellet.
7. Wash pellet with 1 ml of 70 % (v/v) ethanol, spin for an additional 10 min, and aspirate supernatant.
8. Dry the pellet in a rotary evaporator for 5 min (*see* **Note 18**).
9. Resuspend mRNA pellet in 100 µl TE buffer.
10. Aliquot mRNA in volumes of 25 µl each, flash-freeze in liquid nitrogen, and store at −80 °C (*see* **Note 19**).

3.2 In Vitro Translation

1. For gel-based analysis of protein labeling (Subheading 3.3): assemble the in vitro translation mixture for fluorophore-containing "sample" only (Table 3, column 1) in a 1.5 ml

Table 1
PCR amplification mixture

Component	Volume (µl)
Nuclease-free H_2O	80.5
10× PCR buffer	10.0
10 mM dNTP mix	2.0
50 mM $MgCl_2$	4.0
50 µM forward primer	1.0
50 µM reverse primer	1.0
100 ng/µl DNA template	1.0
DNA polymerase (5U/µl)	0.5
Total	100.0

Table 2
In vitro transcription mixture

Component	Volume (μl)
Nuclease-free H_2O	43.3
10× Transcription buffer	10.0
100 mM DTT	10.0
100 mM rNTP mixture	4.0
0.1 U/μl (5′)ppp(5′)G	13.0
20 U/μl RNasin	2.0
PCR product DNA	13.0
SP6 RNA polymerase	3.5
0.5 U/μl pyrophosphatase	1.2
Total	100.0

Table 3
In vitro translation mixture

Component	Sample volume (μl)	Blank volume (μl)
Nuclease-free H_2O	240.0	240.0
10× Translation buffer	50.0	50.0
100 mM DTT	5.0	5.0
200× Protease inhibitors	2.5	2.5
20 U/μl RNasin	2.5	2.5
EGS-C	40.0	40.0
Wheat germ extract	100.0	100.0
mRNA	40.0	40.0
15 pmol/μl NBD-tRNA	20.0	n/a
tRNA buffer	n/a	20.0
Total	500.0	500.0

microfuge tube and incubate the reaction at 26 °C for 40 min (*see* **Note 20**).

2. For spectral analysis of labeled protein (Subheading 3.4): assemble the in vitro translation (Table 3) for both “sample” (with fluorophores) and “blank” (lacking fluorophores, but otherwise identical) in separate 1.5 ml microfuge tubes and incubate the reactions at 26 °C for 40 min (*see* **Note 20**).

3. Translation reactions can be stored on ice while setting up the import reaction (*see* **Note 21**).

3.3 Mitochondrial Import for Sample Characterization

1. Thaw mitochondria (0.5 mg aliquot) on ice for 30 min and dilute to a final concentration of 0.5 mg/ml in ice-cold Import Buffer (Table 4).
2. Prepare 1.0 ml of Import Buffer with respiratory substrate and ATP (886 μl of Import Buffer with 38 μl each of 0.2 M malate, 0.2 M pyruvate, and 0.1 M ATP, pH 7.5).
3. Assemble six 0.3 ml import reactions (Table 5), one set of three reactions for mitochondria with a membrane potential (+Δψ, samples 1–3) and another set of three reactions for those lacking a membrane potential (−Δψ, samples 4–6), and incubate at 26° for 20 min (*see* **Notes 22** and **23**). Save the remaining translation products for gel analysis.

Table 4
Buffers for protein import and spectral measurements

Component	Import buffer	Measurement buffer
H_2O	9.7 ml	9.7 ml
1 M HEPES-KOH, pH 7.5	240 μl	240 μl
2 M KCl	480 μl	480 μl
1 M $MgCl_2$	60 μl	60 μl
1 M KPO_4, pH 7.5	24 μl	24 μl
2 M sucrose	1.5 ml	1.5 ml
BSA (fatty acid-free)	36 mg	n/a
Total	12 ml	12 ml

Table 5
Import reaction mixture: samples for gel-based analysis

Component	Import (+Δψ)	Import (−Δψ)
Import buffer (with malate/pyruvate/ATP)	118.5 μl	118.5 μl
Mitochondria (0.5 mg/ml)	150 μl	150 μl
100 % Ethanol	1.5 μl	n/a
10 mM CCCP	n/a	1.5 μl
NBD-translation reaction	30 μl	30 μl
Total	0.3 ml	0.3 ml

Table 6
Post-import buffers for gel-based analysis

Component	Stop buffer	Hypotonic buffer
H_2O	4.28 ml	9.9 ml
1 M HEPES, pH 7.5	100 μl	100 μl
2 M sucrose	0.625 ml	n/a
Total	5 ml	10 ml

4. Prepare ice-cold Stop Buffer and Hypotonic Buffer (Table 6) for post-import treatments (the following **steps 5–7**).
5. No protease treatment (samples 1 and 4): add 600 μl of Stop Buffer to 300 μl imports and incubate on ice for 20 min.
6. Protease treatment of intact mitochondria (samples 2 and 5): add 600 μl of Stop Buffer to 300 μl imports and add 36 μl of 2.5 mg/ml Proteinase K. Mix well and incubate on ice for 20 min.
7. Protease treatment with hypotonic swelling (samples 3 and 6): divide import reactions into 3 × 100 μl reactions and add to each 900 μl of Hypotonic Buffer. Add to each tube 36 μl of 2.5 mg/ml Proteinase K, mix well, and incubate on ice for 20 min (*see* **Notes 24** and **25**).
8. To all samples (**steps 5–7**), add 10 μl of 100 mM PMSF to quench protease activity and incubate on ice for 5 min.
9. Sediment mitochondria in refrigerated microfuge (15,000 × *g*) for 5 min and aspirate supernatant. Resuspend all samples in a total 21 μl SDS-PAGE Sample Buffer and heat at 60 °C for 10 min (*see* **Note 26**).
10. From the remaining translations, take two 80 μl aliquots. To the first ("−RNase A"), add 1.0 μl H_2O and mix. To the second ("+RNase A"), add 1.0 μl of 3 mg/ml RNase A. Incubate tubes at 26 °C for 10 min. Add 80 μl of ice-cold 50 % TCA to both tubes, incubate for 15 min on ice, pellet precipitated proteins in a microfuge, aspirate supernatant, and wash each pellet in 200 μl of acetone: hydrochloric acid (19:1). Aspirate supernatants, air-dry pellets, and resuspend each pellet in 100 μl SDS-PAGE Sample Buffer.
11. Set up the electrophoresis apparatus according to the manufacturer's instructions.
12. Load fluorescent and (if necessary) radiolabeled molecular weight markers (5 μl each) in lanes 1 and 2 (*see* **Note 27**); load translation reactions ("−RNase A" and "+RNase A", 20 μl each) in lanes 3 and 4; load mitochondria import reactions 1–6 (20 μl each) in the subsequent lanes. *See* Fig. 2a for gel loading schematic.

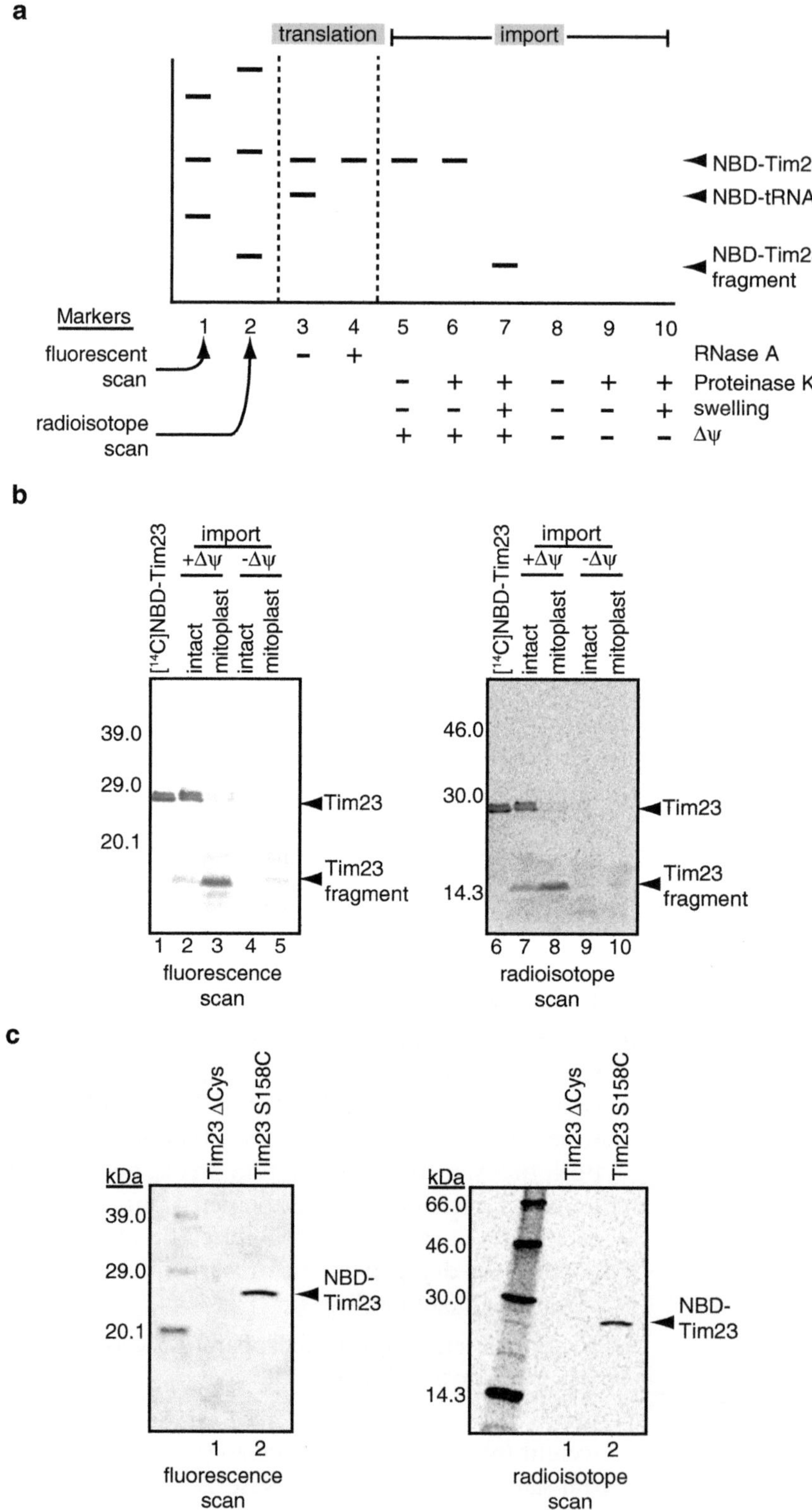

Fig. 2 Assay for probe labeling specificity, integration, and topogenesis of NBD-Tim23. (**a**) Schematic of gel banding pattern expected for labeling and import of Tim23 described in the protocol. Fragmented Tim23 is due

13. Run the gel at 125 V until the bromophenol blue dye front reaches the bottom of the separating gel.
14. Wash the gel in 100 ml of Destain solution for 10 min followed by two 10 min washes in water.
15. Perform in-gel fluorescence scan using the molecular imager to detect the NBD-labeled protein (*see* **Notes 28** and **29**).
16. For radiolabeled samples, dry gel for 40 min at 80 °C and place on a phosphor imaging screen (*see* **Note 30**).
17. Perform radioisotope scan using the molecular imager to detect ^{14}C-labeled protein.
18. For an inner membrane-integrated protein with a large intermembrane space domain like Tim23, a gel loaded as described in this protocol will ideally resemble the banding pattern schematized in Fig. 2a. Results from one such import test of Tim23 labeled at position 131 are shown in Fig. 2b.
19. To demonstrate that labeling occurs specifically at the desired locus (a cysteine in this case), one should perform a negative control by programming a translation with an mRNA transcript devoid of an in-frame cysteine codon (Fig. 2c). Posttranslational ribonuclease treatment allows one to differentiate between the NBD-labeled protein and the NBD-Cys-tRNACys (Figs. 2a and 3).

3.4 Mitochondrial Import and Analytical Fluorescence

1. Thaw mitochondria (0.5 mg aliquot) on ice for 30 min and dilute to a final concentration of 0.167 mg/ml in ice-cold Import Buffer (Table 4).
2. Prepare 4.0 ml of Import Buffer with respiratory substrate and ATP (3.76 ml of Import Buffer with 80 μl each of 0.2 M malate, 0.2 M pyruvate, and 0.1 M ATP, pH 7.5).
3. Assemble six 0.8 ml import reactions (Table 7), three identical "Sample" and three identical "Blank" reactions, and incubate at 26 °C for 20 min.
4. Sediment mitochondria in refrigerated microfuge (10,000 × *g*) for 5 min and aspirate supernatant. Gently resuspend pellets in 0.8 ml of Import Buffer with respiratory substrate and incubate on ice for 15 min. Perform a second centrifugation step,

Fig. 2 (continued) to a subpopulation of mitochondria that do not remain intact. (**b**) Assay for the integration and topogenesis of NBD-Tim23. [^{14}C]NBD-Tim23 K131C was incubated with isolated mitochondria in the presence ("+Δψ") or absence ("−Δψ") of a membrane potential, and left intact ("intact") or subject to hypotonic swelling ("mitoplast") prior to proteinase K treatment. A small fraction of intact samples (lanes 2 and 7) yield a proteolytic fragment, likely due to a subpopulation of mitochondria that are not intact. (**c**) NBD labeling specificity during cell-free translation. Translations were programmed with the indicated mRNA transcript in the presence of NBD-[^{14}C]Cys-tRNA^{Cys}, resolved on SDS-PAGE and subjected to fluorescence and radioisotope scan as shown. Reproduced from [4] with permission from Elsevier

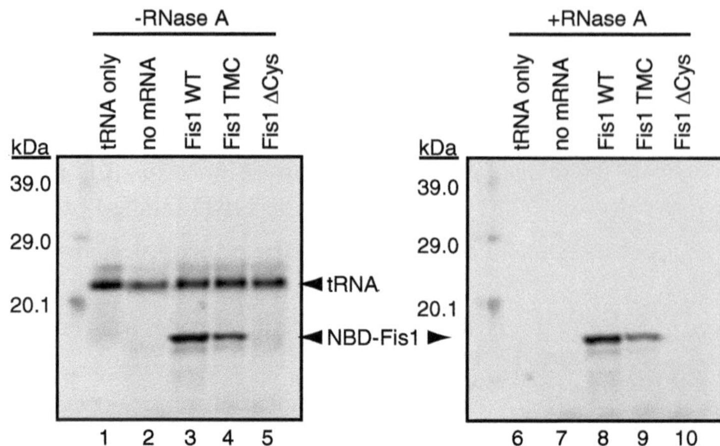

Fig. 3 Assay for probe labeling specificity of Fis1. All translations were programmed with the indicated mRNA transcript in the presence of NBD-Cys-tRNA^{Cys}. Fis1 WT has two native cysteines in the soluble domain, Fis1 TMC has a single cysteine in the transmembrane segment, and Fis1 ΔCys is devoid of cysteines. After translation, half of the samples were treated with RNase A as indicated. Samples were resolved on SDS-PAGE and subjected to fluorescence scans

Table 7
Import reaction mixture: samples for spectral analysis

Component	Sample	Blank
Import buffer (with malate/pyruvate/ATP)	576 μl	576 μl
Mitochondria (0.167 mg/ml)	144 μl	144 μl
Sample translation reaction	80 μl	n/a
Blank translation reaction	n/a	80 μl
Total	0.8 ml	0.8 ml

aspirate supernatant, and resuspend each pellet in 87 μl of Measurement Buffer (Table 4) with respiratory substrate and ATP. Combine all three "Sample" fractions and all three "Blank" fractions to have two uniform samples, one "Sample" and one "Blank" (*see* **Note 31**).

5. Load blank fraction and NBD-sample fraction (each one 260 μl) into two separate quartz microcells preloaded with clean micro stir bars and cap them (*see* **Note 32**). Keep cuvettes on ice during spectral measurements.
6. Using a large magnetic stir bar, stir the sample by moving the internal micro stir bar up and down a total of 15 times per side. Remove all condensation from the sides of the cuvette with a Kimwipe. Insert the cuvette into the fluorimeter chamber.

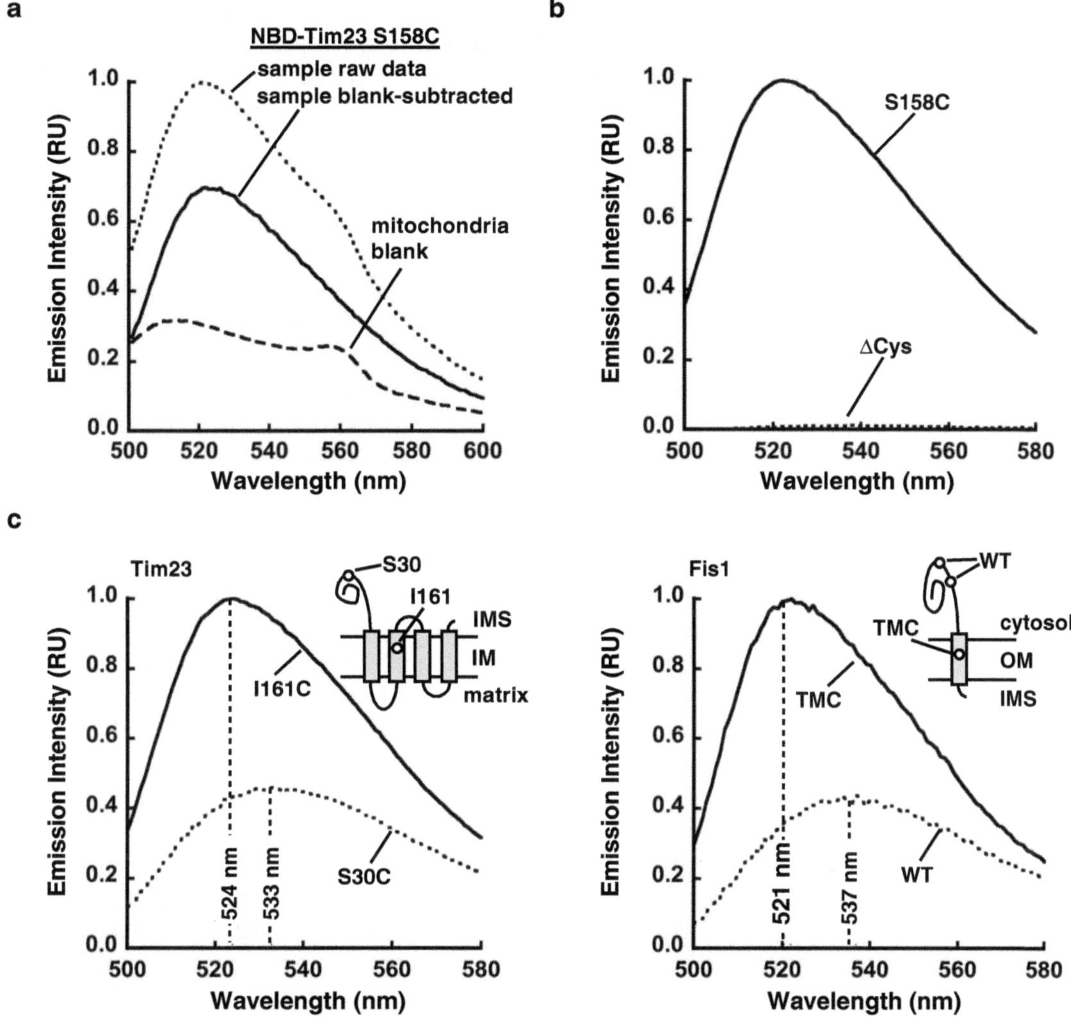

Fig. 4 Emission scans of integrated NBD-Tim23 and NBD-Fis1. (**a**) Obtaining the net NBD signal. Scans of mitochondria containing NBD-Tim23 S158C (sample raw data, *dotted trace*) and unlabeled Tim23 S158C (mitochondria blank, *dashed trace*) with the net signal originating from NBD-Tim23 S158C (*solid trace*). (**b**) Testing the specificity of labeling. Translations containing NBD-[^{14}C]Cys-tRNA^{Cys} were programmed with either the cysteine-less Tim23 ΔCys or Tim23 S158C. The construct with the cysteine probe incorporation site (S158C, *solid trace*) should yield a strong NBD signal; the signal from the cysteine-less construct (ΔCys, *dotted trace*) should be negligible. (**c**) Analyzing the NBD microenvironment by solvent effects. Variants of Tim23 (*left*) and Fis1 (*right*) were labeled with NBD at solvent-exposed or transmembrane sites as indicated in the diagrams (*IM* inner membrane, *OM* outer membrane, *IMS* intermembrane space, *grey boxes* transmembrane helices). In these measurements, the membrane localized probe on Tim23 has a ~2.5-fold higher emission intensity than the aqueous-exposed probe and a 9 nm blue-shifted λ_{max}. The membrane localized probe on Fis1 has a ~2.5-fold higher emission intensity than the aqueous-exposed probes and a 16 nm blue-shifted λ_{max}. Tim23 scans in panels (**b**) and (**c**) reproduced from [4] with permission from Elsevier

7. Perform emission scans of NBD-containing samples by exciting at 470 nm and scanning over emission wavelengths of 490–590 nm (*see* **Note 33**).

8. After performing the scans for the blank (Fig. 4a, dashed trace) and the NBD-containing sample (Fig. 4a, dotted trace), the net signal originating from the NBD-labeled protein is determined by subtracting the blank from the sample (Fig. 4a, solid trace) (*see* **Note 34**).
9. To confirm that the measured emission originates only from the NBD probe at the desired position, one can perform parallel translations in the presence of NBD-Cys-tRNA^{Cys} (one programmed with mRNA encoding a monocysteine construct and one encoding a ΔCys construct) and perform imports into mitochondria. The lack of signal from the ΔCys negative control (Fig. 4b) confirms that the emission does not come from dye spuriously incorporated into the polypeptide or from nonspecifically bound aa-tRNA.
10. NBD probes positioned at different regions of the polypeptide of interest serve as reporters for the polarity and water exposure of the probe microenvironment (Fig. 4c). Using emission scans, the aqueous accessibility can be measured by both the emission intensity (higher in nonpolar environments) and spectral shifts (the wavelength of maximum emission intensity [λ_{max}] decreases [blue-shifts] in nonpolar environments). This analysis provides information on the topology of the labeled protein (i.e., which sites are exposed to the aqueous environment and which sites are exposed to the nonpolar core of the bilayer). Analysis of multiple independent samples in which probes are located sequentially along a structural element (e.g., a channel-facing α-helix) can render high-resolution information regarding the water-facing and bilayer-facing surfaces.

4 Notes

1. Because in vitro transcription and translation steps include tRNA and mRNA, it is imperative that all solutions be prepared under RNase-free conditions with nuclease-free H_2O.
2. As an alternative to the transcription protocol outlined here, kits are commercially available such as the Riboprobe® Combination System (Promega).
3. Plasmid stocks are stored at concentrations of about 100 ng/μl in TE Buffer (*see* Subheading 2.1 Reagent #16) and stored at −20 °C.
4. Primer stocks are stored at concentrations of 50 μM in TE Buffer at −20 °C.
5. We use a Bio-Rad C1000 thermal cycler and low tube strip individual PCR tubes.

6. We synthesize our RNA polymerases in-house. SP6 polymerase is available commercially from vendors such as tRNA Probes and New England Biolabs.
7. As an alternative to the translation protocol outlined here, wheat germ lysate-based commercial kits are available such as the TNT® Coupled Wheat Germ Extract System (Promega).
8. We prepare our own wheat germ lysate in-house following established protocols [15, 16]. Wheat germ extract is available commercially from vendors such as tRNA Probes.
9. The radiolabeled version of this analog, NBD-[^{14}C]Cys-tRNACys (tRNA Probes, #C-35), is required for radioisotope-based characterization (Fig. 2b, c).
10. We use a Bio-Rad Pharos FX Plus Molecular Imager with external lasers. This system allows in-gel imaging of radiolabeled and fluorescent samples. For gel-based imaging of fluorescent samples, alternative systems can be used, provided the excitation source and detectors are compatible with the dye used for labeling. When radiolabeled probes are used, imaging can be accomplished with systems designed for detection of radiolabeled samples (e.g., Bio-Rad PMI system).
11. Our steady state instrument is a Spex Fluorolog 3-22 spectrofluorometer equipped with photon counting electronics and double-grating excitation and emission monochromators. We have empirically found that this instrument offers superior sensitivity and rejection of background signal. These features are critical for the proper analysis of mitochondria-containing samples described here, which display strong light scattering and high background fluorescence (Fig. 4a).
12. Instead of using PCR-amplified DNA fragments as transcription templates, one can use plasmid linearized with an appropriate restriction endonuclease.
13. Reagent concentrations here have been optimized for the Bio-Rad iTaq Polymerase system; optimal concentrations should be tested for enzyme from other sources.
14. PCR amplification efficiency is strongly dependent on magnesium concentration; the optimal [$MgCl_2$] should be empirically tested for each construct.
15. Optimal temperatures and times will vary depending on properties of the construct (e.g., ORF length).
16. Purification of PCR products is done in accordance with the manufacturer's protocol (we use the QIAGEN QIAquik PCR Purification Kit). We typically run PCR DNA on a 1.8 % agarose gel alongside DNA standards to confirm successful amplification of the desired product.

17. To enhance the incorporation of diguanosine triphosphate, add one-tenth the concentration of GTP initially, then supplement the reaction with 4 mM GTP after 1 h.
18. Do not exceed a time of 5 min in the rotary evaporator or the pellet will be very difficult to resuspend.
19. mRNA in TE Buffer can be stored in an ultracold freezer for several months and can withstand multiple freeze/thaw cycles.
20. To enhance readout of endogenous mRNA prior to addition of the label, we incubate reactions at 26 °C for 5 min prior to mRNA and tRNA addition.
21. We prepare all translations immediately before use and do not subject them to freeze/thaw cycles.
22. Import reactions conducted with depolarized inner membranes ("$-\Delta\psi$" samples) serve as negative controls for the integration of inner membrane proteins, because this process requires the presence of an electric field across the membrane (e.g., [17]).
23. During the setup of the import reaction when uncouplers like CCCP are used, we typically preincubate mitochondria with the uncoupler at room temperature for at least 1 min before adding the translation products.
24. Samples need to be split into 3 × 100 μL aliquots because after adding 900 μL Hypotonic Buffer sample volumes would be too large to fit into one microfuge tube; however, these three identical samples will be combined and run on the gel as one uniform sample (*see* **Note 26**). The splitting apart of samples due to maximum volume concerns is completed again in Subheading 3.4, **step 3**, for reactions prepared for spectral analysis.
25. Hypotonic swelling of the osmotically active inner membrane ruptures the outer membrane (producing "mitoplasts"), which exposes protein domains of the intermembrane space to added protease. For proteins such as Tim23 with a large intermembrane space region, the protease protection of the membrane-bound region in mitoplasts serves as a diagnostic indicator of inner membrane integration.
26. For mitoplasts (samples 3 and 6), resuspend each pellet in 7 μl SDS-PAGE Sample Buffer and combine like samples.
27. Fluorescent markers are heated 10 min at 55 °C prior to gel loading. ^{14}C methylated markers are mixed in an equal volume of SDS-PAGE sample buffer (but not heated) prior to loading.
28. On the Pharos FX Plus Imager, NBD is detected using the "FITC" setting (488 nm blue laser source for excitation, emission set to 530 nm).

29. The step of gel preparation that will yield the optimal in-gel fluorescence signal can vary. Scans should be taken directly after electrophoresis, following destain incubation, between water washes, and then following gel drying to obtain the clearest image possible.
30. Sufficient signal from ^{14}C-labeled proteins can generally be obtained by overnight exposure to the phosphor screen.
31. One must resuspend very gently, generally by pipetting up and down a total of 20 times per pellet. In our experience, this mode of sample preparation is sufficient to remove excess dye and off-pathway labeled intermediates. More stringent wash steps may be required for other dyes or constructs.
32. Quartz cuvettes must be cleaned thoroughly between measurements to avoid cross-contamination of fluorophores. We clean our microcells by soaking them in nitric acid for 20 min, washing thoroughly in nuclease-free H_2O, and drying with compressed nitrogen. Stir bars and cuvette caps are washed with a mild detergent and rinsed thoroughly with nuclease-free H_2O.
33. Using a 4 nm bandpass and integration times of 2 s per 1 nm increments will generally give sufficient signal for NBD-labeled proteins in mitochondria.
34. In any aqueous sample, the Raman peak will appear at a wavenumber 3,600 cm^{-1} below that of the incident wavenumber. Moreover, fluorescence signal from mitochondria alone (Fig. 4a, dashed trace) originates from its many endogenous fluorophores (e.g., pyridine and flavin nucleotides). Often the NBD probe of interest in the sample will have very low intensity (e.g., when it is completely solvated with water or near a quenching moiety); thus, the total NBD signal may represent only a small fraction of the total fluorescence signal in the sample.

Acknowledgments

We thank Dr. Doron Rapaport (University of Tuebingen) for the pGEM4-Fis1 plasmid constructs described herein. This work is supported by grants from the American Heart Association (09SDG2380019) and from the National Science Foundation (MCB-1024908) to NA and by an NSF Graduate Research Fellowship to CS.

References

1. Alder NN (2011) Biogenesis of lipids and proteins within biological membranes. In: Yeagle PL (ed) The structure of biological membranes, 3rd edn. CRC, New York, pp 315–377
2. Schmidt O, Pfanner N, Meisinger C (2010) Mitochondrial protein import: from proteomics to functional mechanisms. Nat Rev Mol Cell Biol 11:655–667
3. Distler AM, Kerner J, Hoppel CL (2008) Proteomics of mitochondrial inner and outer membranes. Proteomics 8:4066–4082
4. Alder NN, Jensen RE, Johnson AE (2008) Fluorescence mapping of mitochondrial TIM23 complex reveals a water-facing, substrate-interacting helix surface. Cell 134:439–450
5. Johnson AE, Woodward WR, Herbert E, Menninger JR (1976) Nepsilon-acetyllysine transfer ribonucleic acid: a biologically active analogue of aminoacyl transfer ribonucleic acids. Biochemistry 15:569–575
6. Flanagan JJ, Chen JC, Miao Y, Shao Y, Lin J, Bock PE, Johnson AE (2003) Signal recognition particle binds to ribosome-bound signal sequences with fluorescence-detected subnanomolar affinity that does not diminish as the nascent chain lengthens. J Biol Chem 278:18628–18637
7. Lin PJ, Jongsma CG, Pool MR, Johnson AE (2011) Polytopic membrane protein folding at L17 in the ribosome tunnel initiates cyclical changes at the translocon. J Cell Biol 195:55–70
8. Khushoo A, Yang Z, Johnson AE, Skach WR (2011) Ligand-driven vectorial folding of ribosome-bound human CFTR NBD1. Mol Cell 41:682–692
9. Tamborero S, Vilar M, Martinez-Gil L, Johnson AE, Mingarro I (2011) Membrane insertion and topology of the translocating chain-associating membrane protein (TRAM). J Mol Biol 406:571–582
10. Gubbens J, Kim SJ, Yang Z, Johnson AE, Skach WR (2010) In vitro incorporation of nonnatural amino acids into protein using tRNA(Cys)-derived opal, ochre, and amber suppressor tRNAs. RNA 16:1660–1672
11. Lancet D, Pecht I (1977) Spectroscopic and immunochemical studies with nitrobenzoxadiazolealanine, a fluorescent dinitrophenyl analogue. Biochemistry 16:5150–5157
12. Lin S, Struve WS (1991) Time-resolved fluorescence of nitrobenzoxadiazole-aminohexanoic acid: effect of intermolecular hydrogen-bonding on non-radiative decay. Photochem Photobiol 54:361–365
13. Johnson AE (2005) Fluorescence approaches for determining protein conformations, interactions and mechanisms at membranes. Traffic 6:1078–1092
14. Daum G, Bohni PC, Schatz G (1982) Import of proteins into mitochondria. Cytochrome b2 and cytochrome c peroxidase are located in the intermembrane space of yeast mitochondria. J Biol Chem 257:13028–13033
15. Erickson AH, Blobel G (1983) Cell-free translation of messenger RNA in a wheat germ system. Methods Enzymol 96:38–50
16. Madin K, Sawasaki T, Ogasawara T, Endo Y (2000) A highly efficient and robust cell-free protein synthesis system prepared from wheat embryos: plants apparently contain a suicide system directed at ribosomes. Proc Natl Acad Sci USA 97:559–564
17. Geissler A, Krimmer T, Bomer U, Guiard B, Rassow J, Pfanner N (2000) Membrane potential-driven protein import into mitochondria. The sorting sequence of cytochrome b(2) modulates the deltapsi-dependence of translocation of the matrix-targeting sequence. Mol Biol Cell 11:3977–3991

Chapter 8

Topology Determination of Untagged Membrane Proteins

Iris Nasie, Sonia Steiner-Mordoch, and Shimon Schuldiner

Abstract

The topology of integral membrane proteins with a weak topological tendency can be influenced when fused to tags, such as these used for topological determination or protein purification. Here, we describe a technique for topology determination of an untagged membrane protein. This technique, optimized for bacterial cells, allows the visualization of the protein in the native environment and incorporates the substituted-cysteine accessibility method.

Key words Topology, Untagged, Membrane protein, Cysteine, EmrE, Transport, SCAM

1 Introduction

The atomic resolution structural information on integral membrane proteins is limited, as they are more difficult to express and crystallize than water-soluble proteins [1, 2]. This has necessitated establishing other approaches to define the topology of these proteins, including the genetic fusion of tags [3–8] or reporters [9–12], and the substituted-cysteine accessibility method (SCAM) [13–16].

SCAM has been widely used to determine integral membrane protein topologies. In this technique, one noncritical residue that is thought to be in an intracellular or extracellular loop is mutated by site-directed mutagenesis to a cysteine. Cysteine residues contain sulfhydryl groups that react with a variety of sulfhydryl-specific reagents. By using neutral (membrane permeable) or charged (membrane impermeable) sulfhydryl reagents, SCAM can precisely locate the position of the introduced cysteine residue depending on its orientation—inside or outside the cell. SCAM requires a functional cysteine-less variant where a single cysteine is introduced at a designated location [17].

Genetic fusion of tags is not only used for determining the topology of integral membrane proteins, but also for the facilitation of their purification. Since tags can influence the topology of proteins with a weak topological tendency [18], the topology of these

Doron Rapaport and Johannes M. Herrmann (eds.), *Membrane Biogenesis: Methods and Protocols*,
Methods in Molecular Biology, vol. 1033, DOI 10.1007/978-1-62703-487-6_8,

proteins should be determined in their untagged form, which makes their purification more challenging. The method presented here is based on SCAM and obviates the need for purification of the protein. It was optimized for bacterial cells and used for the topology determination of EmrE, a transporter residing the inner membrane of *Escherichia coli* (*E. coli*) [18]. For this method, we designed single-cysteine mutations positioned in loops that are expected to be on opposite sides of the membrane. Those single-cysteine mutants were expressed in cells and specifically labeled metabolically with [^{35}S]Methionine as described in [19]. The cells were then challenged with a sulfhydryl reagent that is impermeable to the cell membrane (sodium (2-sulfonatoethyl) methanethiosulfonate; MTSES) and can therefore only react with extracellular thiol groups. After the solubilization and denaturation of the protein, the proportion of unreacted thiols was assessed with methyl-polyethylene glycol-maleimide 5,000 (Mal-PEG) [20]. Reaction with Mal-PEG results in a mass addition of 5,000 Da to the protein and therefore in a significant shift of its mobility that can be easily detected by SDS-polyacrylamide gel electrophoresis (PAGE) analysis. The results from the gel were then digitally analyzed: the band intensities were quantified and the ratio of Mal-PEG-reacted thiols to total thiols was calculated (*see* flow chart in Fig. 1a). For comparison, same procedure was done with the tagged form of these same single-cysteine mutants.

2 Materials

2.1 Specific Labeling with [^{35}S]Methionine

1. Desirable gene harboring single-cysteine mutation (in a cysteine-less background) inserted into an expression plasmid containing a phage T7 RNA polymerase promoter (*see* **Note 1**).
2. An *E. coli* strain transformed with the pGP1-2 "induction" plasmid. pGP1-2 encodes for the T7 polymerase under the inducible control of the λ PL promoter and contains the genes for the heat-sensitive λ repressor cI857 and kanamycin resistance (*see* **Note 2**).
3. 2.5 mg/ml Thiamine stock solution in double distilled water (DDW). Filter and store at room temperature (*see* **Note 3**).
4. 70 % (w/v) Glycerol stock solution in DDW. Autoclave and store at room temperature.
5. 10 % (w/v) $MgSO_4$ stock solution in DDW. Autoclave and store at room temperature.
6. 25 mg/ml Kanamycin stock solution in DDW. Filter. Leave one aliquot at 4 °C for current use and store remaining aliquots at −20 °C.

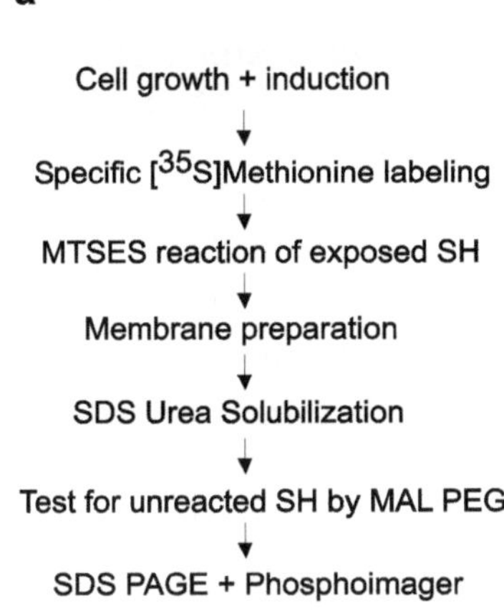

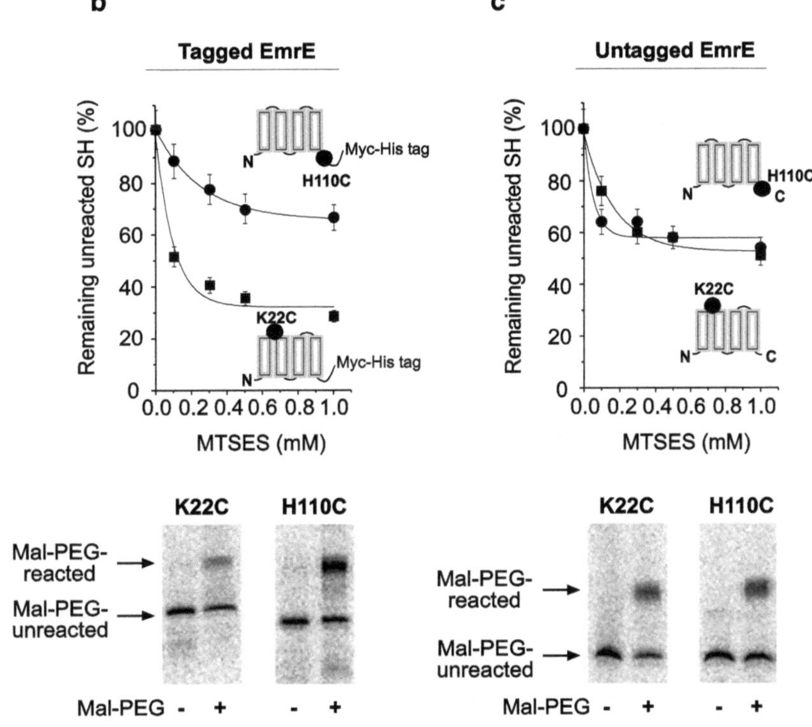

Fig. 1 Tags introduce bias in the topology of EmrE. (**a**) Simplified flow chart of procedure. Tagged (**b**) and untagged (**c**) EmrE-K22C and EmrE-H110C were metabolically labeled with [^{35}S]methionine in *E. coli* cells and treated with MTSES at the indicated concentrations. The unreacted thiols were estimated from the degree of reaction with Mal-PEG. EmrE that reacted with Mal-PEG displays a higher apparent molecular weight (M_r) in SDS-PAGE ((**b**, **c**) *bottom panels*, compare lanes with and without Mal-PEG). The ratios of MTSES-unreacted thiols (intensity of higher M_r bands) over total thiols (intensity of higher M_r bands + intensity of lower M_r bands) were calculated using Image Gauge 3.46 Fujifilm software and are shown in the graphs. The *bottom panels* in (**b**) and (**c**) are samples of the types of changes at one concentration of MTSES: 0.3 mM and 0.1 mM, respectively. Figure originally published in The Journal of Biological Chemistry © Nasie, I., Steiner-Mordoch, S., Gold, A. and Schuldiner S. (2010) Topologically random insertion of EmrE supports a pathway for evolution of inverted repeats in ion-coupled transporters 285, 15234–44. The American Society for Biochemistry and Molecular Biology

7. Minimal Medium A (MMA) + supplements.

   Minimal Medium A (×1)—60 mM K_2HPO_4, 33 mM KH_2PO_4, 7.5 mM $(NH_4)_2SO_4$ (*see* **Note 4**).

   Supplements (final concentrations)—2.5 μg/ml thiamine, 0.01 % $MgSO_4$, 0.5 % glycerol (*see* **Note 5**), 50 μg/ml kanamycin, and a suitable antibiotic for the selection of expression plasmid (we used 100 μg/ml ampicillin for pT7-7).
8. 20 mg/ml Rifampicin stock solution in methanol (*see* **Note 6**).
9. L-[^{35}S]methionine (specific activity, >1,000 Ci/mmol) (Institute of Isotopes Co., Ltd.). Store at 4 °C (*see* **Note 7**).
10. 2 M NaCl stock solution in DDW. Autoclave and store at room temperature.
11. 1 M Tris–HCl pH 7.5 stock solution. For 500 ml: weigh 60.5 g Tris–Base and transfer to a 500 ml measuring cylinder. Add DDW to a volume of 400 ml. Mix on a magnetic stirrer up to dissolving and adjust pH with HCl (*see* **Note 8**). Make up to 500 ml with DDW. Autoclave and store at room temperature.
12. Na buffer: 150 mM NaCl, 15 mM Tris–HCl pH 7.5. Store at room temperature.

2.2 Topology Determination

1. Na buffer: *see* Subheading 2.1 [12].
2. 1 M $MgSO_4$ stock solution in DDW. Autoclave and store at room temperature.
3. 50 mM MTSES (sodium (2-sulfonatoethyl) methanethiosulfonate) (Anatrace, Inc.) stock solution in DDW (*see* **Note 9**).
4. 50 % (w/v) Sucrose stock solution in DDW. Autoclave and store at room temperature.
5. 1 M Tris–HCl pH 8.0 stock solution in DDW. Preparation procedure as in Subheading 2.1 [11].
6. 5 mg/ml Lysozyme stock solution in DDW. Prepare fresh.
7. 0.5 M EDTA stock solution in DDW. Autoclave and store at room temperature.
8. 1.5 μg/ml DNaseI stock solution in 5 mM NaAc, 1 mM $CaCl_2$ pH 4.6–5.0 (titrated by acetic acid). Leave one aliquot at 4 °C for current use and store remaining aliquots at −20 °C.
9. Solubilization solution: 2 % (w/v) SDS, 6 M Urea, 15 mM Tris–HCl, pH 7.5 (*see* **Note 10**).
10. 50 mM Mal-PEG 5,000 (methyl-polyethylene glycol-maleimide; M_R 5,000) (Nektar Transforming Therapeutics, Huntsville, AL) stock solution in DDW. Prepare fresh. Store at 4 °C.

2.3 SDS-Urea Polyacrylamide Gel Electrophoresis

1. Gel buffer: 3 M Tris–HCl, pH 8.45, 0.3 % (w/v) SDS. For 200 ml: Weigh 72.6 g Tris–Base and transfer to a 250 ml measuring cylinder. Add water to a volume of 100 ml. Mix on a magnetic stirrer up to dissolving and adjust pH with HCl (*see* **Note 8**). Make up to 200 ml with water. Add 600 mg SDS and mix gently (*see* **Note 11**). Store at room temperature.
2. 40 % Acrylamide/Bis solution (29:1 acrylamide:Bis). Store at 4 °C (*see* **Note 12**).
3. 10 % (w/v) Ammonium persulfate stock solution in water. Store at 4 °C (*see* **Note 13**).
4. *N*,*N*,*N*′,*N*′-tetramethyl-ethylenediamine (TEMED). Store at 4 °C (*see* **Note 14**).
5. Separating gel: gel buffer (×1) (1 M Tris–HCl, pH 8.45, 0.1 % (w/v) SDS), 16 % acrylamide/Bis solution (29:1 acrylamide:Bis), 10 % (w/v) Glycerol, 6 M Urea.
6. Stacking gel: gel buffer (×1) (1 M Tris–HCl, pH 8.45, 0.1 % (w/v) SDS), 4 % acrylamide/Bis solution (29:1 acrylamide:Bis), 6 M Urea.
7. Cathode buffer (×10): 1 M Tris–Base, 1 M Tricine, pH 8.25, 1 % (w/v) SDS. For 500 ml: Weigh 60.5 g Tris and 89.6 g Tricine and transfer to a 500 ml measuring cylinder. Add water to a volume of 400 ml. Mix on a magnetic stirrer up to dissolving and adjust pH with Tricine (*see* **Note 8**). Make up to 500 ml with water. Add 5 g SDS and mix gently (*see* **Note 11**). Store at room temperature (*see* **Note 15**).
8. Anode buffer (×10): 2 M Tris–HCl, pH 8.9. For 500 ml: Weigh 121 g Tris–Base and transfer to a 500 ml measuring cylinder. Add water to a volume of 400 ml. Mix on a magnetic stirrer up to dissolving and adjust pH with HCl (*see* **Note 8**). Make up to 500 ml with water. Store at room temperature (*see* **Note 15**).
9. Sample buffer (×6): 0.3 M Tris–HCl pH 6.8, 12 % (w/v) SDS, 0.6 % (w/v) bromophenol blue, 0.6 M β-mercaptoethanol, 60 % (v/v) Glycerol. Leave one aliquot at 4 °C for current use and store remaining aliquots at −20 °C (*see* **Note 16**).

3 Methods

3.1 Heat-Induced Over-expression and Specific Labeling with [^{35}S]Methionine

Day 1

Transform expression plasmid containing desirable gene into the pGP1-2 containing strain (*see* **Note 17**). Incubate at 30 °C (*see* **Note 18**).

Day 2

Grow cells at 30 °C in minimal medium A + supplements, over night with shaking (200 rpm) (*see* **Note 19**).

Day 3

1. Dilute cells to $A_{600} = 0.1$ into 50 ml of fresh MMA + supplements (use an Erlenmeyer of 250 ml). Grow at 30 °C.
2. When the culture reaches an $A_{600} = 0.6$, transfer it to a water bath at 42 °C with shaking to induce the T7 polymerase expression. Incubate for 15 min.
3. Add rifampicin (final CONC. 200 μg/ml), and continue incubating with shaking for an additional 10 min at 42 °C (*see* **Note 20**).
4. Shift the culture back to 30 °C for 40 min.
5. Add L-[^{35}S]methionine (final CONC. 10 μCi/ml), and continue incubating at 30 °C for an additional 35 min (*see* **Note 21**).
6. Collect cells by centrifugation (4 °C, 3,200 × *g*, 10 min). Discard supernatant. Put on ice.
7. Wash cells by an addition of 50 ml of ice-cold Na buffer followed by centrifugation (4 °C, 3,200 × *g*, 10 min). Discard supernatant. Put on ice. Repeat washing once more with 10 ml of same solution (*see* **Note 22**).

3.2 Topology Determination (and SDS-Urea PAGE)

Carry out all procedures at room temperature unless otherwise specified.

1. Resuspend pellet with 5.5 ml of Na buffer containing 5 mM $MgSO_4$ (*see* **Note 23**).
2. Divide cells into aliquots of 1 ml (~10 ml of cultured cells per assay), harvest them by centrifugation (4 °C, 21,000 × *g*, 1 min), and resuspend pellet with 200 μl of Na buffer containing 5 mM $MgSO_4$.
3. Add increasing concentrations of MTSES (0, 0.1, 0.3, 0.5, 1.0 mM) and incubate samples at 30 °C for 20 min.
4. Remove MTSES by addition of 1 ml Na buffer and centrifugation (4 °C, 21,000 × *g*, 2 min).
5. Rinse cells three times with 1 ml Na buffer (4 °C, 21,000 × *g*, 2 min).
6. Lysis: resuspend cells pellet with 150 μl of a solution containing 30 % (w/v) sucrose, 30 mM Tris–HCl, pH 8.0, 50 μg/ml lysozyme, and 10 mM EDTA. Incubate at 37 °C for 15 min.
7. Proceed lysis by the addition of 900 μl of DDW containing 0.15 μg/ml DNaseI and 15 mM $MgSO_4$. Incubate at 37 °C for 15 min.

8. Collect membrane fraction by centrifugation (4 °C, 21,000 × *g*, 20 min). Resuspend membranes in 50 μl Na buffer.
9. Measure radioactivity: take 2 μl of membrane sample into a scintillation tube and add 18 μl of DDW. Add scintillation liquid, mix, and read cpm in a beta-counter.
10. For each sample, take a membrane volume corresponding to ~200,000 cpm. For membrane solubilization add 20 μl of solution containing 2 % (w/v) SDS, 6 M urea, 15 mM Tris–HCl, pH 7.5, in the presence of 2.5 mM Mal-PEG 5,000, and incubate with shaking at 30 °C for 1 h.
11. Stop the reaction by the addition of 4 μl sample buffer (×6).
12. Load protein samples on 16 % Tricine SDS-Urea PAGE.
13. Visualize the radioactive bands with a Phosphor-Imager (*see* **Note 24**).
14. Use suitable software to digitally analyze the intensity of the radioactive bands (*see* **Note 25**).
15. Calculate the ratios of MTSES-unreacted/unexposed thiols (intensity of higher molecular weight bands) over total thiols (intensity of higher molecular weight bands + intensity of lower molecular weight bands).

3.3 An Example for Results Obtained for EmrE Using the Technique

EmrE is a transporter residing in the inner membrane of *E. coli* and composed of four transmembrane helices connected by short hydrophilic loops. EmrE has only a weak topological tendency, meaning that it has no specific tendency to assume a topology of N_{in}–C_{in} (both termini are toward the cytoplasm) or N_{out}–C_{out} (both termini are toward the periplasm). We studied its topology using the SCAM technique. For that, we used cysteine (Cys)-less untagged EmrE, and also Cys-less tagged EmrE, as a control (the protein is fused at its C terminus to a Myc epitope followed by a hexa-histidine tag). In each construct, we engineered the single Cys residue to be at position 22 (substitution of lysine 22; loop 1) or at position 110 (substitution of histidine 110; C terminus); these two residues are expected to be on either side of the membrane according to hydropathy analysis. The mutants were expressed in cells and specifically labeled metabolically with [^{35}S] methionine. The cells were then challenged with the impermeant reagent, MTSES. After the solubilization and denaturation of the protein, the available thiols that did not react with the MTSES were assessed with Mal-PEG.

The topological results obtained for EmrE are shown in Fig. 1. In tagged mutants, cysteine in loop 1 at position 22 (EmrE-K22C) but not at the C terminus, position 110 (EmrE-H110C), reacted with MTSES in the intact cell, therefore only a fraction of ~30 % of the protein reacted with Mal-PEG (Fig. 1b). However, with untagged mutants bearing Cys at the same positions, reactivity of both EmrE-K22C and EmrE-H110C was very similar and partial; ~50–60 % of the residues are exposed in each case (Fig. 1c). These results show that the untagged protein displays a dual topology, i.e., approximately half of the protein is N_{in}–C_{in}, whereas the other half is N_{out}–C_{out}, as expected for a protein with a weak topological tendency. In contrast, the Myc-His tag is biasing the protein towards a N_{in}–C_{in} topology.

4 Notes

1. We use the pT7-7 expression plasmid which confers resistance to ampicillin [19].
2. It is also possible to induce gene expression in the BL21(DE3) strain [21], where the gene coding for the T7 polymerase has been inserted into the chromosome under the inducible control of the *lac*UV5 promoter, but we found that the labeling is more efficient when the expression is heat-induced.
3. Thiamine stock solution should be protected from light, since it is light-sensitive.
4. Prepare Minimal medium A (×10). Autoclave and store at room temperature. Right before use, dilute tenfold with DDW and add the supplements.
5. 0.5 % Glucose can be used as carbon source instead of 0.5 % Glycerol. Prepare 20 % (w/v) Glucose stock solution in DDW. Autoclave and store at room temperature.
6. Methanol should be handled with appropriate precautionary measures in accordance with safety procedures. Prepare Rifampicin stock solution fresh. Protect from light since Rifampicin is light-sensitive. Rifampicin inhibits transcription of the *E. coli* RNA polymerase without affecting the phage T7 polymerase which is responsible for the transcription of the genes to be over-expressed. This allows the exclusive labeling of the protein of interest [19].
7. Should be handled with appropriate precautionary measures in accordance with local radioactive safety procedures. The radioactive labeling is for the detection of the unpurified untagged protein. Detection of the protein can also be done by using a suitable antibody instead.
8. Tris can be dissolved faster provided the water is warmed to about 37 °C. However, the downside is that care should be taken to bring the solution to room temperature before adjusting pH.
9. Prepare right before use because MTSES has a very short half life [22].

10. First dissolve the Urea with DDW, then add the Tris–HCl, and eventually add the SDS.
11. SDS has a tendency to create foam when mixed on a magnetic stirrer too vigorously.
12. Harmful substance that should be handled according to its material safety data sheet (MSDS).
13. We find that it is best to use it for no more than a week.
14. Harmful substance that should be handled according to its MSDS. We find that storing at 4 °C reduces its pungent smell.
15. Before running a gel, dilute cathode and anode buffers by 10.
16. SDS precipitates at 4 °C. Therefore, the sample buffer needs to be warmed prior to use.
17. Spread transformed bacteria on a plate containing antibiotics that select for each of the plasmids. We used plates containing ampicillin and kanamycin for the selection of pT7-7 and pGP1-2, respectively.
18. The induction is done at 42 °C which is close to 37 °C. Thus, to avoid preinduction, cells are incubated at 30 °C. Cells exposed to higher temperatures should be discarded.
19. Since growth in a minimal medium is slower, add two to three colonies to each tube that should contain no more than 2 ml, and incubate starters for a minimal time of ~20 h.
20. Add while the Erlenmeyers are still in incubator. To avoid loss of rifampicin, make sure to pipette it close to the Erlenmeyer, as methanol tends to drip.
21. Should be handled with appropriate precautionary measures in accordance with local radioactive safety procedures. Add while the Erlenmeyers are still in incubator.
22. Pellet can be stored at −70 °C. We did not.
23. If pellet was stored at −70 °C, thaw it in 37 °C water bath before resuspending.
24. We use the FLA-3000 Phosphor-Imager; Fujifilm, Tokyo.
25. We use the Image Gauge 3.46 Fujifilm software.

Acknowledgments

SS is Mathilda Marks-Kennedy Professor of Biochemistry at the Hebrew University of Jerusalem. Work in our laboratory is supported by National Institutes of Health Grant NS16708 and Grant 11/08 from the Israel Science Foundation.

References

1. Walian P, Cross TA, Jap BK (2004) Structural genomics of membrane proteins. Genome Biol 5:215
2. White SH (2009) Biophysical dissection of membrane proteins. Nature 459:344–346
3. Chang HC, Bush DR (1997) Topology of NAT2, a prototypical example of a new family of amino acid transporters. J Biol Chem 272: 30552–30557
4. Kast C, Canfield V, Levenson R et al (1995) Membrane topology of P-glycoprotein as determined by epitope insertion: transmembrane organization of the N-terminal domain of mdr3. Biochemistry 34:4402–4411
5. Kast C, Canfield V, Levenson R et al (1996) Transmembrane organization of mouse P-glycoprotein determined by epitope insertion and immunofluorescence. J Biol Chem 271:9240–9248
6. Kast C, Gros P (1998) Epitope insertion favors a six transmembrane domain model for the carboxy-terminal portion of the multidrug resistance-associated protein. Biochemistry 37:2305–2313
7. McKenna E, Hardy D, Kaback HR (1992) Insertional mutagenesis of hydrophilic domains in the lactose permease of Escherichia coli. Proc Natl Acad Sci USA 89:11954–11958
8. Pan CJ, Lei KJ, Annabi B et al (1998) Transmembrane topology of glucose-6-phosphatase. J Biol Chem 273:6144–6148
9. Manoil C, Beckwith J (1986) A genetic approach to analyzing membrane protein topology. Science 233:1403–1408
10. Drew D, Sjostrand D, Nilsson J et al (2002) Rapid topology mapping of Escherichia coli inner-membrane proteins by prediction and PhoA/GFP fusion analysis. Proc Natl Acad Sci USA 99:2690–2695
11. Feilmeier BJ, Iseminger G, Schroeder D et al (2000) Green fluorescent protein functions as a reporter for protein localization in Escherichia coli. J Bacteriol 182:4068–4076
12. Rothman A, Padan E, Schuldiner S (1996) Topological analysis of NhaA, a Na+/H+ antiporter from Escherichia coli. J Biol Chem 271:32288–32292
13. Fujinaga J, Tang XB, Casey JR (1999) Topology of the membrane domain of human erythrocyte anion exchange protein, AE1. J Biol Chem 274:6626–6633
14. Tang XB, Fujinaga J, Kopito R et al (1998) Topology of the region surrounding Glu681 of human AE1 protein, the erythrocyte anion exchanger. J Biol Chem 273:22545–22553
15. Zhu Q, Lee DW, Casey JR (2003) Novel topology in C-terminal region of the human plasma membrane anion exchanger, AE1. J Biol Chem 278:3112–3120
16. Ninio S, Elbaz Y, Schuldiner S (2004) The membrane topology of EmrE - a small multidrug transporter from Escherichia coli. FEBS Lett 562:193–196
17. Zhu Q, Casey JR (2007) Topology of transmembrane proteins by scanning cysteine accessibility mutagenesis methodology. Methods 41:439–450
18. Nasie I, Steiner-Mordoch S, Gold A et al (2010) Topologically random insertion of EmrE supports a pathway for evolution of inverted repeats in ion-coupled transporters. J Biol Chem 285:15234–15244
19. Tabor S, Richardson CC (1985) A bacteriophage T7 RNA polymerase/promoter system for controlled exclusive expression of specific genes. Proc Natl Acad Sci USA 82: 1074–1078
20. Li J, Xu Q, Cortes DM et al (2002) Reactions of cysteines substituted in the amphipathic N-terminal tail of a bacterial potassium channel with hydrophilic and hydrophobic maleimides. Proc Natl Acad Sci USA 99: 11605–11610
21. Studier FW, Moffatt BA (1986) Use of bacteriophage T7 RNA polymerase to direct selective high-level expression of cloned genes. J Mol Biol 189:113–130
22. https://www.nacalai.co.jp/ss/Contact/pdf/2011-12_Anatrace_products_catalog.pdf. Accessed 22 May 2012

Chapter 9

self-assembling GFP: A Versatile Tool for Plant (Membrane) Protein Analyses

Katharina Wiesemann, Lucia E. Groß, Manuel Sommer, Enrico Schleiff, and Maik S. Sommer

Abstract

The investigation of cellular processes on the molecular level is important to understand the functional network within plant cells. *self-assembling* GFP has evolved to be a versatile tool for (membrane) protein analyses. Based on the autocatalytical reassembling property of the nonfluorescent strands 1–10 and 11, protein distribution and membrane protein topology can be analyzed in vivo. Here, we provide basic protocols to determine membrane protein topology in *Arabidopsis thaliana* protoplasts.

Key words Membrane proteins, Topology, *self-assembling* GFP, Plant protoplasts, *Arabidopsis thaliana*

1 Introduction

Since its discovery in 1962 [1], the green fluorescent protein (GFP) became an indispensable tool in modern research, especially in the field of cell biology [2]. Being initially used to investigate the spatiotemporal expression of proteins in living cells [3], it was later on engineered, e.g., to visualize the intracellular fate and dynamics of proteins and to trace protein–protein interactions, folding, transport and topology [4–8].

The protein itself is comprised of 238 amino acids forming an approx. 26 kDa large 11-stranded β-barrel with a central coaxial α-helix [9]. The fluorophore is formed by spontaneous autocatalytical cyclization and oxidation of the amino acid triad serine 65, tyrosine 66 and glycine 67 inside of the barrel, emitting bright green fluorescence when being exhibited to blue light [10]. It was recently found that the recombinantly expressed nonfluorescent β-strands 1–10 and the β-strand 11 of GFP have the capacity to spontaneously auto-/self-assemble, forming the functional fluorophore ([11]; *see* Fig. 1).

Doron Rapaport and Johannes M. Herrmann (eds.), *Membrane Biogenesis: Methods and Protocols*,
Methods in Molecular Biology, vol. 1033, DOI 10.1007/978-1-62703-487-6_9,

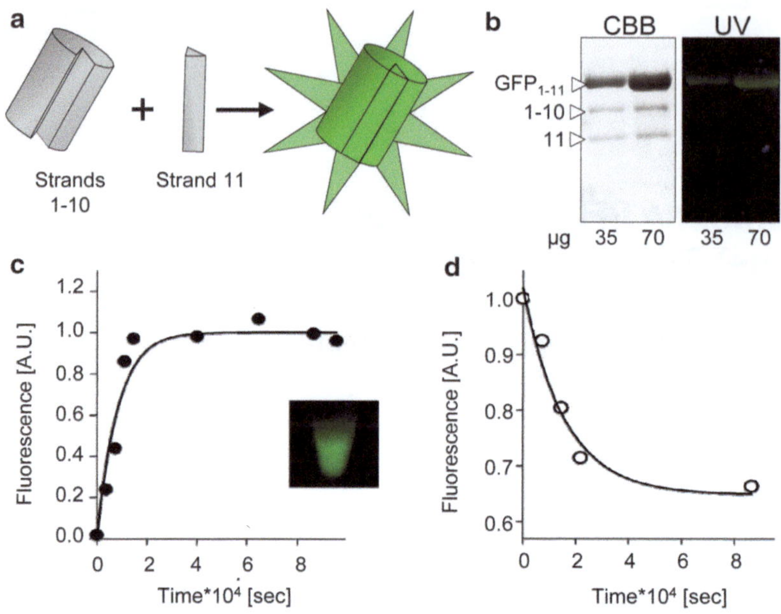

Fig. 1 *self-assembling* GFP. (**a**) Schematic depiction of *sa*GFP. Nonfluorescent strands 1–10 and strand 11 (*grey*) can self-assemble to form functional GFP (*green*). (**b**) Indicated amounts of recombinantly co-expressed and purified strands 1–10 and strand 11 were analyzed via SDS-PAGE. The analysis reveals that reconstituted assembled *sa*GFP is highly stable and resistant to chemical denaturants such as 8 M urea present in the SDS sample buffer (compare bands of strands 1–10 and strand 11 to band of reconstituted GFP (GFP_{1-11})). *CBB* Coomassie brilliant blue, *UV* ultraviolet light. (**c**, **d**) Assembly and disassembly of *sa*GFP were measured. Recombinant strand 11 (4 or 10 μM) was mixed with molar excess (16, 50, 100, or 200 μM) of strands 1–10 and incubated at room temperature. The assembly of 4 μM strand 11 with 16 μM of strands 1–10 is shown exemplarily. (**c**) Reassembling was tracked by measuring the increase of fluorescence (see tube) over time. Fluorescence was excited at 488 nm and emission was detected at 495–530 nm. *Line* represents the least square fit to an exponential equation. (**d**) Assembled *sa*GFP (after 24 h) was diluted 100-fold with reaction buffer and stability of GFP was examined for several hours. Dissociation of 10 μM strand 11 incubated with 200 μM of strands 1–10 is shown exemplarily. The values were analyzed by least square fit to an exponential decay equation

self-assembling GFP (short *sa*GFP) became a versatile tool for membrane protein analyses. Both fragments can be independently targeted to the various cellular sub-compartments, when being fused to appropriate topogenic signals (exemplified for plant cells in Fig. 2), where they self-assemble and fluoresce (e.g., [12–14]).

Fig. 2 (continued) 1–10 reporter constructs with various strand 11 fused control proteins of known localization. Cytosol: *atHsp18-511C* (CYT_{11}; line 1), plastidic intermembrane space (cIMS): *at*Tic22$_{11C}$ ($cIMS_{11}$; line 2), chloroplast stroma: RBC3$_{11C}$ (STR_{11}; line 3), peroxisomes (PEX): *at*PKT3$_{11C}$ (PEX_{11}; line 4), mitochondrial intermembrane space (mIMS): *at*VDAC3$_{11N}$ ($mIMS_{11}$; line 5), mitochondrial matrix (MTX): p(ATPb)$_{11C}$ (MTX_{11}; line 6). Localization was assessed via overlay (Merge, column 3) of the GFP signal (GFP, column 1) and the chlorophyll autofluorescence (AF, lines 1–6, column 2). Mitochondria were additionally stained with MitoTracker (lines 5–6, column 2). Protoplast integrity was monitored by differential interference contrast (DIC, column 4). Scale bars represent 10 μm. For detailed descriptions of the co-transfected strands 1–10 reporter constructs *see* Table 1 and references therein

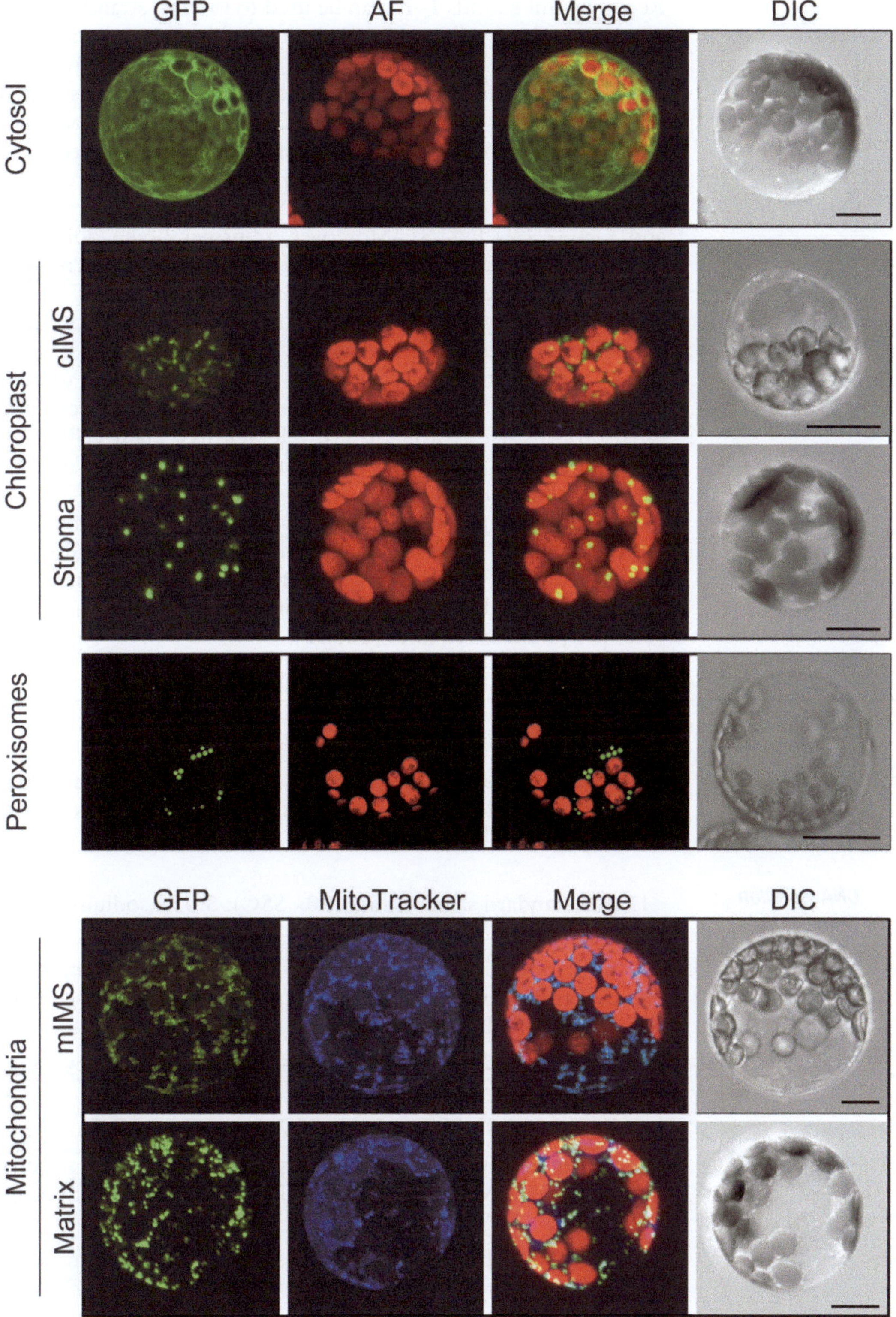

Fig. 2 Strands 1–10 and strand 11 fusion proteins can be efficiently targeted to the various organelles in plant protoplasts. Strands 1–10 and strand 11 fused proteins targeting to specific cellular sub-compartments were co-transfected into *Arabidopsis thaliana* mesophyll protoplasts. Each line shows co-expression of strands

Recombinant strands 1–10 can be used to localize strand 11 fusion proteins in fixed cells, similar to immunofluorescence microscopy [15]. Membrane protein interactions can be detected in living cells when co-expressing potential interactors either fused to strand 11 or strand 1–10 [16]. Recently, *sa*GFP was adapted to assess the topologies of α-helical- or β-barrel-shaped membrane proteins in living plant cells [12–14], algae [17], and human parasites [6], as exemplified here for the chloroplast outer envelope proteins Oep7 and Oep37 (*see* Fig. 3). Moreover, transmembrane domains of single and multi membrane spanning proteins were determined in vivo, by co-expressing truncated, 11-tagged versions of the proteins lacking single domains with strands 1–10 either being targeted in *cis* or *trans* to the respective membrane [6, 12].

In this chapter, we provide the basic protocols for topology and assessment of membrane proteins in freshly isolated leaf mesophyll protoplasts. Protoplasts are cell wall-free plant cells, which retain their cellular identity and differentiation and have the capability to take up DNA. They are physiologically versatile cells that offer the unique possibility to study a broad spectrum of cell biological processes [18]. The simplistic applicability of GFP gives reason to expect that *sa*GFP complementation ability can be used in almost any other cell type as well, as shown already, e.g., for yeast [19] human cell lines [20, 21].

2 Materials

All buffers and solutions are prepared in deionized water and either autoclaved or filtered. Prepare and store stock solutions at room temperature, buffers at 4 °C.

2.1 DNA Isolation

1. 20× standard saline citrate (20× SSC): 3.0 M sodium chloride (NaCl), 0.3 M sodium citrate (NaCitrate).
2. Buffer 1: 0.05 M 2-Amino-2-hydroxymethyl-propane-1,3-diol (Tris base), 0.05 M EDTA, 15 % (w/v) sucrose. Adjust pH with hydrogen chloride (HCl) to 8.0 (*see* **Note 1**).
3. Buffer 2: 0.2 M NaOH, 1 % (w/v) sodium dodecyl sulfate (SDS).
4. Buffer 3: 0.01 M Tris–HCl pH 7.6, 0.001 M EDTA.
5. Buffer 4: 0.05 M 3-(*N*-morpholino) propane sulfonate (MOPS), 5 M lithium chloride (LiCl). Adjust pH with NaOH to 8.0.
6. RNase mix: RNase A/RNase T1 500 U/mL in water. Aliquot and store at −20 °C.
7. PCI: Phenol/trichloromethane ($CHCl_3$)/isoamyl alcohol (ratio 25:24:1).

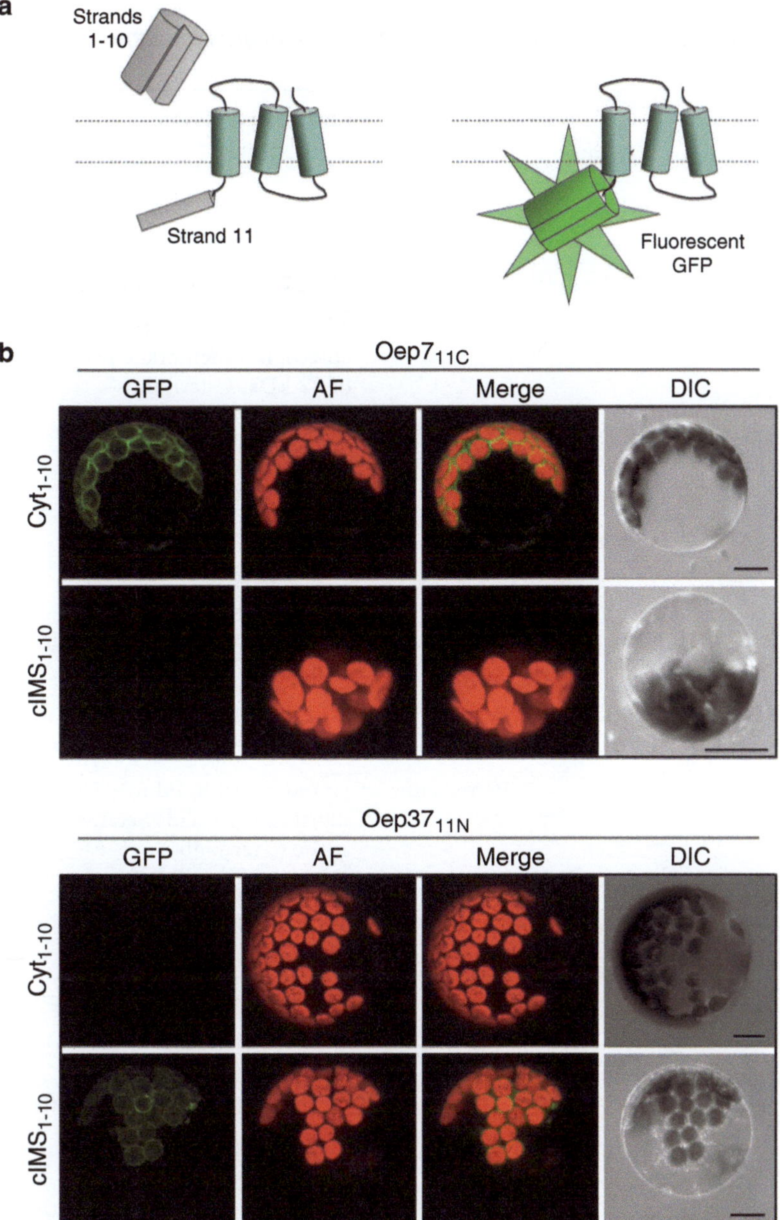

Fig. 3 Topology assessment using *sa*GFP fusion proteins. (**a**) The orientation of an entire membrane protein can be assigned by simplistic fusion of the protein to strand 11 of *sa*GFP. According to the orientation (in or outward facing), a GFP signal should be only observable if strands 1–10 (reporter) are expressed in the adjacent compartment. (**b**) Topology assessment of Oep7 and Oep37. *Arabidopsis thaliana* mesophyll protoplasts co-expressing either the chloroplast outer envelope proteins $Oep7_{11C}$ (lines 1 and 2) or $Oep37_{11N}$ (lines 3 and 4) together with the cytosolic (CYT_{1-10}, lines 1 and 3) or plastidic intermembrane space ($cIMS_{1-10}$, lines 2 and 4) reporter constructs (*see* Table 1). The topology of the test proteins is assessed by the GFP fluorescence (column 1) resulting from the assembly of the terminally fused strand 11 with strand 1–10 fused reporters (**a**). Using *sa*GFP it was possible to confirm the established C_{IN}–N_{OUT} topology of single membrane spanning Oep7 [13, 31, 32]. Moreover, the previously predicted C_{IN}–N_{IN} topology of β-barrel-shaped Oep37 [33, 34] was revised as the *sa*GFP analyses revealed a converse topology of the protein [14]. *Left* to *right*: GFP fluorescence (*GFP*), chlorophyll autofluorescence (*AF*), overlay (*Merge*), and differential interference contrast (*DIC*). Scale bars represent 10 μm

Table 1
Strand 11 control and 1–10 reporter fusion proteins for *Arabidopsis thaliana*

Name	Gene	Protein	Localization	Remarks; position of strand(s) 11 or 1–10	References
CYT_{11}	AT2g19310	*at*Hsp18.5	Cytoplasm	Heat-shock-protein of 18.5 kDa; C-terminus	[13]
CYT_{1-10}	–	*sa*GFP1–10	Cytoplasm	Strands 1–10 of *sa*GFP; unfused	[11, 13]
$cIMS_{11}$	AT4g33350	*at*Tic22-IV	cIMS	TRANSLOCASE of the inner chloroplast membrane protein of 22 kDa; C-terminus	[13]
$cIMS_{1-10}$	AT4g31780	*at*Mgd1	CIM, outer leaflet	Mono-galactosyl-diacylglycerol synthase 1, of the cIEM; C-terminus	[13, 24]
STR_{11}	AT1g12410	*at*Clp2	Stroma	ATP-dependent caseinolytic protease (Clp) subunit 2; C-terminus	[25]
STR_{1-10}	AT5g38410	*at*RBC3	Stroma	Precursor of the small subunit of RUBISCO; C-terminus	[12]
PEX_{11}	AT2G33150	*at*PKT3	Peroxisome	3-Ketoacyl-CoA thiolase 2 with N-terminal PTS2; C-terminus	[26]
PEX_{1-10}	–	$1\text{–}10_{SKL}$	Peroxisome	Strands 1–10 fused to PTS1 signal (amino acids Ser-Lys-Leu at the C-terminus)	[27]
$mIMS_{11}$	AT5g15090	*at*VDAC3	MOM	Voltage-dependent anion channel isoform 3, N-terminus	[14, 28]
$mIMS_{1-10}$	AT3g46560	*at*Tim50	MIM	Translocase of the inner mitochondrial membrane protein of 50 kDa, N-terminus	[13, 29]
MTX_{11}	AT5g08670	*at*ATPb	Matrix	Targeting peptide of the precursor of the F1-ATP synthase, beta subunit; C-terminus	[14, 30]
MTX_{1-10}	AT5g08670	*at*ATPb	Matrix	Targeting peptide of the precursor of the F1-ATP synthase, beta subunit; C-terminus	[14, 30]

CYT cytoplasm, *MIM/MOM* mitochondrial inner/outer envelope membrane, *mIMS* mitochondrial intermembrane space, *MTX* mitochondrial matrix, *CIM/COM* chloroplast inner/outer envelope membrane, *cIMS* chloroplast intermembrane space, *STR* chloroplast stroma, *PEX* peroxisomes

8. 3 M potassium acetate (KAc). Adjust pH to 5.2 with acetic acid (*see* **Note 2**).
9. 3 M sodium acetate (NaAc) pH 5.2. Adjust pH to 5.2 with acetic acid.
10. Trichloromethane.

11. Ethanol (EtOH), 70 % (v/v) and 96 % (v/v).
12. 10× Proteinase K: 0.1 mg Proteinase K in 1 % (w/v) SDS, 0.01 M EDTA, 1× SSC (*see* **Note 3**). Aliquot and store at −20 °C.

2.2 Protoplast Isolation and Transformation

1. BSA: 10 % (w/v) albumin fraction V (BSA) in water.
2. Washing solution: 50 % (v/v) MCP, 0.1 % (w/v) BSA.
3. MCP: 0.02 M 2-(*N*-morpholino)ethane-sulfonate (MES), 0.5 M sorbitol, 0.001 M $CaCl_2$. Adjust pH to 5.6 with potassium hydroxide (KOH).
4. MCP-Percoll 100: 0.02 M MES, 0.5 M sorbitol, 0.001 M $CaCl_2$ in 100 % (v/v) Percoll®. Store at 4 °C.
5. MCP-Percoll 25: Mix 12.5 mL Percoll® with 37.5 mL of MCP. Store at 4 °C.
6. MMg: 0.005 M MES-KOH pH 5.6, 0.4 M sorbitol, 0.015 M $MgCl_2$.
7. PEG-Ca solution: 40 % (w/v) polyethylene glycol (PEG) 4,000, 0.1 M $CaCl_2$, 0.4 M sorbitol.
8. K3: 0.02 M MES/KOH pH 5,6 0.4 M sucrose, 0.001 M $CaCl_2$, 1× MS salts (Murashige and Skoog medium incl. vitamins, Duchefa Biochemistry).
9. Enzyme Mix: 1 % (w/v) Cellulase, 0.3 % (w/v) Macerozyme in MCP. Prepare freshly.
10. K240 sandpaper.
11. 75 μm nylon mesh.

3 Methods

3.1 Plasmids and Construct Design

To transiently transfect isolated *Arabidopsis thaliana* leaf protoplasts, we use modified pAVA vectors (*see* **Note 4**). Proteins are expressed under the mid-strong constitutive 35S promoter of the cauliflower mosaic virus [22]. Three vectors were created, one for the expression of C-terminal fusion proteins with saGFP1–10 reporter and two for N- or C-terminal tagging of proteins with the saGFP11 (*see* Fig. 4). Steric hindrance of the GFP fragments and the fusion proteins to ensure for proper folding is minimized by using a short linker peptide, e.g., $(GSSS)_2$ [13, 23]. Proteins can either be directed to a cellular (sub-) compartment by fusion to a full-length protein or the appropriate topogenic signal (exemplified in Table 1, *see* **Note 5**). If the intracellular localization and therewith the topogenic signals are not known, it is recommended to tag the proteins at either the N- or the C-terminus.

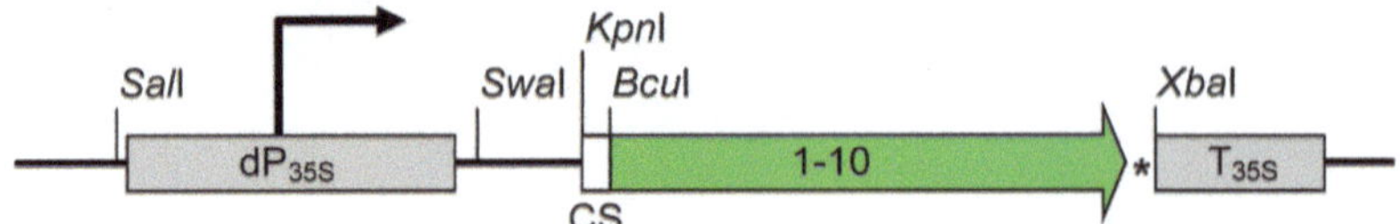

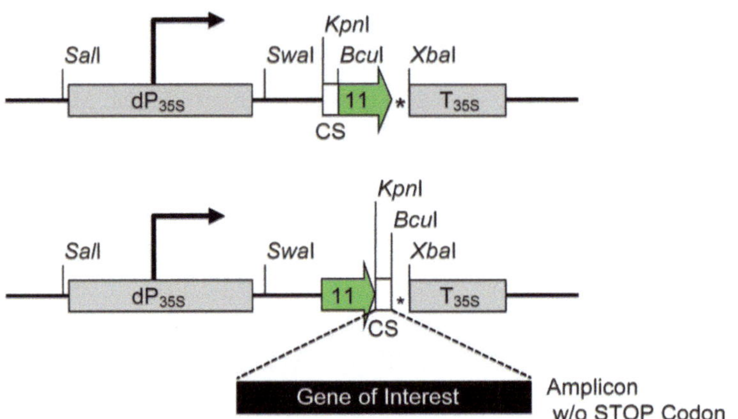

Fig. 4 Construct design. The *sa*GFP vector system is based on the pBlueScript derivate pAVA [22]. The vector encodes the *bla* gene for ampicillin resistance. Genes are expressed under control of the constitutively active 35S promoter (P_{35S}) of the cauliflower mosaic virus. The promoter can be substituted by a specific or an inducible one via *SalI* and *SwaI*. Genes coding the protein of interest (without stop codon) can be inserted using *KpnI* and *BcuI*. The coding regions for strand 11 or strands 1–10 of *sa*GFP are separated by a short linker and precede or succeed the cloning site (*CS*), respectively. Vectors can be obtained upon request. The permission to use the *sa*GFP fragments has to be obtained from G.S. Waldo (Los Alamos). Modified from [23]

3.2 Isolation of Ultra-pure Plasmid DNA

1. Start with 50–100 mL (*see* **Note 6**) overnight grown bacterial culture, expressing the plasmid of interest.
2. Harvest cells for 20 min at 5,000 × *g*, room temperature; Remove supernatant.
3. Resuspend cells in 3 mL ice-cold buffer 1.
4. Mix lysate with 7 mL of freshly prepared buffer 2, shake carefully (4–5 times, upside down) and incubate at room temperature for 10 min.
5. Add 3.5 mL ice-cold 3 M KAc pH 5.2; Keep sample on ice for 20 min.
6. Centrifuge for 20 min at 5,000 × *g*, 4 °C.
7. Filtrate the supernatant through cotton wool (using a small plastic funnel).

8. Add 0.7 volume of isopropyl alcohol, mix, and incubate for at least 30 min at −20 °C.
9. Precipitate DNA by centrifuging 20 min at 15,000 × *g*, 4 °C.
10. Pour off the supernatant, dry the pellet and resuspend DNA in 0.5 mL of buffer 3.
11. Rinse tubes once by adding 0.5 mL of buffer 4 and pool with the resuspended DNA.
12. Incubate for 30–60 min on ice, then centrifuge for 10 min at 6,000 × *g*, 4 °C.
13. Transfer the supernatant into a new tube and add 0.1 volume of 3 M NaAc pH 5.2 and 2.5 volumes of cold abs. EtOH, vortex and precipitate the DNA for at least 30 min (or overnight) at −20 °C.
14. Pellet DNA by centrifugation for 20 min at 15,000 × *g*, 4 °C; Discard the supernatant, dry the pellet and resuspend DNA in 0.25 mL buffer 3.
15. Add 0.01 mL of RNase mix, mix and incubate for 30 min at 37 °C.
16. Add 0.03 mL of 10X Proteinase K, mix and incubate at 37 °C for further 15 min.
17. Add 0.35 mL PCI, vortex and spin for 2 min at 20,000 × *g*, room temperature.
18. Transfer the upper, DNA-containing aqueous phase (~0.3 mL) into a new tube and add 1 volume of $CHCl_3$. Vortex and centrifuge for 1 min at 20,000 × *g*, room temperature.
19. Transfer supernatant into a new tube, add 1 volume of 3 M NaAc pH 5.2 and 2.5 volumes cold abs. EtOH.
20. Precipitate DNA by at least 1 hour incubation (or overnight) at −20 °C.
21. Centrifuge for 15 min at 20,000 × *g*, 4 °C.
22. Discard the supernatant and wash DNA with 0.5 mL ice-cold 70 % (v/v) EtOH.
23. Centrifuge again for 5 min at 20,000 × *g*, 4 °C.
24. Remove the supernatant, dry pellet in a Speedvac for 15 min at 42 °C and resuspend DNA 0.05–0.1 mL of buffer 4 by gentle agitation at 37 °C (30 min).
25. Determine DNA concentration and adjust to 1 mg/mL.

3.3 Isolation of *Arabidopsis thaliana* Protoplasts

Carry out all steps at room temperature unless it is specified otherwise.

Protoplasts are fragile. Avoid sheer-forces by extended pipetting. Use pipette tips with a large orifice. Figure 5 shows a schematic depiction of all steps.

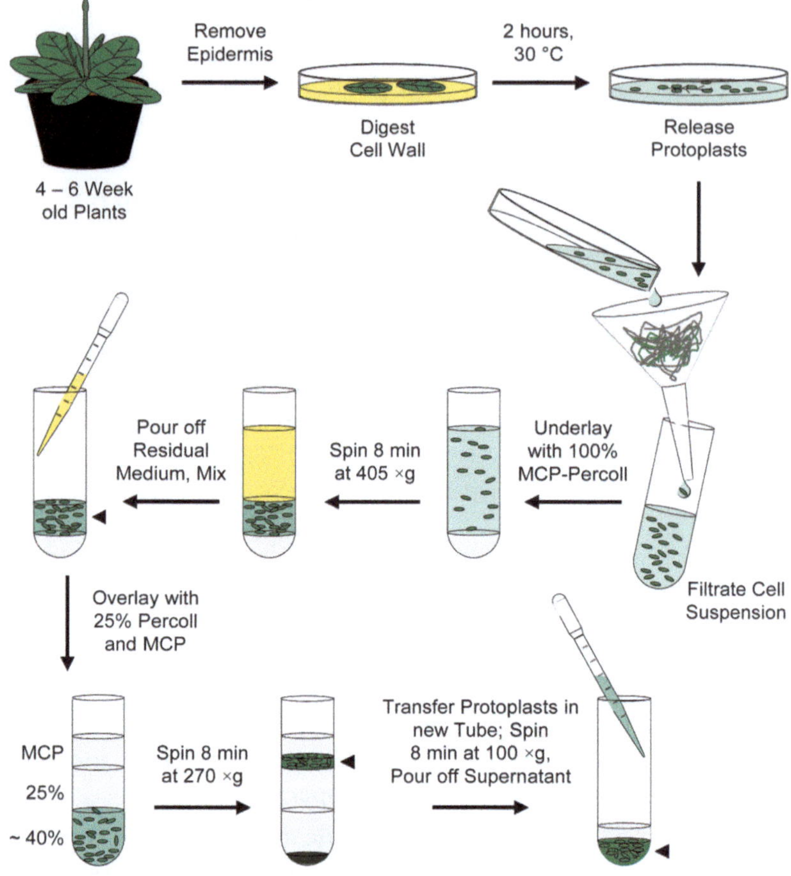

Fig. 5 Schematic depiction of the protoplast isolation procedure. Isolation of leaf mesophyll protoplasts from 4- to 6-week-old *Arabidopsis thaliana* plants. For details see text

1. Cut leaves of 4–6-week-old *Arabidopsis thaliana* plants (*see* **Note** 7).
2. Remove epidermis by roughening the bottom side of each leaf carefully with a K240 sandpaper (rub 1–2 times until cell sap leaks).
3. Place leaves bottom-side down into a Petri dish containing 20 mL washing solution.
4. After preparation of a sufficient amount of leaves carefully remove washing solution by pipetting and add 20 mL of enzyme mix.
5. Incubate for 2 h at 30 °C to digest the cell's cell wall.
6. Release protoplasts by soft swiveling of the dish.
7. Filter released protoplasts through a 75 μm nylon mesh.
8. Underlay filtrate with 2.5 mL of 100 % (v/v) MCP-Percoll and centrifuge for 8 min at 405 × *g*, RT (without brake); Remove approximately 20 mL of the clear supernatant.

9. Mix the remaining (protoplast containing) fraction gently with the Percoll cushion to yield approx. 40 % (v/v) Percoll.
10. Subsequently overlay with 7.5 mL 25 % (v/v) Percoll in MCP and 5 mL MCP.
11. Centrifuge for 8 min at 270×*g*, RT (without brake); Transfer protoplasts (green band between MCP and 25 % (v/v) Percoll) into a new tube.
12. Pellet protoplasts by centrifugation for 8 min at 100×*g*, Remove supernatant.
13. Resuspend protoplasts in a small volume of MMg; Count cells using a Fuchs-Rosenthal cell counter and adjust to a cell number of 10^6 cells per mL MMg.

3.4 PEG-Mediated Transfection of Protoplasts and Microscopic Analyses

For a single transfection reaction:

1. Mix 0.1 mL (10^5) protoplasts with 5–20 μg of highly purified pDNA (*see* **Note 8**) and 0.1 mL PEG-Ca.
2. Incubate for 20 min at RT.
3. Stop reaction by adding 1 mL K3 and gently mix. Protoplast sediment over time!
4. Analyze protoplasts 8–12 h post-transfection (*see* **Notes 9** and **10**). Fluorescence is excited at 488 nm (*sa*GFP) and 514 nm (chlorophyll autofluorescence of the chloroplasts) and detected using a confocal laser-scanning microscope at 505–525 nm or 650–750 nm, respectively (*see* **Note 11**). Pay attention; Do not squeeze cover slip on microscope slides after application of the sample.
5. Verify (full-length) expression of *sa*GFP constructs if necessary via SDS-PAGE and Western Blotting using specific strand 11 or strands 1–10 antibodies (*see* **Note 12**). Integration of membrane proteins can be tested by carbonate or high-salt treatment [14].

4 Notes

1. Use a 0.5 M EDTA stock solution. EDTA only dissolves at a pH of 8.0. To dissolve EDTA add NaOH pellets until the solution clears up. *Note*: Check pH using pH stripes, since EDTA corrodes the pH-electrode.
2. *Caution*: More than half of the final buffer volume of acidic acid is needed to adjust pH.
3. SSC buffer is required to ensure optimal proteolytical activity of Proteinase K.
4. *Crucial*: The transformation efficiency of protoplasts strongly depends on the quality of the used DNA. The given protocol

(*see* Subheading 3.2), although tedious, yields good amounts of highly pure plasmid DNA for best transformation efficiencies. When using commercial kits, it is recommended applying the Proteinase K treatment, followed by the subsequent steps.

5. Mitochondrial and plastid proteins usually possess cleavable N-terminal targeting signals. Hence, C-terminal fusion is advisable. Contrarily, we observed that proteins with a C-terminal signal, such as a peroxisomal targeting sequence (PTS1, the amino acids serine, lysine and leucine), "prefers" to be N-terminally labeled. If there are more than two signals or signals at either side of the protein (e.g., an ER-resident proteins), strand 11 can be placed in between the signal and the mature domain.
6. Choose culture volume in dependence of the copy number of the plasmid.
7. Plants were grown under short-day conditions (8/16 h day/night cycle with 120 μmol/m^2 s). To yield approximately 1×10^6 protoplasts use >10 leafs.
8. For co-expression of several constructs mix DNA beforehand. When titrating DNA amounts of certain constructs, adjust overall DNA amount with empty vector to 10 μg.
9. To determine the transformation efficiency, transform protoplasts with 10 μg of a plasmid expressing, e.g., cytoplasmic GFP. The efficiency is estimated by visual inspection (No. of cells which show GFP fluorescence vs. No. of protoplasts which do not).
10. Although all expressed under the same promoter, constructs sometimes have (even drastically) different expression rates. These differences can be overcome by varying the ratio between the constructs to be transfected simultaneously. This might be required, if the fusion proteins have to be intracellularly distributed. To overcome artifacts, for instance, because of incomplete translocation, when being imported into sub-organellar compartments, the ideal DNA amount used for transfection has to be elucidated empirically for each construct combination. In some cases strong expression of one of the fusion partners might block the transport route of the other partner and lead to subsequent miss-targeting and aggregation. In those cases, the analyses should begin latest 8 h post-transfection.
11. We use a TCS SP5 confocal laser-scanning microscope (Leica) with a HCX PL APO CS 40× 1.25 NA 1.25 oil objective. Its unique opto-acoustical beam splitter is ideal for the analyses of plant material, since it allows the clear separation from GFP and chloroplast chlorophyll auto fluorescence. Moreover, the analyses can be combined with a broad range of fluorescent probes and proteins (such as Mito or ER Tracker®).

12. Strands 1–10 of GFP can be detected using a *mouse* monoclonal antibody against GFP (Roche). For detection of strand 11 fusion proteins, antibodies were raised against the peptide RDHMVLHEYVNAAGIT (Peptide Specialty Laboratories).

Acknowledgments

We thank Geoffrey S. Waldo (Los Alamos National Laboratory, Los Alamos, NM) for providing templates for the self-assembling GFP. This work was supported by Deutsche Forschungsgemeinschaft (SFB807 P17).

References

1. Shimomura O, Johnson FH, Saiga Y (1962) Extraction, purification and properties of aequorin, a bioluminescent protein from the luminous hydromedusan, Aequorea. J Cell Comp Physiol 59:223–239
2. Remington SJ (2011) Green fluorescent protein: a perspective. Protein Sci 20:1509–1519
3. Chalfie M, Tu Y, Euskirchen G et al (1994) Green fluorescent protein as a marker for gene expression. Science 263:802–805
4. Thomas JD, Daniel RA, Errington J et al (2001) Export of active green fluorescent protein to the periplasm by the twin-arginine translocase (Tat) pathway in Escherichia coli. Mol Microbiol 39:47–53
5. Patterson GH, Lippincott-Schwartz J (2002) A photoactivatable GFP for selective photolabeling of proteins and cells. Science 297:1873–1877
6. van Dooren GG, Tomova C, Agrawal S et al (2008) Toxoplasma gondii Tic20 is essential for apicoplast protein import. Proc Natl Acad Sci USA 105:13574–13579
7. Fiebiger E, Story C, Ploegh HL et al (2002) Visualization of the ER-to-cytosol dislocation reaction of a type I membrane protein. EMBO J 21:1041–1053
8. Ciruela F (2008) Fluorescence-based methods in the study of protein-protein interactions in living cells. Curr Opin Biotechnol 19:338–343
9. Ormo M, Cubitt AB, Kallio K et al (1996) Crystal structure of the Aequorea victoria green fluorescent protein. Science 273:1392–1395
10. Cody CW, Prasher DC, Westler WM et al (1993) Chemical structure of the hexapeptide chromophore of the Aequorea green-fluorescent protein. Biochemistry 32: 1212–1218
11. Cabantous S, Terwilliger TC, Waldo GS (2005) Protein tagging and detection with engineered self-assembling fragments of green fluorescent protein. Nat Biotechnol 23:102–107
12. Machettira AB, Gross LE, Sommer MS et al (2011) The localization of Tic20 proteins in Arabidopsis thaliana is not restricted to the inner envelope membrane of chloroplasts. Plant Mol Biol 77:381–390
13. Sommer MS, Daum B, Gross LE et al (2011) Chloroplast Omp85 proteins change orientation during evolution. Proc Natl Acad Sci USA 108:13841–13846
14. Ulrich T, Gross LE, Sommer MS et al (2012) Chloroplast beta-barrel proteins are assembled into the mitochondrial outer membrane in a process that depends on the TOM and TOB complexes. J Biol Chem 287:27467–27479
15. Kaddoum L, Magdeleine E, Waldo GS et al (2010) One-step split GFP staining for sensitive protein detection and localization in mammalian cells. Biotechniques 49:727–728, 730, 732 passim
16. Hempel F, Bullmann L, Lau J et al (2009) ERAD-derived preprotein transport across the second outermost plastid membrane of diatoms. Mol Biol Evol 26:1781–1790
17. Bullmann L, Haarmann R, Mirus O et al (2010) Filling the gap, evolutionarily conserved Omp85 in plastids of chromalveolates. J Biol Chem 285:6848–6856
18. Sheen J (2001) Signal transduction in maize and Arabidopsis mesophyll protoplasts. Plant Physiol 127:1466–1475
19. Ferrara F, Listwan P, Waldo GS et al (2011) Fluorescent labeling of antibody fragments using split GFP. PLoS One 6:e25727
20. Chaudhary A, Ganguly K, Cabantous S et al (2012) The Brucella TIR-like protein TcpB interacts with the death domain of MyD88. Biochem Biophys Res Commun 417:299–304

21. Chun W, Waldo GS, Johnson GV (2011) Split GFP complementation assay for quantitative measurement of tau aggregation in situ. Methods Mol Biol 670:109–123
22. von Arnim AG, Deng XW, Stacey MG (1998) Cloning vectors for the expression of green fluorescent protein fusion proteins in transgenic plants. Gene 221:35–43
23. Gross LE, Machettira AB, Rudolf M et al (2011) GFP-based in vivo protein topology determination in plant protoplasts. J Endocytobiosis Cell Res 21:89–97
24. Kobayashi K, Nakamura Y, Ohta H (2009) Type A and type B monogalactosyldiacylglycerol synthases are spatially and functionally separated in the plastids of higher plants. Plant Physiol Biochem 47:518–525
25. Sun Q, Zybailov B, Majeran W et al (2009) PPDB, the plant proteomics database at Cornell. Nucleic Acids Res 37:D969–D974
26. Johnson TL, Olsen LJ (2003) Import of the peroxisomal targeting signal type 2 protein 3-ketoacyl-coenzyme a thiolase into glyoxysomes. Plant Physiol 133:1991–1999
27. Bionda T, Tillmann B, Simm S et al (2010) Chloroplast import signals: the length requirement for translocation in vitro and in vivo. J Mol Biol 402:510–523
28. Clausen C, Ilkavets I, Thomson R et al (2004) Intracellular localization of VDAC proteins in plants. Planta 220:30–37
29. Geissler A, Chacinska A, Truscott KN et al (2002) The mitochondrial presequence translocase: an essential role of Tim50 in directing preproteins to the import channel. Cell 111:507–518
30. Jansch L, Kruft V, Schmitz UK et al (1996) New insights into the composition, molecular mass and stoichiometry of the protein complexes of plant mitochondria. Plant J 9: 357–368
31. Salomon M, Fischer K, Flugge UI et al (1990) Sequence analysis and protein import studies of an outer chloroplast envelope polypeptide. Proc Natl Acad Sci USA 87: 5778–5782
32. Schleiff E, Tien R, Salomon M et al (2001) Lipid composition of outer leaflet of chloroplast outer envelope determines topology of OEP7. Mol Biol Cell 12:4090–4102
33. Goetze TA, Philippar K, Ilkavets I et al (2006) OEP37 is a new member of the chloroplast outer membrane ion channels. J Biol Chem 281:17989–17998
34. Schleiff E, Eichacker LA, Eckart K et al (2003) Prediction of the plant beta-barrel proteome: a case study of the chloroplast outer envelope. Protein Sci 12:748–759

Part III

Studying Protein–Protein and Protein–Lipids Interactions Within Membranes

Chapter 10

The Use of Cardiolipin-Containing Liposomes as a Model System to Study the Interaction Between Proteins and the Inner Mitochondrial Membrane

Milit Marom and Abdussalam Azem

Abstract

The interaction of proteins with biological membranes is a key factor in their biogenesis and proper function. Hence, unraveling the properties of this interaction is very important and constitutes an essential step in deciphering the structural and functional characteristics of a membrane protein. Here we describe the use of cardiolipin-containing liposomes to analyze the interaction of the import protein Tim44 with the inner mitochondrial membrane. Using this system we showed that Tim44 is peripherally attached to the membrane and we detected the membrane binding site of the protein. The cardiolipin-containing liposomes serve as an excellent in vitro model system to the inner mitochondrial membrane and thus provide a good tool to analyze the interaction of various mitochondrial proteins with the inner membrane.

Key words Liposomes, Tim44, Cardiolipin, Mitochondria, Inner mitochondrial membrane, TIM23 import complex

1 Introduction

Liposomes are spherical particles which can resemble natural membranes. This similarity is manifested by their bilayer structure, which forms spontaneously when dry phospholipids are introduced to an aqueous media. Liposomes were first characterized in the mid 1960s when it was observed using electron microscopy that phospholipids form concentric multilamellar spherulites in the presence of water or salt solutions [1]. Since this initial observation liposomes were the subject of extensive research and today liposomes have many applications in medicine and biology.

The high homology between liposomes and biological membranes makes them an excellent model system for membrane research. Liposomes are widely used in the analysis of membrane proteins (i.e., integral membrane proteins, channel forming proteins, and peripheral membrane proteins), as model systems for

Doron Rapaport and Johannes M. Herrmann (eds.), *Membrane Biogenesis: Methods and Protocols*,
Methods in Molecular Biology, vol. 1033, DOI 10.1007/978-1-62703-487-6_10,

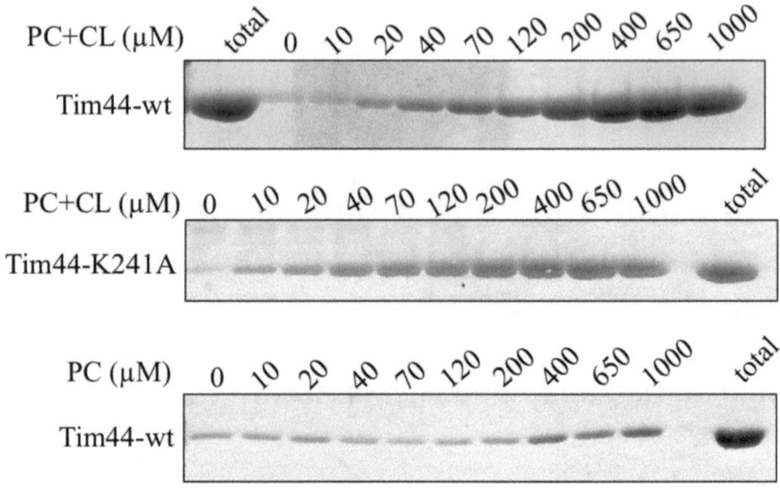

Fig. 1 SDS-PAGE analysis of the binding of Tim44 to liposomes. 1 μM of purified Tim44-wt or the mutant Tim44-K241A (in which lysine 241 was changed to alanine) were incubated for 1 h with increasing concentrations of sucrose-loaded liposomes composed of either a mixture of PC and CL (5:1) or PC alone. The liposomes were pelleted by centrifugation, and 40 μL of SDS sample buffer was added to solubilize the bound proteins. 20 μL of the bound fraction was loaded on 14 % SDS-PAGE and stained with Coomassie Blue. Lipid concentration is indicated above each *lane*. Total represents the total amount of protein added

membrane fusion, and for the characterization of various physical properties of membranes (for example, lipid rafts and phase separation) [2–5]. Here we demonstrate the use of liposomes as a model system for the mitochondrial inner membrane. Cardiolipin is the characteristic phospholipid of the inner mitochondrial membrane which constitutes about 20 % of the total inner mitochondrial membrane phospholipids [6]. Accordingly, the liposomes we used in our experiments were composed of 20 % cardiolipin (CL) and 80 % phosphatidylcholine (PC). Tim44 is an essential member of the mitochondrial TIM23 import complex. Its function in the import process entails its localization next to the inner mitochondrial membrane. Our initial aim was to analyze whether Tim44 is an integral membrane protein or is it peripherally attached to the membrane [7–9]. As can be observed in Fig. 1, Tim44 showed specific binding to cardiolipin-containing liposomes, whereas only minor binding is observed to phosphatidylcholine only liposomes. To identify specific residues of Tim44 that mediate its binding to cardiolipin, we created several Tim44 mutants in which we replaced a positively charged residue with Alanine. The mutant Tim44-K241A (in which lysine 241 was changed to alanine) is one of these mutants, and as can be seen in Fig. 1, it binds the cardiolipin-containing liposomes similarly to the native protein. None of the mutants we created in this area showed reduced binding to the cardiolipin-containing liposomes, suggesting that the binding to the membrane is not mediated via a single amino acid but rather it is the accumulative result of several interactions. A further truncation analysis of the protein enabled us to identify the major binding site of Tim44 to the membrane [10]. Thus, using the cardiolipin-containing liposomes we were able to determine the binding site of Tim44 to the mitochondrial inner membrane. Our results were corroborated with molecular dynamics simulations which provided additional validity for the use of liposomes as a model system for the mitochondrial inner membrane [10]. Furthermore, another study that analyzed the binding of Tim44 to the membrane used a structural approach and reached similar results, thus supporting again the use of this system as a model [11]. In addition to Tim44, which was the first protein of the TIM23 complex that was shown to have a high affinity to cardiolipin-containing membranes [12], a large number of mitochondrial proteins bind specifically to cardiolipin and this binding is vital for their biogenesis and proper function [13–15]. Cardiolipin-containing liposomes thus may serve as an excellent system to study those interactions in vitro. Moreover, liposomes with various phospholipid compositions can be used as a model for different membranes (various organelle membranes, the membranes of different organisms) and widen the scope of interactions that can be analyzed utilizing this system.

The binding assay to the liposomes was used successfully in previous studies [16, 17] and is relatively simple and straightforward. This simplicity arises from the peripheral binding of the protein to the membrane which does not require major alterations in the membrane structure and deep penetration into the bilayer.

2 Materials

All solutions should be prepared using double distilled water. All chemicals should be handled and disposed of according to appropriate safety procedures.

2.1 Liposome Preparation

1. Synthetic dioleoylphosphatidylcholine (PC).
2. Cardiolipin from bovine heart (CL).
3. Chloroform.
4. Rotatory evaporation system.
5. Buffer A: 10 mM Na-Hepes (pH 7.4), 500 mM sucrose.
6. Mini extruder.
7. 0.1 μm polycarbonate membranes.
8. Filter supports.
9. Buffer B: 10 mM Na-Hepes (pH 7.4), 100 mM KCl, 100 mM NaCl.

2.2 Phospholipids Calibration Curve

1. Synthetic dioleoylphosphatidylcholine.
2. Chloroform.
3. Ammonium Ferrothiocyanate solution—dissolve 27.3 g $FeCl_3(H_2O)_6$ and 30.4 g Ammonium Thiocyanate in 1 L double distilled water (*see* **Note 1**).

2.3 Measuring the Liposomes Concentration

1. Buffer B: as above.
2. Ammonium Ferrothiocyanate solution.
3. Chloroform.

2.4 Tim44 Binding Assay

1. Purified Tim44 as reported previously [18].
2. Buffer C: 20 mM Na-Hepes (pH 7.4), 200 mM NaCl.
3. Sample buffer: 2 % SDS, 62.5 mM Tris–HCl pH 6.8, 20 % Glycerol, 5 % β-mercaptoethanol, 0.05 % Bromophenol blue.
4. Ammonium persulfate (APS): 10 % solution in water.
5. 40 % Acrylamide:bisacrylamide (29:1) solution.
6. *N*,*N*,*N*′,*N*′-tetramethylethane-1,2-diamine (TEMED).
7. Resolving gel buffer: 1.5 M Tris–HCl pH 8.8, 0.4 % SDS.
8. Stacking gel buffer: 0.5 M Tris–HCl pH 6.8, 0.4 % SDS.
9. SDS-PAGE running buffer: 0.025 M Tris, 0.192 M glycine, 0.1 % SDS.
10. Coomassie dye solution: 0.12 % Coomassie R250, 40 % Ethanol, 10 % acetic acid.

3 Methods

3.1 Liposome Preparation

Carry out all procedures at room temperature unless otherwise specified.

1. Weigh 5 mg of PC (for PC only liposomes) or 4.8 mg PC and 1.2 mg CL (for PC + CL liposomes) (*see* **Note 2**), into a round bottomed flask.
2. In a fume hood add 1 mL Chloroform into each flask and mix gently until a homogenous solution is obtained. Since chloroform is volatile, hold the flask cap with your finger to prevent it from falling.
3. Remove the chloroform using rotary evaporation under reduced pressure in order to obtain a thin lipid film on the sides of the flask (*see* **Note 3**).
4. Add 1.1 mL Buffer A (swelling solution) and vortex rigorously for 30 s.
5. Perform five freeze–thaw cycles using a 40 °C water bath and a Dewar with liquid nitrogen (*see* **Note 4**).

Table 1
Samples for the phospholipid calibration curve

PC dissolved in chloroform (0.1 mg/mL) [mL]	0	0.1	0.2	0.3	0.4	0.5	0.6	0.7	0.8	0.9	1
Ammonium ferrothiocyanate [mL]	2	2	2	2	2	2	2	2	2	2	2
Chloroform [mL]	2	1.9	1.8	1.7	1.6	1.5	1.4	1.3	1.2	1.1	1

6. Using a mini extruder, extrude 11 times through 0.1 μm polycarbonate membranes (*see* **Note 5**) until homogeneously sized sucrose-loaded unilamellar liposomes are obtained.
7. Dilute the liposome mixture to a final volume of 5 mL with Buffer B.
8. Divide the liposomes into five 1.5 mL centrifuge tubes, each containing 1 mL liposome solution. Centrifuge at 137,000 × *g* for 1 h at 22 °C (*see* **Note 6**).
9. Remove the supernatant and add 100 μL Buffer B to each pellet (total 500 μL). Incubate for 15 min at room temperature.
10. Resuspend the pellets by pipetting and combine the aliquots. The liposomes can be saved at 4°C for approximately 1 week.
11. Determine the liposomes concentration as described in the following sections.

3.2 Phospholipids Calibration Curve

The phospholipids quantification is based on a colorimetric assay which detects the complex formation between phospholipids and ammonium ferrothiocyanate [19].

Carry out all procedures at room temperature unless otherwise specified.

1. Dissolve 10 mg PC in 100 mL Chloroform (concentration—0.1 mg/mL).
2. Prepare 11 samples, in glass tubes, as indicated in Table 1 (*see* **Note 7**).
3. The samples will contain two phases—an upper water phase and a lower chloroform phase. Vortex each sample for 1 min in order to mix the phases.
4. Wait a few minutes until the two phases separate. By using a glass pipette remove the upper phase.
5. Insert the lower phase into a quartz cuvette and read absorption at 488 nm. Use the sample without PC as a blank.
6. Prepare a calibration curve of O.D. vs. lipid concentration in μM.

3.3 Measuring the Liposomes Concentration

1. In a glass tube mix 2 mL of Ammonium Ferrothiocyanate solution with 10 μL of liposomes (vortex the liposomes before adding), or Buffer B (as a blank). Allow the mixture a few seconds to react.

Table 2
Samples for the Tim44 binding assay

Sample	Buffer B (μL)	Tim44 10 μM (μL)	Liposomes 1 mM (μL)	Liposomes 5 mM (μL)	Final liposomes concentration (μM)
1	180	20	–	–	0
2	178	20	2	–	10
3	176	20	4	–	20
4	172	20	8	–	40
5	166	20	14	–	70
6	156	20	24	–	120
7	172	20	–	8	200
8	164	20	–	16	400
9	154	20	–	26	650
10	140	20	–	40	1,000

2. Add 2 mL chloroform and vortex.
3. Wait a few minutes until the two phases separate.
4. Remove the upper phase using a glass pipette and measure the O.D. of the lower phase at 488 nm in a quartz cuvette.
5. Using the calibration curve, calculate the liposomes concentration. Note that the concentration has to be multiplied by 200 since the liposomes are diluted.

3.4 Tim44 Binding Assay to Liposomes

The binding experiment is carried out by incubating recombinant Tim44 with increasing concentrations of liposomes. Upon centrifugation, bound proteins will sediment together with the liposomes, while unbound proteins will remain in the supernatant. The first step of each experiment is to remove any protein aggregates that can interfere with the binding assay (since they precipitate even without liposomes).

1. For every experiment (ten samples) prepare a solution containing 250 μL of 10 μM Tim44 in Buffer C (*see* **Note 8**).
2. Incubate for 15 min at room temperature.
3. Centrifuge at 137,000×*g* for 1 h at 22 °C (*see* **Note 9**). Continue with the supernatant for the binding experiment.
4. Prepare a "total" protein sample for direct loading to the gel—20 μL Tim44 from 10 μM stock (from **step 3**) with 20 μL sample buffer.
5. Prepare solutions as described in Table 2, in 250 μL polycarbonate centrifuge tubes (*see* **Note 10**).

6. Incubate the samples of **step 5** for 1 h at room temperature.
7. Separate bound Tim44 from unbound protein by centrifugation at 137,000 × *g* for 1 h at 22 °C.
8. Remove the supernatant and add 40 μL sample buffer.
9. Incubate the pellet with the sample buffer for 2 h at room temperature or overnight at 4 °C (*see* **Note 11**).
10. After incubation, resuspend the pellet by pipetting and transfer into Eppendorf tubes.
11. Load 20 μL of each sample and 20 μL of the total sample onto a 14 % polyacrylamide gel.
12. Stain the gel with Coomassie dye (see, for example, Fig. 1).

4 Notes

1. $FeCl_3(H_2O)_6$ and Ammonium Thiocyanate should be saved in a desiccator. Ammonium Ferrothiocyanate solution is stable for months in room temperature.
2. It is possible to use other phospholipids but you have to check their phase transition temperature and perform the extrusion step above this temperature. We use a PC:CL ratio of 80 %:20 % (w/w) since our liposomes are a model for the inner mitochondrial membrane in which CL is approximately 20 % of the total phospholipids [6]. When preparing PC and PC + CL liposomes, it is advisable to weigh first the PC in order to prevent contamination with CL. Since the lipids are sticky, use two spatulas when weighing. With one spatula pick up a small amount of material and with the second spatula gently place the lipids on the weighing plate.
3. We keep the lipid film in 4 °C up to 1 day.
4. This step is performed since it induces the formation of unilamellar vesicles and it improves the homogeneity of the size distribution of the liposomes after extrusion. Start the freeze-thaw cycles with freezing—prepare a Dewar with liquid nitrogen, hold the upper part of the flask with a long tweezer, and gently place the lower part of the flask in the Dewar. The upper part of the flask has to be above the liquid surface in order to prevent the liquid from entering the flask. When transferring to the water bath, submerge only the bottom of the flask into the water to prevent the liquid from entering the flask.
5. Extrusion yields liposomes having a diameter near the pore size of the membrane used. In general, the more extrusion cycles performed the more homogenous the size distribution of the sample. 11 Extrusion cycles provide a rather narrow distribution around the membrane pore size which is sufficient

for our experiments. It is important to finish the extrusion on the alternate syringe (i.e., extrude an uneven number of times) in order to reduce contamination by larger particles. A membrane pore size of lower than 0.2 μm is needed for obtaining unilamellar liposomes. When performing the extrusion, it is important to pay attention to the amount of pressure needed to force the liposomes through the membrane—a sudden decrease in pressure may indicate a tear in the membrane.

6. Since the liposomes pellet, at low lipid concentrations, is rather difficult to observe, we mark the external side of the tubes on which the pellet is expected to stick. In order to clean the tubes, immerse them in ethanol and mix. Next, wash with deionized water and dry.
7. When working with chloroform work with glass materials because plastic can be dissolved by the chloroform.
8. We prepare the protein solution in the centrifuge tube.
9. Since the pellet can be hard to observe mark the external side of the tube on which the pellet will stick.
10. Add components in the following order: Buffer B, Tim44, and liposomes. Vortex the liposomes before their addition since they tend to accumulate in the bottom of the tube. The first sample is a background sample of Tim44 without liposomes. Again, mark the external side of the tube on which the pellet will stick.
11. If you choose to incubate overnight close the tubes with parafilm in order to avoid reduction of volume and contaminations.

Acknowledgment

This work was supported by the German-Israeli Foundation for Scientific Research and Development (GIF-1012/08) and Israel Science Foundation (452/09).

References

1. Bangham AD, Horne RW (1964) Negative staining of phospholipids and their structural modification by surface-active agents as observed in the electron microscope. J Mol Biol 8:660–668
2. Marsden HR, Tomatsu I, Kros A (2011) Model systems for membrane fusion. Chem Soc Rev 40:1572–1585
3. Wesolowska O, Michalak K, Maniewska J et al (2009) Giant unilamellar vesicles—a perfect tool to visualize phase separation and lipid rafts in model systems. Acta Biochim Pol 56:33–39
4. Sanchez SA, Tricerri MA, Ossato G et al (2010) Lipid packing determines protein-membrane interactions: challenges for apolipoprotein A-I and high density lipoproteins. Biochim Biophys Acta 1798:1399–1408
5. Walde P, Cosentino K, Engel H et al (2010) Giant vesicles: preparations and applications. Chembiochem 11:848–865
6. Osman C, Voelker DR, Langer T (2011) Making heads or tails of phospholipids in mitochondria. J Cell Biol 192:7–16
7. Horst M, Jeno P, Kronidou NG et al (1993) Protein import into yeast mitochondria: the inner membrane import site protein ISP45 is the MPI1 gene product. EMBO J 12: 3035–3041

8. Blom J, Kubrich M, Rassow J et al (1993) The essential yeast protein MIM44 (encoded by MPI1) is involved in an early step of preprotein translocation across the mitochondrial inner membrane. Mol Cell Biol 13:7364–7371
9. Schneider HC, Berthold J, Bauer MF et al (1994) Mitochondrial Hsp70/MIM44 complex facilitates protein import. Nature 371: 768–774
10. Marom M, Safonov R, Amram S et al (2009) Interaction of the Tim44 C-terminal domain with negatively charged phospholipids. Biochemistry 48:11185–11195
11. Cui W, Josyula R, Li J et al (2011) Membrane binding mechanism of yeast mitochondrial peripheral membrane protein TIM44. Protein Pept Lett 18:718–725
12. Weiss C, Oppliger W, Vergeres G et al (1999) Domain structure and lipid interaction of recombinant yeast Tim44. Proc Natl Acad Sci USA 96:8890–8894
13. Claypool SM (2009) Cardiolipin, a critical determinant of mitochondrial carrier protein assembly and function. Biochim Biophys Acta 1788:2059–2068
14. Schlattner U, Tokarska-Schlattner M, Ramirez S et al (2009) Mitochondrial kinases and their molecular interaction with cardiolipin. Biochim Biophys Acta 1788:2032–2047
15. Ott M, Zhivotovsky B, Orrenius S (2007) Role of cardiolipin in cytochrome c release from mitochondria. Cell Death Differ 14:1243–1247
16. Vergeres G, Manenti S, Weber T et al (1995) The myristoyl moiety of myristoylated alanine-rich C kinase substrate (MARCKS) and MARCKS-related protein is embedded in the membrane. J Biol Chem 270:19879–19887
17. Buser CA, Sigal CT, Resh MD et al (1994) Membrane binding of myristylated peptides corresponding to the NH2 terminus of Src. Biochemistry 33:13093–13101
18. Slutsky-Leiderman O, Marom M, Iosefson O et al (2007) The interplay between components of the mitochondrial protein translocation motor studied using purified components. J Biol Chem 282:33935–33942
19. Charles J, Stewart M (1979) Colorimetric determination of phospholipids with ammonium ferrothiocyanate. Anal Biochem 104: 10–14

Chapter 11

Analysis of the Interaction Between Membrane Proteins and Soluble Binding Partners by Surface Plasmon Resonance

Zht Cheng Wu, Jeanine de Keyzer, Ilja Kusters, and Arnold J.M. Driessen

Abstract

The interaction between membrane proteins and their (protein) ligands is conventionally investigated by nonequilibrium methods such as co-sedimentation or pull-down assays. Surface Plasmon Resonance can be used to monitor such binding events in real-time using isolated membranes immobilized to a surface providing insights in the kinetics of binding under equilibrium conditions. This application provides a fast, automated way to detect interacting species and to determine the kinetics and affinity (K_d) of the interaction.

Key words SPR, Membrane proteins, Ribosomes binding, SecA binding, SecYEG, IMVs immobilization, L1 chip

1 Introduction

Studies on the interaction between soluble proteins and membrane proteins are complicated by the need to maintain the membrane protein in a functional state typically the membrane-embedded state. Conventional methods to monitor such interactions include nonequilibrium methods such as co-sedimentation or pull-down assays using either proteoliposomes with the reconstituted membrane protein or isolated membrane vesicles. Such methods have the disadvantage that they do not monitor the kinetics of the interaction, and also low affinity interactions are difficult to capture. More recently developed (equilibrium) methods often involve (fluorescent) labeling of the proteins and require the membrane protein to be either in detergent solution or reconstituted in a lipid bilayer. These modifications are not always desirable, as is illustrated by our recent observation that the interaction between the membrane protein complex SecYEG and ribosomes is strongly dependent on the molecular environment of SecYEG [1].

Doron Rapaport and Johannes M. Herrmann (eds.), *Membrane Biogenesis: Methods and Protocols*,
Methods in Molecular Biology, vol. 1033, DOI 10.1007/978-1-62703-487-6_11, © Springer Science+Business Media, LLC 2013

Whereas with the membrane-embedded SecYEG complex barely interacts nonprogrammed ribosomes, this interaction is remarkably stimulated suggesting that detergent augments particular nonnative interactions.

Here, we describe how surface plasmon resonance (SPR) can be used to monitor the interaction between soluble and membrane proteins without the need for labeling or membrane solubilization. SPR follows the formation and dissociation of the biomolecular complexes in real-time. The interactions are monitored on a sensor surface by detecting changes in the refractive index near this surface. One of the binding partner (ligand) is attached to the sensor surface that forms one wall of a flow cell. Through this flow cell passes a continuous flow of running buffer into which the other binding partner (analyte) is injected. Binding of the analyte to the immobilized ligand results in an increase in the refractive index at the sensor surface, which is detected and plotted in a so-called sensogram as resonance units (RUs) against time. This application not only provides a fast, automated way to detect interacting species, but also can determine the kinetics and the affinity of the interaction. The sensor surface generally consists of a glass coated with gold (or another inert metal) which is further modified to allow the immobilization of biomolecules, for example by covalent linkage of a carboxymethylated dextran layer. A decade ago, Biacore introduced the L1 chip, on which the carboxymethylated dextran layer is modified with lipophilic groups. Since membrane vesicles and (proteo-)liposomes can attach to this surface directly while maintaining their lipid bilayer structure, the L1 chip allows for the analysis of the interaction between soluble analytes and membrane proteins in their native environment, the lipid bilayer.

We have developed a robust method in which we use the L1 chip to compare the binding of a purified, soluble, analyte to *E. coli* inner membranes with or without over-expression of the membrane protein of interest [1, 2]. The membranes are isolated from *E. coli* using a sucrose density centrifugation protocol and immobilized without further modification on the L1 chip. The multiflow cell set-up of the Biacore system allows simultaneous monitoring of analyte binding to the membranes with and without over-expression. Subtraction of the background (no over-expression) signal, which includes the effect of differences in the refractive index between the running and injection buffer (bulk effect) and the nonspecific binding of the analyte, results in direct detection of the specific binding to the ligand. The method will be described here using the bacterial Sec-translocase as an example. The Sec-translocase mediates the transport of proteins across and the insertion of membranes proteins into the cytoplasmic membrane of *E. coli* [3]. We will show how SPR can be used to analyze the interaction of the membrane-embedded core of the translocase, the protein conducting channel SecYEG, with its cytosolic binding partners: the ATPase SecA and membrane protein translating ribosomes. However, the method can readily be adapted to other membrane proteins that interact with a soluble partner protein.

2 Materials

Prepare all solutions in autoclaved ultrapure water (using purifying deionized water to obtain a resistance of 18 MΩ cm at room temperature).

2.1 Membrane Isolation Components

1. LB (Luria Bertani) medium: 10 g tryptone, 5 g yeast extract, and 10 g NaCl are dissolved in 950 mL water. The pH is adjusted to 7.0 using 5 N NaOH and additional water is added to obtain a final volume of 1 L.
2. 100 mg/mL ampicillin: 1 g ampicillin is dissolved in 10 mL water and filter sterilized (*see* **Note 1**).
3. 1 M isopropyl-β-D-1-thiogalactopyranoside (IPTG): 2.38 g IPTG is dissolved in 10 mL water and filter sterilized (*see* **Note 1**).
4. 100 mM phenylmethylsulfonyl fluoride (PMSF): 1.74 g PMSF is dissolved in 10 mL ethanol (*see* **Note 1**).
5. 1 M $MgSO_4$: 1.20 g $MgSO_4$ is dissolved in 10 mL water.
6. DNAse (Sigma, St. Louis, MO).
7. 50 mM Tris–HCl pH 8.0: 1.51 g Tris is dissolved in 200 mL water. The pH is adjusted to 8.0 using 5 N HCl and water is added to obtain a final volume of 250 mL. Cool to 4 °C before use.
8. 50 mM Tris–HCl pH 8.0, 20 (or 36, 45, 51, 55) % (w/v) sucrose: Water is added to 0.3 g Tris and 10 g (or 18, 22.5, 25.5, 27.5 g) sucrose to a total volume of 20 mL. The pH is adjusted to 8.0 using 5 N HCl and water is added to obtain a final volume of 50 mL. Cool to 4 °C before use.
9. 50 mM Tris–HCl pH 8.0, 20 % glycerol: 0.151 g Tris and 4.6 mL 87 % glycerol are dissolved in 15 mL water. The pH is adjusted to 8.0 using 5 N HCl and water is added to obtain a final volume of 25 mL. Cool to 4 °C before use.
10. One Shot cell disrupter; Constant Systems, Daventry, UK or equivalent.

2.2 SecA Isolation Components

1. LB and ampicillin: *see* Subheading 2.1.
2. Buffer D (20 mM Hepes-KOH pH 6.5, 10 % glycerol): 2.38 g Hepes and 57.5 mL 87 % glycerol are dissolved in 450 mL of water and adjust pH to 6.5 using 5 N KOH. Add water to a final volume of 500 mL and cool to 4 °C before use.

3. Buffer E (Buffer D supplemented with 1 M NaCl): 1.19 g HEPES, 14.61 g NaCl, and 2.9 mL 87 % glycerol are dissolved in 200 mL water. The pH is adjusted to 6.5 with 5 N KOH. Add water to a final volume of 250 mL and cool to 4 °C before use.
4. 5 mL HiTrap SP HP column (GE Healthcare).
5. Äkta FPLC (GE Healthcare) or equivalent.

2.3 Crude Ribosome and Ribosome Nascent Chain Isolation Components

1. LB and ampicillin: *see* Subheading 2.1.
2. Buffer R: 50 mM Tris–HCl pH 7.5, 150 mM KCl, and 10 mM $MgCl_2$. 1.51 g Tris, 2.8 g KCl, and 0.51 g $MgCl_2$ are dissolved in 200 mL water. The pH is adjusted using 5 N HCl and water is added to obtain a final volume of 250 mL. Autoclave and cool to 4 °C before use.
3. Buffer Z: 50 mM Tris–HCl pH 7.5, 1 M sucrose, 150 mM KCl, and 10 mM $MgCl_2$. 85.6 g Sucrose, 1.51 g Tris, 2.8 g KCl, and 0.51 g $MgCl_2$ are dissolved in 150 mL water. The pH is adjusted using 5 N HCl and water is added to a final volume of 250 mL. Autoclave for 15 min and cool to 4 °C before use.
4. Buffer Y: 25 mM HEPES/NaOH pH 7.0, 100 mM $CaCl_2$. 0.6 g HEPES. 1.47 g $CaCl_2$ are dissolved in 80 mL water. The pH is adjusted using 5 N NaOH and water is added to a final volume of 100 mL. Autoclave and cool to 4 °C before use.
5. RNase free-DNase (Fluka 25,770 U/mL).
6. Cell lysing solution: 1 g of lysozyme (Merck 50,000 U/mg) is dissolved in 10 mL of water yielding a stock of 10 mg/mL.
7. Streptactin beads (IBA).
8. Beads Regeneration buffer: Dilute 5 mL of 10× regeneration buffer (IBA) into 45 mL water.
9. Column washing buffer: Dilute 5 mL of 10× washing buffer (IBA) into 45 mL water.
10. Elution buffer: 2.5 mM desthiobiotin in Buffer R. Add 5 mg of desthiobiotin (IBA) in 10 mL Buffer R.
11. Poly-prep chromatography column (Biorad).
12. Millipore Amicon Ultra 4 or Ultra-15 centrifugation units (Millipore).
13. Streptactin-AP conjugate (IBA).

2.4 SPR Components

All solutions are freshly prepared, filtered, and degased (*see* **Note 2**) before use and should not be reused (unless indicated otherwise).

1. Buffer A: 50 mM Tris–HCl pH 8.0, 50 mM KCl, 5 mM $MgCl_2$, and 1 mM DTT. Transfer 3.03 g Tris, 1.864 g KCl, 0.51 g $MgCl_2$, and 0.077 g DTT to a volumetric cylinder and add water to a volume of 450 mL. Mix and adjust pH with 5 N HCl and add water to a final volume of 500 mL.

2. Buffer B: 50 mM Tris–HCl pH 8.0, 150 mM KCl, 5 mM $MgCl_2$, 1 mM DTT, and 0.5 mg/mL BSA. 1.864 g KCl and 125 mg BSA are added to 250 mL Buffer A. The BSA should only be added when the degasing is almost completed (*see* **Note 3**).
3. 100 mM sodium carbonate pH 10. 1.06 g of sodium carbonate is dissolved in 95 mL water. The pH is adjusted using 5 N HCl and water is added to 100 mL (*see* **Note 4**).
4. 0.5 % SDS solution: 0.5 g SDS is dissolved in 100 mL water (*see* **Note 5**).
5. 40 mM octyl glucoside: 0.117 g of octyl β-D-glucopyranoside is dissolved in 10 mL water (*see* **Note 4**).
6. L1 Chip (GE healthcare).
7. Biacore 2000 or equivalent (GE healthcare).

3 Methods

3.1 Inner Membrane Vesicle Isolation

1. Transform *E. coli* SF100 [4] with empty vector pET302 [5] or SecYEG over-expression plasmid pET610 [6].
2. Plate on LB agar supplemented with 100 μg/mL ampicillin and grow overnight at 37 °C.
3. Inoculate 100 mL LB supplemented with 100 μg/mL ampicillin and 0.5 % (w/v) glucose with a single colony and grow overnight at 37 °C with shaking.
4. Dilute the overnight culture 50 times in 1 L LB supplemented with 100 μg/mL ampicillin and grow at 37 °C with shaking until OD_{660} of about 0.6.
5. Induce the culture with 0.5 mM isopropyl-d-thiogalactopyranoside (IPTG) and continue growth for an additional 2 h.
6. Harvest cells by centrifugation (6,000 × *g*, 15 min, 4 °C) and resuspended in 7 mL 50 mM Tris–HCl (pH 8.0), 20 % (w/v) sucrose.
7. Snap-freeze the cells in liquid nitrogen and store at −20 or −80 °C.
8. To prevent proteolysis, all subsequent steps should be performed at 4 °C.
9. Slowly thaw the cells (*see* **Note 6**) and add 0.5 mM phenylmethylsulfonyl fluoride (PMSF), 1 mM $MgSO_4$ and 1 mg/mL DNAse (Sigma, St. Louis, MO).
10. Lyse the cells by passing twice through an One Shot cell disrupter at 8,000 lb/in^2.
11. After the first lysis step, increase the PMSF concentration to 1 mM.

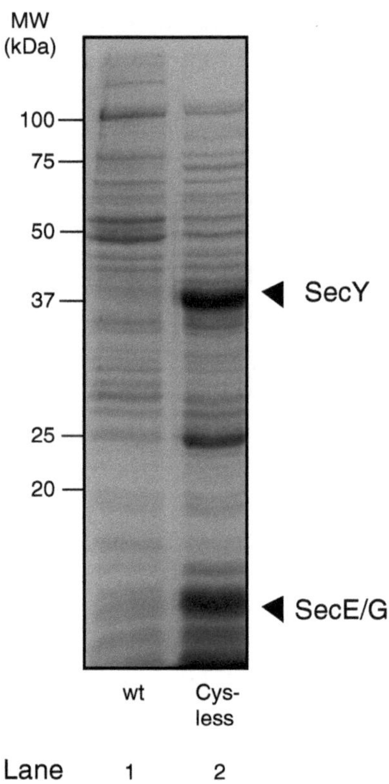

Fig. 1 Over-expression of Cys-less SecYEG: *E. coli* IMVs with and without over-expressed Cys-less SecYEG were analyzed on SDS-PAGE followed by Coomassie brilliant blue staining

12. Remove unbroken cells by centrifugation (6,000 × *g*, 15 min) and collect the membranes by ultracentrifugation (125,000 × *g*, 90 min).
13. Resuspend the membrane pellet in 1 mL 50 mM Tris–HCl pH 8.0.
14. Separate outer and inner membranes on a sucrose step gradient: Divide the resuspended membranes over two Beckman TLA110 tubes (4 mL) containing layers of 55 % (0.55 mL), 51 % (1 mL), 45 % (0.45 mL), and 36 % (0.45 mL) (w/v) sucrose in 50 mM Tris–HCl pH 8.0, respectively and spin for 30 min at 4 at 90,000 rpm (440,000 × *g*) in a TLA110 rotor.
15. Collect the brownish inner membrane fraction from the 45 % sucrose layer, dilute at least fivefold with 50 mM Tris–HCl pH 8.0, and recollected by centrifugation (440,000 × *g*, 30 min).
16. The pellet is resuspended in 50 mM Tris–HCl pH 8.0, 20 % glycerol, frozen in liquid nitrogen, and stored in small aliquots at −80 °C.
17. Over-expression of SecYEG is verified by SDS-PAGE gel followed by Coomassie brilliant blue staining as shown in Fig. 1.

3.2 SecA Isolation

This procedure is adapted from the method described by Kusters et al. [7].

1. Transform *E. coli* DH5α [8] with plasmid pMKL18 [unpublished, gift of R. Freudl (Institut für Bio- und Geowissenschaften, IBG-1: Biotechnologie; Forschungszentrum Jülich GmbH, D-52425 Jülich, Germany); SecA gene cloned in pUC19 vector, expression of SecA, *amp*].
2. Plate on LB agar supplemented with 100 μg/mL ampicillin and incubate overnight at 37 °C.
3. Inoculate 1 L LB supplemented with 100 μg/mL ampicillin with a single colony and grow overnight at 37 °C with shaking (*see* **Note 7**).
4. Collect the cells by centrifugation for 15 min at 6,000 × *g*, 4 °C in a Beckman Avanti-J-26XP centrifuge or equivalent.
5. Wash the cells by resuspending the pellet in 200 mL Buffer D and collect by centrifugation for 15 min at 6,000 × *g*, 4 °C.
6. Resuspend the cell pellet in 2–3 pellet volumes of Buffer D, transfer to a 15 mL Falcon tube, freeze in liquid nitrogen, and store at −80 °C (*see* **Note 8**).
7. Defrost the cells by placing the Falcon tube in ice-water.
8. Lyse the defrosted cells by sonication, using a tip sonicator for 15 cycles of 30 s on and 30 s off. Place the tube in an ice-water bath during sonication to prevent overheating of the sample.
9. Remove membranes and unbroken cells by ultracentrifugation at 440,000 × *g* for 30 min at 4 °C in a Beckman Optima Max XP centrifuge or equivalent and collect the supernatant (called cell free extract, CFE) in fresh falcon tube.
10. Adjust the NaCl concentration of the CFE to 100 mM by mixing with an appropriate volume of Buffer D.
11. The CFE is loaded on a 5 mL HiTrap SP HP column, equilibrated with Buffer D with 100 mM NaCl (*see* **Note 9**) at a flow rate of 0.5 mL/min.
12. The column is washed with 5 column volumes Buffer D with 100 mM NaCl (*see* **Note 9**) at a flow rate of 1 mL/min.
13. Bound SecA is eluted in 10 column volumes with a linear gradient of 0.1–0.5 M NaCl in Buffer D (*see* **Note 9**) at a flow rate of 1 mL/min and collected in 1 mL fractions.
14. 5 μL of the fractions are analyzed by 10 % sodium dodecyl sulfate–polyacrylamide gel electrophoresis (SDS-PAGE).
15. For storage at −80 °C, dilute SecA containing fractions at least twofold (below 200 mM NaCl) using Buffer D (*see* **Note 10**).

3.3 Ribosome Isolation

1. Plate *E. coli* strain MRE600 [9] on LB agar and grow overnight at 37 °C (200 rpm, New Brunswick Scientific Excella E25).
2. The next day inoculate 25 mL LB with a single colony and grow overnight at 37 °C with shaking.
3. Dilute the overnight culture 100 times into 1 L LB prewarmed at 37 °C and grow cells at 37 °C until OD_{660} 0.6 is reached.
4. Cool the cells on ice and harvest them by centrifugation using a JLA 8.1000 rotor (9,500 ×*g* at 4 °C, 10 min) in a Beckman Avanti-J-26XP centrifuge or equivalent.
5. Resuspend the cell pellet in 10 mL ice-cold Buffer R using a 20 mL glass syringe (Fortuna® Optima®, Germany) or a 10 mL pipet and transfer the cells into a 15 mL falcon tube.
6. Add 1 mL cell lysing solution to the cells and put the falcon tube directly in the −80 °C freezer (without liquid N_2 snap freezing) for at least 30 min (*see* **Note 11**). Slowly thaw the cells by placing the tubes in ice-water. Add 20 μL RNase free-DNase to reduce the viscosity of the cell lysate and repeat the freeze–thawing procedure once more.
7. Remove the cell debris by two low spin centrifugation steps (30,000 ×*g*, 4 °C, 30 min) using a Beckman Optima MAX XP Ultracentrifuge or equivalent.
8. Lay the cleared cell lysate (about 7 mL) on 50 mL Buffer Z (*see* **Note 12**) in a Ti45 centrifuge tube and centrifuge at 112,000 ×*g* for 17 h at 4 °C in a Beckman Optima L-90K ultracentrifuge or equivalent.
9. Dissolve the translucent ribosome pellet in 500 μL ice-cold Buffer R by gentle shaking (*see* **Note 13**).
10. The concentration can be determined spectrophotometrically at a wavelength 260 using extinction coefficient of 4.2×10^7 [10]. The ribosome solution is diluted 100 times in order to get an accurate reading.
11. Store the ribosomes in small aliquots at −80 °C for further use.

3.4 Ribosome Nascent Chain Isolation

This procedure is adapted from the method described by Rutkowska et al. and Evans et al. [11, 12]. The method involves the in vivo expression of in this case a truncate of the membrane protein FtsQ followed by the stalling motif of the *E. coli* secretion monitor protein SecM [13] that halts translation without promoting release of the nascent chain. The construct is preceded by a triple STREP-tag to allow separation of RNCs from non-translating ribosomes [11, 14]. All procedures are performed at 4 °C unless indicated otherwise.

1. Transform *E. coli* BL21(DE3)Δtig [11] with pUK19strep$_3$Fts-QSecM [14] or an equivalent nascent chain construct.
2. Plate on LB agar containing 100 μg/mL ampicillin and incubate overnight at 37 °C.

3. Inoculate 25 mL LB containing 100 μg/mL ampicillin with a single colony and incubate at 30 °C for overnight with shaking.
4. The next day, dilute the overnight culture 100 times into 1 L LB prewarmed at 30 °C and grow cells at 30 °C until OD_{660} 0.5 is reached.
5. Induce the cells with 500 μM IPTG for 30 min.
6. Rapidly cool the cells by transferring the culture to a 1 L centrifugation bucket, adding five or six ice cubes made from Buffer R and cooling the bucket in ice-water.
7. Centrifuge using a JLA 8.1000 rotor (9,500 × *g* at 4 °C, 10 min) in a Beckman Avanti-J-26XP centrifuge or equivalent.
8. Resuspend the cell pellet in 10 mL ice-cold Buffer R.
9. Lyse the cells and isolate crude ribosomes by following **steps 6–9** from the ribosome isolation procedure in Subheading 3.3.
10. While dissolving the RNC pellet, pack 2 mL 50 % StrepTactin beads (IBA) in a poly-prep chromatography column.
11. Equilibrate the column with 3 column volumes of cold Buffer R.
12. Load the dissolved pellet onto the column.
13. Wash the column with 1 column volume of cold Buffer R.
14. Wash the column 2 column volumes of Buffer R containing 0.5 M KCl.
15. Wash the column five times column volumes of Buffer R (*see* **Note 14**).
16. Elute the RNCs with three times column volumes of elution buffer and collecting all in a same falcon tube.
17. Concentrate the eluate using a Millipore Amicon Ultra 4 or Ultra-15 centrifugational units using Beckman allegraX-15R (1,455 × *g*, 4 °C) or equivalent to a final volume of approximately 300 μL (*see* **Note 15**).
18. The concentration can be determined spectrophotometrically at a wavelength 260 using extinction coefficient of 4.2×10^7 [10]. The ribosome solution should be diluted 100 times for accurate reading.
19. Store the ribosomes at −80 °C for further use.
20. The presence of ribosomes bearing the nascent chain can be verified using immunoblot using Streptactin-AP conjugate (IBA) as shown in Fig. 2.
21. The packed column can be regenerated by adding 2 column volumes of regeneration buffer until the beads become completely orange, followed by thorough washing with buffer until the color vanishes.

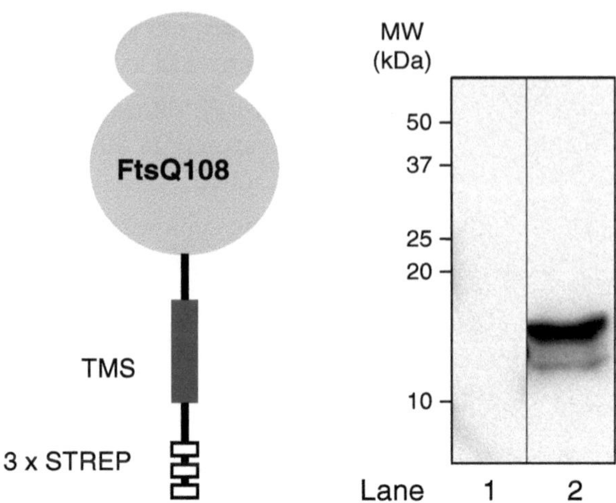

Fig. 2 Detection of FtsQ108 nascent chain ribosome by Western Blot: control non-translating ribosome (*lane 1*) and FtsQ108 (*lane 2*). The antibody used detects the STREP-tag. This research was originally published in the Journal of Biological Chemistry. Zht Cheng Wu, de Keyzer, J., Kedrov, A., Driessen, A.J.M. (2012) Competitive Binding of SecA and Ribosomes to the SecYEG translocon. *J. Biol. Chem.* 287: 7885–7895 © the American Society for Biochemistry and Molecular Biology

3.5 Immobilization of IMVs

3.5.1 Homogenization of the IMVs Before the Immobilization

1. Dilute the IMVs to a final concentration of 2 mg/mL in a final volume of 130 µL.
2. Place the nitrocellulose membrane (Avestin, Ottawa, Ontario Canada diameter: 19 mm, pore size: 200 nm) in between the two sealed blocks of the extruder (Avestin, Germany) and tighten by screwing.
3. Rinse the syringes with milliQ water and extrude the water back-and-forth to eliminate air bubbles (*see* **Note 16**).
4. Push out the water completely from one side and load the open syringe with degassed Buffer A.
5. Extrude with Buffer A several times and avoid the generation of bubbles.
6. Empty the syringe by disconnecting one end. Push all Buffer A out.
7. Pipette 130 µL 2 mg/mL IMVs directly in the opening of the syringe (*see* **Note 17**).
8. From the starting side where the syringe connects, push 11 times before sample collection (*see* **Note 18**).
9. After extrusion, samples are placed on ice before loading.

3.5.2 Immobilization of IMVs on L1 Chip

All steps are performed at 25 °C and with degassed buffers.

1. Dock the L1 chip into the Biacore 2000 or equivalent and prime the system with Buffer A.
2. Start a sensogram and inject 15–20 μL non-over-expression IMVs into the control channel (*see* **Note 19**) at a flow speed 1 μL/min.
3. Change the flow path to the second channel and inject 15–20 μL of SecYEG over-expression IMVs at a flow speed 1 μL/min.
4. Repeat **step 3** for every sample, if multiple types of IMVs are analyzed.
5. Change the flow path to all channels and set the flow at 20 μL/min.
6. Inject 15 μL 100 mM sodium carbonate to remove loosely bound IMVs and peripherally associated proteins.
7. Determine the level of IMV loading in each channel from the binding response (RU) and, if necessary, adjust loading by injecting 5–15 μL IMVs into channels with lower response unit (*see* **Note 20**).
8. Inject 15 μL 100 mM sodium carbonate to remove loosely bound IMVs and peripherally associated proteins.
9. Stop the sensorgram and prime system with Buffer B.

3.6 Binding Analysis

All steps are performed at 25 °C and with degassed buffers.

1. Start a new sensorgram and set the flow speed at 20 μL/min.
2. Equilibrate the system with Buffer B for 15 min (*see* **Note 21**).
3. Dilute the ligand to the desired concentration with Buffer B (approximately 250 μL) (*see* **Note 22**).
4. Inject 15 μL 100 mM sodium carbonate to prepare the binding surface.
5. Select K-inject option to inject 200 μL sample (*see* **Notes 23** and **24**).
6. Regenerate binding surface by injecting 15 μL 100 mM carbonate. The response should go back to the initial value observed in step (*see* **Note 25**).
7. Repeat **steps 3–6** if multiple sample concentrations are tested.
8. Examples of binding curve (with or without subtracting the control channels) are shown in Fig. 3.

3.7 Cleaning and Storage of the Chip

1. Start a new sensorgram and prime with MilliQ water.
2. Inject 20 μL 0.5 % SDS and 20 μL 40 mM octylglucoside sequentially (*see* **Note 26**).

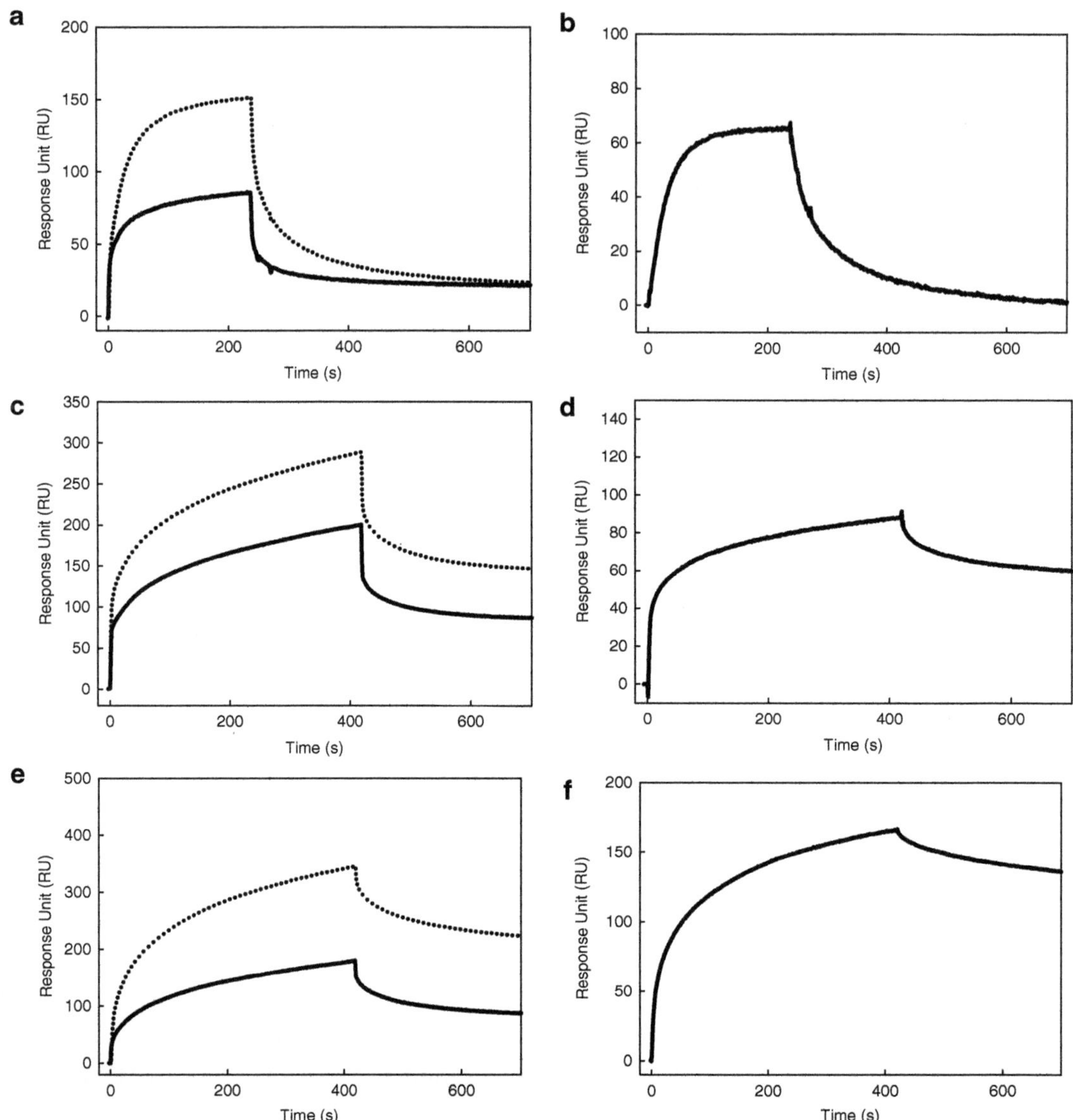

Fig. 3 Binding of SecA, ribosomes, and RNCs to Cys-less SecYEG present in surface immobilized *E. coli* IMVs. (**a**) Binding of SecA (24 nM) in control (*solid line*) and in over-expressed (*dotted line*) channel. (**b**) Specific binding of SecA to Cys-less SecYEG after subtracting the control from the over-expressed channel. (**c**) Binding of ribosomes (27 nM) in the control (*solid line*) and over-expressed (*dotted line*) channel. (**d**) Specific binding of ribosomes to Cys-less SecYEG after subtracting the control from the over-expressed channel. (**e**) Binding of FtsQ108 RNCs (27 nM) in the control (*solid line*) and over-expressed (*dotted line*) channel. (**f**) Specific binding of FtsQ108 RNCs to Cys-less SecYEG after subtracting the control from the over-expressed channel. This research was originally published in the Journal of Biological Chemistry. Zht Cheng Wu, de Keyzer, J., Kedrov, A., Driessen, A.J.M. (2012) Competitive Binding of SecA and Ribosomes to the SecYEG translocon. *J. Biol. Chem.* 287: 7885–7895 © the American Society for Biochemistry and Molecular Biology

3. Repeat **step 2**.
4. Undock the chip and store under N_2 gas in a 50 mL falcon tube at 4 °C (*see* **Note 27**).

5. Switch the system to the standby mode if it will be used shortly or enter the shutdown mode if not needed for a longer period of time (*see* **Note 28**).

3.8 Data Analysis

1. The obtained binding curves from injections of multiple ligand concentrations can be used to fit using "Bia-evaluation" software to obtain the affinity of the interaction.
2. An alternative way of determining the affinity is to use binding curves that have reached saturation plateau and plot the height of the curve. The K_d can be obtained using nonlinear regression fitting.

4 Notes

1. These solutions can be stored for several months at −20 °C.
2. The degasing procedure can be done using a Heto SUE30 water pump or a DIVAC 2.4 L dry pump. The device must have connected to a filtration device that contains a nitrocellulose membrane (0.45 μM Millipore®) and a stirring bar is placed in the collection container. Proper degasing is crucial due to the fact that air bubbles could disturb the reading of the binding during the detection. The degasing is completed when no bubbles are formed on the magnetic stirring bar.
3. To prevent foaming, the BSA is only added 10 min before the end of degasing procedure. The speed of stirring must be a minimum to avoid foaming.
4. Buffer C requires degassing and should be stored at room temperature until its further use. However, this buffer should be freshly prepared daily.
5. Degassing is not required but the solutions need to be filtered though a Whatman FP 30/0.2 μm filter disc before use and can be reused if stored at −20 °C.
6. Thawing can be done in a water-filled beaker at room temperature.
7. SecA is constitutively expressed from plasmid pMKL18 and high SecA over-expression inhibits growth. Therefore, cells are grown without induction.
8. Cells can be stored frozen in Buffer D for several months.
9. The NaCl concentration of Buffer D can be adjusted by mixing with the appropriate volume of Buffer E.
10. In the diluted elution buffer, SecA can be stored at −80 °C for at least 2 years.
11. At this point, it is possible to stop the experiment by storing the cells at −80 °C for further use.

12. The usage of Beckmann Ti45 tubes is advised and the volume of Buffer Z must be above 80 % of tube's maximum holding capacity.
13. Avoid strong mechanical stress while dissolving the ribosomal pellet. Placing the tubes in an ice box on a shaking device is recommended. The tube containing the pellet should be in a tilted angle so that the pellet is submerged in buffer. Ribosomes can be pipetted up and down gently before storing at −80 °C in small aliquots.
14. Before eluting the RNCs, make sure that the absorbance (at 260 nm) of the flow through is less than 0.01. If it is high, more washing is needed in order to remove the untagged ribosomes.
15. Concentrate the RNCs to a final volume that does not get under 200 μL. Check the volume of the sample in every 15 min centrifugation time and empty the catch compartment once a while to speed up the process. Never concentrate the RNCs sample to a volume that protein can form aggregates.
16. Push the water-filled syringe (with no bubbles in the syringe) to the other open-connector to drain bubble in the membrane compartment. Fill the other syringe with water and connect it to the open-connector. Try to push back-and-forth to see if any bubbles are generated during water extrusion.
17. While pipetting IMVs into the syringe, just pull the plunger to transfer the liquid into the syringe. Before connecting the syringe to the extruder, push some liquid out in order to obtain an air-free seal.
18. From the starting side, push the syringes ten times up and forth and perform the eleventh time by disassembling the syringe from the opposite side. The content is directly extruded into an Eppendorf tube on the eleventh push.
19. The loading of the control IMVs should always be in channel 1 or channel 3 (if more than one control is desired) so that direct subtraction of the over-expressed channel can be visualized during injection of the analytes.
20. All channels have to be loaded with similar amount of RU. In general, a good loading should be in between 4,000 and 5,500 RU. If certain IMVs are hard to immobilize, all other IMVs from other channels should be adjusted to the lowest loading level.
21. Since Buffer B has a higher salt concentration that Buffer A and contains BSA, the SPR response curves may drift at the start of the sensogram. It is recommended to wait until no significant fluctuation occurs and the drift is limited to around 3 RU/min. The presence of BSA reduces unspecific binding.

22. The volume necessary for injections is larger than indicated in particular when using k-inject. Always check how much the system consumes to prevent the introduction of air.
23. The volume of injection depends on the type of experiment. If complete saturation is required, a larger volume needs to be prepared for the injection.
24. Normally, the flow cell is vigorously washed with running buffer after injection has been completed. This often causes instabilities in the response, obscuring the dissociation phase. In the k-inject mode, the washing step can be delayed for a period of time that can be specified manually (typically 5 min) allowing accurate analysis of the dissociation phase.
25. The carbonate wash is needed to prepare the binding surface for a second analyte injection. If washing with carbonate does not remove all previously bound analyte, other regeneration conditions (for example: high salt, pH change, etc.) need to be explored.
26. To prevent damage of the chip, the contact time with the detergents should not exceed 1 min.
27. Use N_2 gas to chase out the air present in the tube and quickly screw the cap on.
28. The system can either switched into the standby mode for 4 days or proceed with the shutdown procedure.

Acknowledgments

This work was supported by the Chemical Sciences division of the Netherlands Foundation for Scientific Research (CW-NWO), the Organization for Fundamental Research on Matter (FOM), and the Foundation for Life Sciences (ALW), which are all financially supported by the Netherlands Organization for Scientific Research (NWO).

References

1. Wu ZC, de Keyzer J, Kedrov A, Driessen AJM (2012) Competitive binding of the SecA ATPase and ribosomes to the SecYEG translocon. J Biol Chem 287:7885–7895
2. de Keyzer J, van der Does C, Kloosterman TG, Driessen AJM (2003) Direct demonstration of ATP-dependent release of SecA from a translocating preprotein by surface plasmon resonance. J Biol Chem 278:29581–29586
3. du Plessis DJ, Nouwen N, Driessen AJM (2011) The Sec translocase. Biochim Biophys Acta 1808:851–865
4. Baneyx F, Georgiou G (1990) *In vivo* degradation of secreted fusion proteins by the *Escherichia coli* outer membrane protease OmpT. J Bacteriol 172:491–494
5. van der Does C, Manting EH, Kaufmann A, Lutz M, Driessen AJM (1998) Interaction between SecA and SecYEG in micellar solution and formation of the membrane-inserted state. Biochemistry 37:201–210
6. Kaufmann A, Manting EH, Veenendaal AK, Driessen AJM, Van der Does C (1999) Cysteine-directed cross-linking demonstrates

that helix 3 of SecE is close to helix 2 of SecY and helix 3 of a neighboring SecE. Biochemistry 38:9115–9125

7. Kusters I, van den Bogaart G, Kedrov A, Krasnikov V, Fulyani F, Poolman B, Driessen AJM (2011) Quaternary structure of SecA in solution and bound to SecYEG probed at the single molecule level. Structure 19:430–439
8. Hanahan D (1983) Studies on transformation of *Escherichia coli* with plasmids. J Mol Biol 166:557–580
9. Salaj-Smic E (1978) Colicinogeny of *Escherichia coli* MRE 600. Antimicrob Agents Chemother 14:797–799
10. Nilsson M, Bülow L, Wahlund KG (1997) Use of flow-field fractionation for the rapid quantitation of ribosome and ribosomal subunits in *Escherichia coli* at different protein production conditions. Biotechnol Bioeng 54:461–467
11. Rutkowska A, Beerbaum M, Rajagopalan N, Fiaux J, Schmieder P, Kramer G, Oschkinat H, Bukau B (2009) Large-scale purification of ribosome-nascent chain complexes for biochemical and structural studies. FEBS Lett 583:2407–2413
12. Evans MS, Ugrinov KG, Frese MA, Clark PL (2005) Homogeneous stalled ribosome nascent chain complexes produced *in vivo* or *in vitro*. Nat Methods 2:757–762
13. Nakatogawa H, Murakami A, Ito K (2004) Control of SecA and SecM translation by protein secretion. Curr Opin Microbiol 7:145–150
14. Schaffitzel C, Ban B (2007) Generation of ribosome nascent chain complexes for structural and functional studies. J Struct Biol 159: 302–310

Chapter 12

Peptide Interaction with and Insertion into Membranes

Ron Saar-Dover, Avraham Ashkenazi, and Yechiel Shai

Abstract

Natural and synthetic membrane active peptides as well as fragments from membrane proteins interact with membranes. In several cases, such interactions cause the insertion of the peptides to the membrane and their assembly within the lipid bilayer. Here we present spectroscopic approaches utilizing NBD and rhodamine fluorescently labeled peptides to measure peptide–membrane interaction and peptide–peptide interaction within the membrane. The usage of the physical properties of NBD and rhodamine in solution and in membranes provides useful information on the interplay between peptides and lipids.

Key words Peptide–membrane interaction, Fluorescence spectroscopy, FRET, NBD, Rhodamine, Liposomes, ATR-FTIR

1 Introduction

The molecular mechanism of protein–membrane interaction is often studied by using peptide fragments from an intact protein or short membrane interacting peptides. This approach significantly increases our knowledge of the mechanism of action of different toxins, hormones, and integral membrane proteins. The activity of a membrane protein-derived peptides, natural host-defense peptides, and their de novo designed derivatives depend on (1) peptide properties (structure, charge, and hydrophobicity) and (2) the characteristics of the membrane. The main steps involved in peptide–membrane interaction include a binding step, insertion, and final organization within the membrane milieu [1], and each step can be investigated using different methods. For example, surface plasmon resonance (SPR) is an accurate method to measure peptide binding to membrane multilayer [2]. The secondary structure and the orientation of the peptide within the membrane can be determined using attenuated total reflectance Fourier transform infrared (ATR-FTIR) spectroscopy [3, 4], scanning calorimetry measurements [5, 6], and solid state NMR [7, 8]. Here, we present fluorescence spectroscopic techniques for studying peptide–membrane interaction.

Doron Rapaport and Johannes M. Herrmann (eds.), *Membrane Biogenesis: Methods and Protocols*, Methods in Molecular Biology, vol. 1033, DOI 10.1007/978-1-62703-487-6_12,

The fluorescence approach is relatively simple, highly sensitive, noninvasive, and allows time-scale investigation.

Tryptophan fluorescence has been used as a natural and intrinsic spectroscopic probe for studying membrane binding of tryptophan containing peptides [9, 10]. Tryptophan fluorescence is sensitive to the lipid environment and increases upon interaction with membranes. In addition, it can be introduced into some classes of antimicrobial peptides (AMPs) with minimal perturbation of the structure [9]. However, in many cases the quantum yield of tryptophan is not sufficient to follow the initial steps in the binding process, especially if the binding is weak. In addition, it is strongly affected by the surrounding amino acids. Therefore, results may be sometimes confusing [1].

In order to overcome these limitations, conjugation of the lipid environmentally sensitive NBD (7-nitrobenz-2-oxa-1, 3-diazole-4-yl) fluorophore to peptides is often used. NBD fluorescence can increase up to approximately tenfold upon interaction with membranes. Its high excitation wavelength (467 nm) and the high quantum yield reduce significantly the contribution of light scattering. NBD-labeled peptides exhibit fluorescence emission maxima around 540 nm in hydrophilic solution [1]. However, upon addition of lipid vesicles, relocation of the NBD group into a more hydrophobic environment results in an increase in its fluorescence intensity and a blue shift of the emission maxima [11]. The first property is used to determine the binding constant of the peptide to the membrane. The second property is exploited to evaluate the depth of penetration [4]. The advantage of the use of the NBD moiety is that it allows the use of experimental conditions in which the lipid:peptide molar ratio ranges from <100:1 up to >15,000:1. The addition of NBD does not change the biological function of most of the peptide, as was found for different AMPs such as pardaxin [12], dermaseptins [13], cecropins [14], and cathelicidin LL-37 [15]. However, pre-examination must be done for each newly investigated peptide.

Some peptides (such as the human cathelicidin LL-37) oligomerize in solution due to hydrophobic interactions, therefore they can reach the membrane as monomers and oligomers [15]. To evaluate whether the peptides remain self-associated within membranes and to compare their oligomerization state, the fluorescence of Rhodamine (Rho)-labeled peptides is monitored in the presence of phospholipid vesicles. When Rho-labeled monomers are associated and their rhodamine fluorescent probes are in close proximity, the result is self-quenching of the fluorescence emission. Unlike NBD, rhodamine fluorescence is only slightly sensitive to the dielectric constant of the medium and therefore is not significantly affected from the interaction with the lipids. Another way to investigate peptide–peptide interactions within the membrane is to measure the Förster resonance energy transfer (FRET) efficiency between donor and acceptor fluorescence probes. For that purpose NBD and Rhodamine can be used as a pair due to an overlap in their wavelength spectra (described in details in Subheading 3.3.2).

Note that all methods described below are usually performed at room temperature for peptides labeled at the N-terminus, but can be applied for C-terminal-labeled peptides. In addition, different lipid vesicles can be used, with a general rule that small vesicles (such as small unilamellar vesicles-SUVs) have lower light scattering and therefore allow higher signal to noise ratio. However, utilization of larger vesicles (large unilamellar vesicles-LUVs) is favorable due to a larger surface area of peptide–membrane interaction and reduced membrane curvature.

2 Materials

Stock lipids and peptides or any other dissolved materials should be kept lyophilized in sealed tubes to avoid oxidation. Preferentially, inert gases such as nitrogen or argon can be gently streamed before sealing. We use commercially available lipids (Avanti Polar Lipids; Sigma Aldrich) for preparation of membrane vesicles (see detailed methods in [16]). Peptides were synthesized by an Fmoc solid-phase method [17] on Rink amide-4-methylbenzhydrylamine hydrochloride salt (MBHA) resin, by using an Applied Biosystems (Foster City, CA) 433A automatic peptide synthesizer. Fluorescent labeling with 4-chloro-7-nitrobenz-2-oxa-1, 3-diazole fluoride (NBD-F) or 5-(and-6)-carboxytetra-methylrhodamine succinimidyl ester (Rhodamine) was followed by peptide cleavage from the resin and purification by reverse phase high-performance liquid chromatography (RP-HPLC) (see detailed method in [15, 18]).

1. LUVs suspension stock at total lipid concentration of 12.5 mM.
2. Fluorescently labeled peptide solution (*see* **Note 1**).
3. Distilled water or Milli-Q reagent grade water.
4. Phosphate buffered saline (PBS pH 7.4).
5. Ethanol 70 % (v/v).
6. A 5 × 5-mm quartz cuvette.

3 Methods

3.1 Binding of NBD-Labeled Peptide to Membranes

The binding constant of a peptide is calculated from a titration of lipid vesicles (SUVs or LUVs) into NBD-peptide solution. To achieve an accurate result, at least 20 points should be recorded for each curve. An accepted dilution factor of the peptide solution is up to 10 % and therefore no more than 40 μL of vesicle solution should be added.

1. Dissolve NBD-peptide in PBS to a final concentration of 0.1 μM (*see* **Note 2**).
2. Set the spectrophotometer to spectrum mode with excitation wavelength of 467 nm and emission wavelength of 500–600 nm. In our device, SLM-Aminco Bowman series 2-luminescence spectrophotometer (SLM-Aminco, Rochester, NY, USA), slits are usually set to 5–10 nm. Using wider slits will improve sensitivity but can increase background noise and therefore should be individually determined for each peptide.
3. Add 400 μL peptide solution to a pre-cleaned cuvette (magnetic stirrer can be used) and read the signal output. Measure again every few minutes until no change is detected. This is the basal signal of the labeled peptide in the absence of membranes.
4. Add 1 μL from the LUVs suspension stock to the cuvette to reach an initial peptide/lipid ratio of 1:312 and read again. Remeasure the signal intensity every 1 min until no change in the signal is detected. This will indicate that binding has reached equilibrium.
5. Repeat **step 4** successively until no change in the pick maxima (around 530 nm) can be detected.
6. Clean the cuvette by washing it three times with 70 % (v/v) ethanol. Wash ethanol traces using DDW.
7. To account for background, the emissions of both PBS and vesicles alone at the same wavelength should be monitored and subtracted (repeat **steps 3–6**).

3.1.1 Binding Curves and Affinity Calculation

The binding curves are commonly presented as the difference in fluorescence intensity ($\Delta F_{530\mathrm{nm}}$) vs. the lipid concentration (an example is given in Fig. 1). The affinity constant (K_a) is determined by a steady state affinity model using nonlinear least squares. The nonlinear least squares fitting is done using the following equation:

$$\Upsilon(x) = \frac{K_a X F_{max}}{1 + K_a X} \tag{1}$$

where X is the lipid concentration, $\Upsilon(x)$ is the fluorescence emission, F_{max} is the maximal difference in the emission of NBD-labeled peptide before and after the addition of lipids (it represents the maximum peptide bound to lipid), and K_a is the affinity constant [M^{-1}].

3.2 Peptide Insertion into the Membrane

3.2.1 Blue Shift

The emission maximum of NBD in PBS is approximately 540 nm. Following binding to lipid vesicles, NBD-labeled peptides exhibit blue shifts in the emission maxima towards 528–533 nm, which reflects a situation in which the NBD group is located at or near the surface of the membrane. Typical NBD emission values for

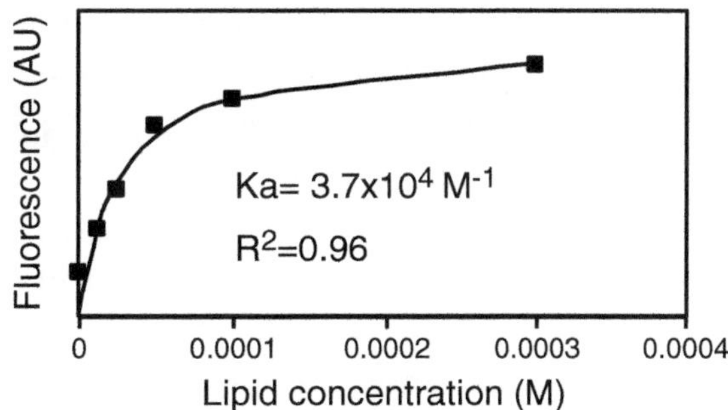

Fig. 1 The binding affinity of human immunodeficiency virus (HIV) gp41 transmembrane domain (TMD) peptides to liposomes. The titration curve was obtained by sequential titration of NBD conjugated peptides (NBD-gp41 TMD) in PBS with increasing amount of phosphatidylcholine:cholesterol (9:1) LUVs. In a typical experiment, a fixed concentration of NBD-gp41 TMD (0.2 μM) was added to 400 μL of PBS and LUVs were added until a plateau was reached. The values were corrected by subtracting the emission of the LUVs at the given concentration. The affinity constant (K_a) was determined by nonlinear least squares analysis. The correlation coefficient is also indicated. Adapted with permission from [19]. Copyright (2012) American Chemical Society

surface localized peptides are 530 nm and for membrane-inserted peptides is 520 nm. Hence, determining the emission maximum can provide information on the location of the NBD probe with respect to the membrane.

1. Dissolve NBD-peptide in PBS to a final concentration of 1 μM.
2. Set the spectrophotometer to scanning mode with excitation wavelength to 467 nm and emission wavelength spectra of 500–600 nm.
3. Add 400 μL peptide solution to a pre-cleaned cuvette (magnetic stirrer can be used) and read the signal output. This is the basal signal of the labeled peptide.
4. Add LUVs to the cuvette to a final concentration of 100 μM.
5. Monitor the shift in the wavelength at the signal pick. To verify that peptide penetration has reached equilibrium, measure again every few minutes until no change is detected.
6. To account for background, the emissions of both PBS and vesicles alone at the same wavelength should be monitored and subtracted (repeat **steps 3–5** while omitting the peptide).

3.2.2 Cleavage of NBD-Labeled Peptides by Proteinase K for Analyzing Peptide Insertion into the Membrane

To verify peptide insertion into the membrane, proteinase K enzymatic cleavage assay is performed. Proteinase K digests peptides and is active in most buffer solutions. It cleaves the exposed part of the peptide, but the enzyme cannot reach the membrane buried region. Following its addition to the cuvette, cleaved labeled fragments are released into the aqueous environment and a substantial reduction in the NBD fluorescence is apparent.

1. Dissolve NBD-peptide in PBS to a final concentration of 0.1–1 μM.
2. Set the spectrophotometer to kinetic mode with excitation wavelength of 647 nm and emission wavelength of 530 nm.
3. Add 400 μL NBD-peptide solution to a pre-cleaned cuvette (magnetic stirrer can be used) and read the signal output until equilibrium is reached.
4. Add LUVs to 100 μM final concentration. Monitor the change in the signal intensity until equilibrium is reached. This is the basal signal for this assay.
5. Add 125 μg/mL of proteinase K and track the change in the signal.
6. To account for background, the emissions of PBS, LUVs, and proteinase K alone at the same wavelength should be monitored and subtracted (repeat **steps 3–5**).
7. In order to evaluate the degree of peptide insertion, calculate the percentage of fluorescence change upon addition of proteinase K to the cuvette.

3.3 Peptide–Peptide Interaction in the Membrane

3.3.1 Rhodamine Quenching for Analyzing Peptides Self-Assembly

1. Dissolve Rho-peptide in PBS to a final concentration of 1 μM.
2. Set the spectrophotometer to kinetic mode with excitation wavelength of 530 nm and emission wavelength of 580 nm.
3. Add 400 μL peptide solution to a pre-cleaned cuvette (magnetic stirrer can be used) and read the signal output until equilibrium is reached. This is the basal signal of the labeled peptide.
4. Add LUVs at a final lipid concentration that was found to bind all soluble peptide (from Subheading 3.1).
5. Monitor the change in the signal intensity until equilibrium is reached (time scale normally varies from few seconds up to 1 h).
6. To account for background, the emissions of both PBS and vesicles alone at the same wavelength should be monitored and subtracted (repeat **steps 3–5**).
7. Add 125 μg/mL of proteinase K and track the change in the signal. The kinetic of recovery of in the fluorescent signal can give qualitative information about the strength of peptides interaction and their accessibility to proteinase K (as described in Subheading 3.2.2).
8. In order to quantify the effect of membranes on the self-interactions of the peptides, calculate the percentage of fluorescence change upon addition of vesicles.

3.3.2 Measurements of FRET

The FRET experiments are performed by using NBD- and Rho-labeled peptides as fluorescence donor and acceptor, respectively [19]. The wavelength of the maximal emission of NBD is at 525–530 nm (in a hydrophobic environment). This wavelength overlaps with the excitation wavelength of Rho: the Förster distance (R_0), at which the FRET efficiency is 50 %, of the NBD-Rho pair is about 56 Å. Thus, when NBD and Rho are in close proximity, the excitation of NBD at 467 nm results in a decrease in the maximal emission of NBD at ~528 nm and an increase in the maximal emission of Rho at ~580 nm (an example is given in Fig. 2).

1. Set the spectrophotometer to scan mode with excitation wavelength of 467 nm and emission wavelength range of 500–600 nm.
2. Add 400 μL of LUVs (final concentration ranges between 100 and 200 μM lipids) to a pre-cleaned cuvette (magnetic stirrer can be used) and read the signal output until equilibrium is reached.
3. Add NBD-peptide at final concentration of 0.4 μM and read the signal output until binding equilibrium is reached.
4. Add Rho-peptide in several sequential doses ranging from 0.0125 to 0.4 μM.
5. Repeat the procedure using the Rho-peptide (acceptor) alone and subtract the signal from the results above. This corrects for the contribution of the emission of the acceptor as a result of direct excitation.
6. Repeat the procedure using the addition of unlabeled peptide instead of the Rho-peptide as described in Subheading 3.3.3. Then, subtract the signal from the results above. This corrects for the possible changes in NBD fluorescence due to membrane insertion of the peptide complex.
7. To account for background, the emissions of both PBS and LUVs alone at the same wavelength should be monitored and subtracted.
8. The percentage of transfer efficiency (E) is determined by measuring the decrease in the quantum yield of the donor, as a result of the energy transfer to the acceptor. It is given by the following equation:

$$E(\%) = \frac{(1 - I_{\mathrm{da}})}{I_{\mathrm{d}}} \times 100 \qquad (2)$$

where I_{da} is the donor intensity (NBD-peptide) in the presence of the acceptor and I_{d} is the donor intensity in the absence of the acceptor.

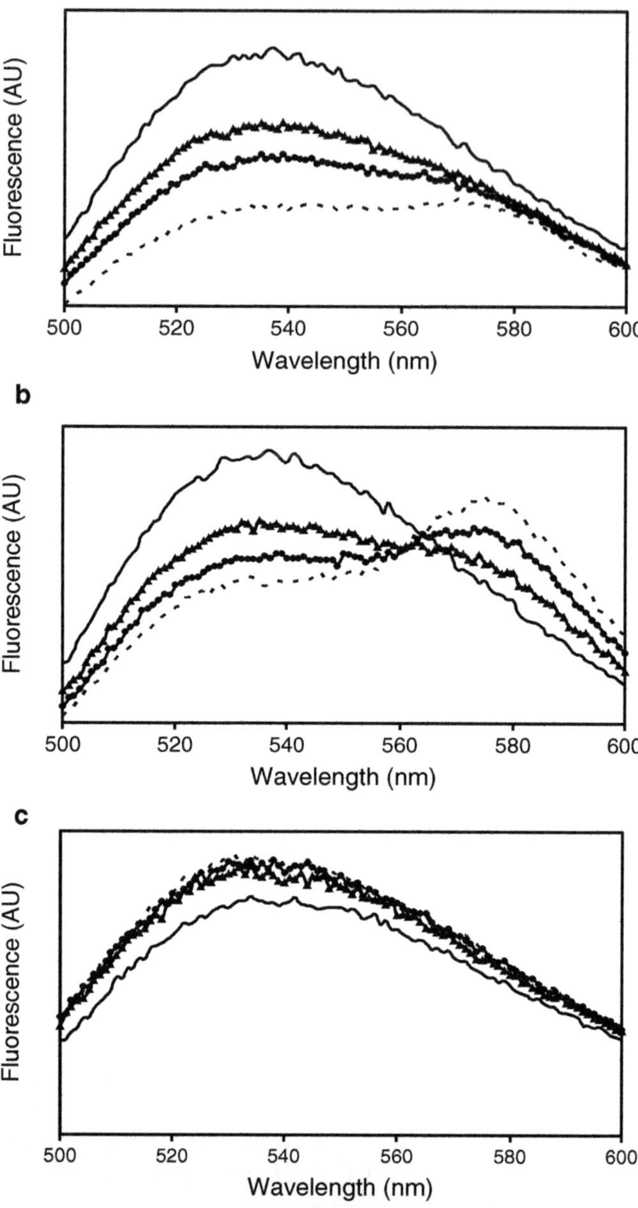

Fig. 2 FRET measurements reveal a specific interaction of the HIV gp41 TMD and the T-cell receptor alpha (TCRα) TMD. Fluorescence spectra were obtained at room temperature, with excitation set at 467 nm (5-nm slit) and emission scan at 500–600 nm (10-nm slits). NBD-labeled TCRα TMD peptide was added first from a stock solution in DMSO (final concentration 0.1 μM and a maximum of 0.25 % (v/v) DMSO) to a dispersion of phosphatidylcholine LUV (100 μM) in PBS. This was followed by the addition of: (**a**) Rho-labeled gp41 TMD peptide; (**b**) Rho-labeled TCRα TMD peptide and (**c**) Rho-labeled TAR/PS TMD control peptide. Fluorescence spectra were obtained in four different ratios of Rho-peptide:NBD-peptide: 0:4 (*black line*), 1:4 (*black triangle*), 2:4 (*black circle*), and 3:4 (*dashed line*). In panel (**a**), there is a decrease in the emission of the NBD donor, but with no typical increase in the emission of the rhodamine acceptor. This is possibly due to dequenching of the rhodamine signal by the oligomerization of the TMD in the membrane. The figure was reproduced from [20]

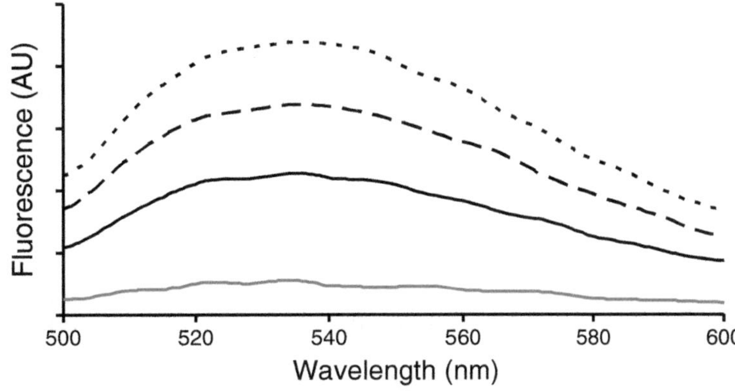

Fig. 3 Changes in the environment of membrane-bound NBD-labeled HIV gp41 fusion peptide (NBD-gp41 FP) upon addition of TCRα TMD. Fluorescence spectra were obtained at room temperature, with excitation set at 467 nm (10-nm slit) and an emission scan at 500–600 nm (10-nm slits). In a typical experiment, a NBD-gp41 FP was added first to a dispersion of phosphatidylcholine LUV (100 μM) in PBS (*bottom line*). This was followed by the addition of unlabeled TCRα TMD peptide in several sequential doses. Fluorescence spectra were obtained in three different NBD-gp41 FP:TCRα TMD ratio ranged from 1:0 to 1:3 (*bottom* to *top*). NBD-gp41 FP exhibits a blue shift concomitant with an increase in its fluorescence intensities around 530 nm, which indicates a lipid environmental change of the labeled peptide upon interaction with the unlabeled peptide. Adapted with permission from [21]. Copyright (2008) American Chemical Society

3.3.3 Measuring Peptide–Peptide Interaction That Leads to Membrane Insertion

The fluorescence experiments are performed using NBD-labeled peptides in the presence of liposomes. The interactive unlabeled peptide is added in sequential doses and the emission of NBD is measured. The increase in the fluorescence emission maxima of the NBD-labeled peptide, concomitant with a blue shift, indicates that the peptide–peptide interaction leads to changes in the lipid environment of the NBD probe probably by penetrating into the membrane (an example is given in Fig. 3).

1. Set the spectrophotometer to scan mode with excitation wavelength of 467 nm and emission wavelength range of 500–600 nm.
2. Add 400 μL of LUVs (100 μM lipids) to a pre-cleaned cuvette (magnetic stirrer can be used) and read the signal output until equilibrium is reached.
3. Add NBD-peptide at final concentration of 0.4 μM and read the signal output until binding equilibrium is reached.
4. Add unlabeled peptide in several sequential doses ranging from 0.0125 to 0.4 μM.
5. To account for background, the emissions of both PBS and LUVs alone at the same wavelength should be monitored and subtracted.

4 Notes

1. Peptide concentration [M] is determined spectroscopic using the Beer–Lambert equation:

$$A = \varepsilon \cdot c \cdot L \tag{3}$$

where A is the actual absorbance at 467 nm (NBD) and at 530 nm (rhodamine). The molar absorption coefficient (ε) of NBD is 16,000 [cm^{-1} M^{-1}], and that of rhodamine is 38,000 [cm^{-1} M^{-1}]. L is the cuvette path length in centimeter.

2. Peptide concentration for the experiment should be at the low micromolar range to reach very low peptide/lipid ratio and is also depended on the labeling and purification quality. We generally use a concentration range of 0.1–1 μM of labeled peptides.

Acknowledgments

This work was supported by the Israel Science Foundation, German-Israel Foundation (GIF), Israel Ministry of Health, Benoziyo Center for Neurological Diseases, and European Community Grant No. 278998. Yechiel Shai is the incumbent of the Harold S. and Harriet B. Brady Professorial Chair in Cancer Research.

References

1. Shai Y (1999) Mechanism of the binding, insertion and destabilization of phospholipid bilayer membranes by alpha-helical antimicrobial and cell non-selective membrane-lytic peptides. Biochim Biophys Acta 1462:55–70
2. Papo N, Shai Y (2003) Exploring peptide membrane interaction using surface plasmon resonance: differentiation between pore formation versus membrane disruption by lytic peptides. Biochemistry 42:458–466
3. Oren Z, Hong J, Shai Y (1999) A comparative study on the structure and function of a cytolytic alpha-helical peptide and its antimicrobial beta-sheet diastereomer. Eur J Biochem 259: 360–369
4. Merklinger E, Gofman Y, Kedrov A, Driessen AJ, Ben-Tal N, Shai Y, Rapaport D (2012) Membrane integration of a mitochondrial signal-anchored protein does not require additional proteinaceous factors. Biochem J 442:381–389
5. Freire E (1995) Differential scanning calorimetry. Methods Mol Biol 40:191–218
6. Epand RM, Rotem S, Mor A, Berno B, Epand RF (2008) Bacterial membranes as predictors of antimicrobial potency. J Am Chem Soc 130:14346–14352
7. Oren Z, Ramesh J, Avrahami D, Suryaprakash N, Shai Y, Jelinek R (2002) Structures and mode of membrane interaction of a short alpha helical lytic peptide and its diastereomer determined by NMR, FTIR, and fluorescence spectroscopy. Eur J Biochem 269:3869–3880
8. Grieco P, Carotenuto A, Auriemma L, Saviello MR, Campiglia P, Gomez-Monterrey IM, Marcellini L, Luca V, Barra D, Novellino E, Mangoni ML (2012) The effect of d-amino acid substitution on the selectivity of temporin L towards target cells: identification of a potent anti-Candida peptide. Biochim Biophys Acta 1828(2):652–660
9. Mansson R, Bysell H, Hansson P, Schmidtchen A, Malmsten M (2011) Effects of peptide secondary structure on the interaction with oppositely charged microgels. Biomacromolecules 12:419–424

10. Gable JE, Schlamadinger DE, Cogen AL, Gallo RL, Kim JE (2009) Fluorescence and UV resonance Raman study of peptide-vesicle interactions of human cathelicidin LL-37 and its F6W and F17W mutants. Biochemistry 48:11264–11272
11. Chattopadhyay A, London E (1987) Parallax method for direct measurement of membrane penetration depth utilizing fluorescence quenching by spin-labeled phospholipids. Biochemistry 26:39–45
12. Rapaport D, Shai Y (1992) Aggregation and organization of pardaxin in phospholipid membranes. A fluorescence energy transfer study. J Biol Chem 267:6502–6509
13. Pouny Y, Rapaport D, Mor A, Nicolas P, Shai Y (1992) Interaction of antimicrobial dermaseptin and its fluorescently labeled analogues with phospholipid membranes. Biochemistry 31:12416–12423
14. Gazit E, Lee WJ, Brey PT, Shai Y (1994) Mode of action of the antibacterial cecropin B2: a spectrofluorometric study. Biochemistry 33:10681–10692
15. Oren Z, Lerman JC, Gudmundsson GH, Agerberth B, Shai Y (1999) Structure and organization of the human antimicrobial peptide LL-37 in phospholipid membranes: relevance to the molecular basis for its non-cell-selective activity. Biochem J 341(Pt 3):501–513
16. Hetru C, Letellier L, Oren Z, Hoffmann JA, Shai Y (2000) Androctonin, a hydrophilic disulphide-bridged non-haemolytic antimicrobial peptide: a plausible mode of action. Biochem J 345(Pt 3):653–664
17. Merrifield RB, Vizioli LD, Boman HG (1982) Synthesis of the antibacterial peptide cecropin A (1–33). Biochemistry 21:5020–5031
18. Avrahami D, Oren Z, Shai Y (2001) Effect of multiple aliphatic amino acids substitutions on the structure, function, and mode of action of diastereomeric membrane active peptides. Biochemistry 40:12591–12603
19. Reuven EM, Dadon Y, Viard M, Manukovsky N, Blumenthal R, Shai Y (2012) HIV-1 gp41 transmembrane domain interacts with the fusion peptide: implication in lipid mixing and inhibition of virus-cell fusion. Biochemistry 51:2867–2878
20. Cohen T, Cohen SJ, Antonovsky N, Cohen IR, Shai Y (2010) HIV-1 gp41 and TCRalpha trans-membrane domains share a motif exploited by the HIV virus to modulate T-cell proliferation. PLoS Pathog 6:e1001085
21. Cohen T, Pevsner-Fischer M, Cohen N, Cohen IR, Shai Y (2008) Characterization of the interacting domain of the HIV-1 fusion peptide with the transmembrane domain of the T-cell receptor. Biochemistry 47: 4826–4833

Chapter 13

Scanning Fluorescence Correlation Spectroscopy in Model Membrane Systems

Joseph D. Unsay and Ana J. García-Sáez

Abstract

Fluorescence correlation spectroscopy (FCS) is an emerging technique employed in biophysical studies that exploits the temporal autocorrelation of fluorescence intensity fluctuations measured in a tiny volume (in the order of fL). The autocorrelation curve derived from the fluctuations can then be fitted with diffusion models to obtain parameters such as diffusion time and number of particles in the diffusion volume/area. Application of FCS to membranes allows studying membrane component dynamics, which includes mobility and interactions between the components. However, FCS encounters several difficulties like accurate positioning and stability of the setup when applied to membranes.

Here, we describe the theoretical basis of point FCS as well as the scanning FCS (SFCS) approach, which is a practical way to address the challenges of FCS with membranes. We also list materials necessary for FCS experiments on two model membrane systems: (1) supported lipid bilayers and (2) giant unilamellar vesicles. Finally, we present simple protocols for the preparation of these model membrane systems, calibration of the microscope setup for FCS, and acquisition and analysis of point FCS and SFCS data so that diffusion coefficients and concentrations of fluorescent probes within lipid membranes can be calculated.

Key words Fluorescence correlation spectroscopy, Supported lipid bilayers, Giant unilamellar vesicles, Scanning fluorescence correlation spectroscopy, Diffusion coefficient

1 Introduction

One of the most exciting features of living systems is the presence of membranes, which not only confine cellular compartments but also control the exchange of energy, matter, and information between the inside of the cell and its external environment. Biological membranes are organized as sheet-like structures with complex lipid and protein compositions. Proteins are responsible for many of the functions of cellular membranes (receptors, transporters, enzymes, adhesion, etc.). The lipids are responsible for making the semipermeable barrier, dividing the inner aqueous compartment and external aqueous surrounding. They also regulate the physical and chemical properties of the membrane and play a role in signaling [1, 2]. Together with structural conformation, the

Doron Rapaport and Johannes M. Herrmann (eds.), *Membrane Biogenesis: Methods and Protocols*, Methods in Molecular Biology, vol. 1033, DOI 10.1007/978-1-62703-487-6_13,

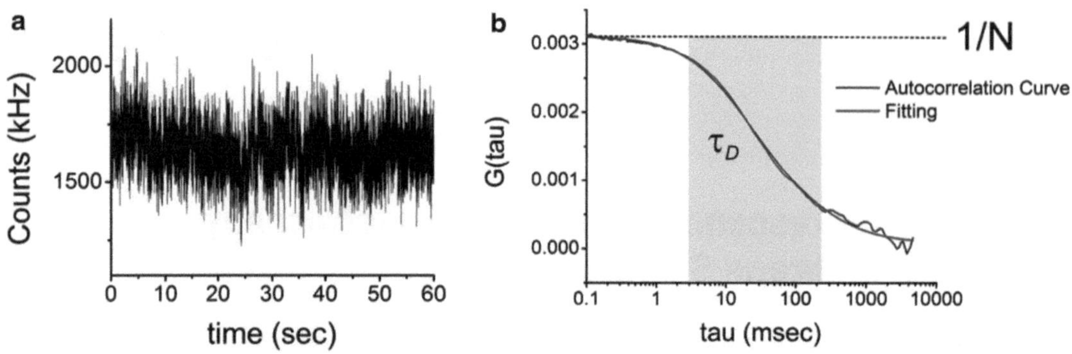

Fig. 1 Principle of FCS. (**a**) Fluorescence fluctuations are recorded by avalanche photodiodes (APDs). (**b**) The fluctuations are temporally autocorrelated to measure self-similarity of the signal over time and then fitted with a model function. Depending on the model function used, one can infer physical parameters (number of particles, N and diffusion time, τ_D) from the fitted curve. In this example, an SLB made up of dioleoylphosphatidylcholine (DOPC): sphingomyelin (SM): cholesterol (Chol) at 2:2:1 ratio is used. From the fluorescence fluctuations, one can observe some instability in the system, which results to poor fitting at some regions of the autocorrelation curve

study of mobility and dynamic properties of these components are essential to understand their function in the membrane.

Studying biological membranes often encounters challenges because of the difficulty in studying its components. Membrane proteins are relatively difficult to purify and reconstitute into lipid bilayers [3–5]. Published reviews and practical guides have already tried to address these difficulties by proposing different strategies [6–10]. Several approaches have been developed to study membranes and their components. Among others, these include bulk techniques like fluorescence measurements and nuclear magnetic resonance spectroscopy [11, 12], surface techniques like atomic force microscopy [13, 14], and advanced optical microscopy that can be applied to the single molecule regime [15]. These techniques became more important with the emerging paradigm of the existence of lipid rafts in which fluid/fluid phase separation in membranes allows clustering of lipids and proteins to perform specific functions [16–18]. To understand these lipid rafts, one needs to have information on both structural and functional dynamics.

Fluorescence correlation spectroscopy (FCS) is a powerful technique to study dynamic intracellular processes [19–21] and is currently gaining ground in applicability to membrane systems as evidenced by recent studies [22–27]. In the next section, we explain the basis of FCS measurements and describe in detail the different physical parameters that one can derive from this technique.

1.1 Fluorescence Correlation Spectroscopy

FCS is a technique that utilizes fluorescence fluctuations rather than absolute fluorescence intensity (Fig. 1a). These fluctuations arise from the diffusion of individual molecules within a sub-micrometer detection volume (in the order of fL), which is usually the focal volume of a confocal microscope [28]. The autocorrelation analysis of these fluctuations (Fig. 1b) allows extracting local concentrations and translational (or rotational) diffusion coefficients. From these parameters one can determine among others chemical rate constants, association and dissociation constants, and structural dynamics [29, 30].

The temporal autocorrelation analysis measures self-similarity of the signal with itself over time. It is calculated with the following expression:

$$G(\tau) = \frac{\langle \delta F(t) \quad \delta F(t+\tau) \rangle}{\langle F(t) \rangle^2} \tag{1}$$

where G is the autocorrelation function, F is the fluorescence intensity as a function of time t, and τ is the correlation time. The angular brackets refer to time averaging, so that $\delta F(t) = F(t) - \langle F(t) \rangle$.

Physical parameters can be obtained from the autocorrelation curve by fitting model equations that take into account the excitation profile and model the fluorescence fluctuations inside the detection volume. This model also considers the size and shape of the focal volume, the molecular brightness of the fluorophore, and its concentration as a function of position and time. Detailed derivation of the fitting functions can be found in previous publications [29, 31]. Table 1 shows some model functions that are used depending on the processes that take place in the sample under study. Such processes include probe diffusion and photophysical processes. Examples of photophysical processes which may affect the autocorrelation curve include blinking and triplet state. Blinking occurs when molecules have a dark and bright state due to the chemical environment, for example, fluorescent proteins like GFP are affected by pH changes. Excited molecules can also transition from singlet excited state to triplet state before relaxing to ground state by emitting a photon and thus remain dark at the time scale of a few microseconds.

Pure Brownian diffusion of fluorescent particles within lipid membranes occurs in two dimensions and the focal plane of the objective has to be focused to contain the membrane. For 2D diffusion in a plane perpendicular to the optical axis at $z = 0$, the autocorrelation function is expressed as:

$$G_{2D}(\tau) = \frac{1}{N}\left(1 + \frac{\tau}{\tau_D}\right)^{-1} \tag{2}$$

Here N is the average number of fluorescent particles in the detection area. This can be calculated from the amplitude of the autocorrelation curve since:

$$G_{2D}(0) = \frac{1}{N} \tag{3}$$

Table 1
Model functions for fitting of FCS data

Type of diffusion	Model function
3D diffusion	$G_{3D}(\tau)=\frac{1}{N}\left(1+\frac{\tau}{\tau_D}\right)^{-1}\frac{1}{\sqrt{1+\frac{\tau}{S^2\tau_D}}}$
2D diffusion	$G_{2D}(\tau)=\frac{1}{N}\left(1+\frac{\tau}{\tau_D}\right)^{-1}$
2D diffusion for two components	$G_{2D+2C}(\tau)=\frac{1}{N_{\text{total}}}\frac{q_f^2\Upsilon_f G_{Df}(\tau)+q_s^2\Upsilon_s G_{Ds}(\tau)}{\left(q_f\Upsilon_f+q_s\Upsilon_s\right)^2}$
2D diffusion with triplet	$G_{2D+T}(\tau)=\left[1+\frac{T}{1-T}\exp\left(-\frac{\tau}{\tau_T}\right)\right]G_{2D}(\tau)$
2D diffusion with blinking	$G_{2D+bl}(\tau)=\left[1+\frac{c_{\text{dark}}}{c_{\text{bright}}}\exp(-k_{bl}\tau)\right]G_{2D}(\tau)$
2D with elliptical-Gaussian profile	$G_{2DG}(\tau)=\frac{1}{N}\left(1+\frac{\tau}{\tau_D}\right)^{-\frac{1}{2}}\frac{1}{\sqrt{1+\frac{\tau}{S^2\tau_D}}}$

Here G is the autocorrelation curve, N is the average number of particles in the detection volume, τ is the lag time, τ_D is the diffusion time, τ_T corresponds to the triplet time. In addition w_0 is the waist of the detection volume, q is the molecular brightness of the f (fast) and s (slow) diffusing components, Υ refers to their molar fraction, T is the fraction of the fluorophores in the triplet state within the detection volume, c is the concentration of the dark and bright species, k_{bl} is the blinking rate, S is the structure parameter, and $S = w_z/w_0$. The terms introduced to correct for two components, triplet and blinking, are also valid for 3D diffusion

The number of particles can be used to calculate the area concentration of the particle in the focal area through the formula: $N = C\pi w_0^2$, with C being the area concentration and w_0, the waist radius of the laser focus, which is an estimate of the size of the laser in the horizontal plane, assuming the laser has a circular cross section.

The other parameter in Eq. 2, diffusion time, τ_D, can be understood as the average amount of time a fluorescent particle stays in the detection volume. This can be inferred from the decay of the autocorrelation curve. It is related to the diffusion coefficient, D, through the expression:

$$\tau_D = \frac{w_0^2}{4D} \tag{4}$$

The diffusion coefficient is a mobility parameter that is dependent on the particle size, the viscosity of the medium, and temperature of the system.

Table 2
Diffusion coefficient of fluorophores commonly used as references in FCS measurements

Fluorophore	Diffusion coefficient (μm²/s)	References
Rhodamine 6G	426, 414 ± 1*	[65, 66]
Alexa488	435	[65]
eGFP	95	[65]
Alexa543	341	[65]
Atto655	426 ± 8*	[60]

Diffusion coefficients calculated in water solutions at 22.5 °C, except those marked with *, which were obtained at 25 °C

To determine the waist radius of the focus, w_0, one can measure the diffusion time of a fluorescent dye whose diffusion coefficient has already been characterized in other studies (*see* Table 2). In that case, the diffusion coefficient of the particles of interest can be estimated by FCS, relative to the diffusion coefficient of the standard used.

However, several challenges arise from using single point FCS, especially when applied to membranes. These challenges include (1) axial positioning and other distortions of the focal volume, (2) stability of the measuring system, and (3) photobleaching [22, 32]. When there are axial (z-axis) positioning errors, the laser is not focused on the membrane and this increases the detection volume, creating systematic errors which cannot be corrected by averaging different measurements. Getting the correct axial position is thus mandatory. Other distortions also arise from proximity to the coverslip surface, differences in the refractive index of the membrane and the water environment, astigmatism, and optical saturation, which then necessitates the use of similar conditions during calibration of the setup [32]. Secondly, because diffusion in membranes is slower than in solution, one needs to have longer measurement times to get statistically accurate results [33]. The current practice involves measuring 10,000 times longer than the expected diffusion time, which is usually in the millisecond time scale for species diffusing in a membrane. This longer measuring time can be limited by stability of the setup since the laser may be slightly displaced, which can result in apparent second component or anomalous diffusion. Furthermore, thermal membrane undulations may lead to defocusing over time. Finally, because fluorophores diffuse slowly, they will have longer residence times in the focal volume, which increases the probability of photobleaching. Usually, this can be corrected by lowering the laser power or taking this into account and use of functions to correct for the effect [22].

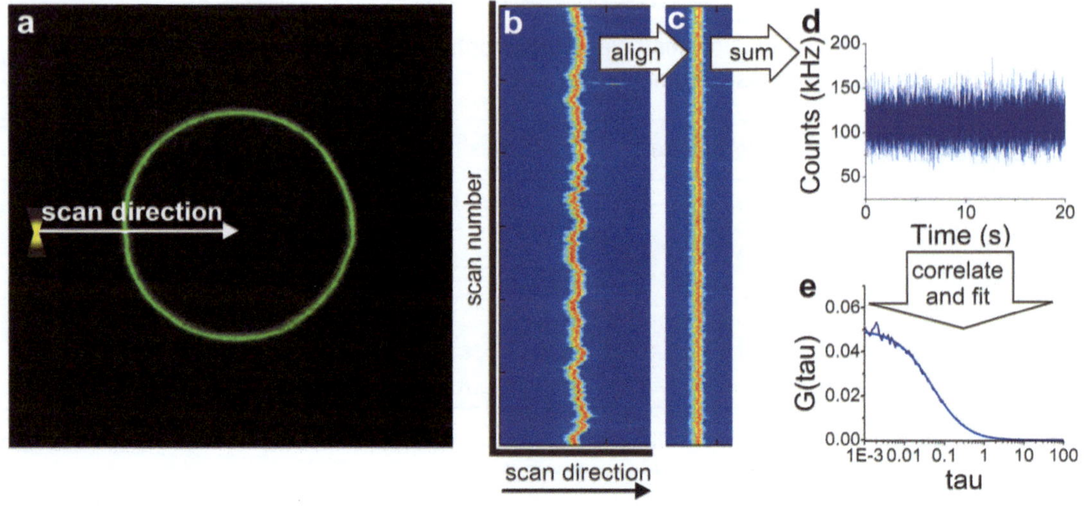

Fig. 2 Principle of scanning FCS perpendicular to the membrane plane. (**a**) The detection volume is scanned repeatedly through the membrane. (**b**) Line scans are arranged beneath each other to produce a pseudo-image where the membrane is clearly visible. (**c**) Membrane movements are corrected by aligning the membrane for all the scans. (**d**) For each line scan, the photons detected during the transient of the membrane are summed up to result into one point of the fluorescence intensity trace. (**e**) The intensity trace can be correlated to result in the auto-correlation curve and then fitted with diffusion models. In this case, we use MatLab to implement all operations

One approach to correcting distortions and axial positioning problems uses the z-scanning ability of current systems to measure FCS in different *z*-positions with respect to the membrane (z-scan FCS) [34]. By checking the dependence of diffusion time and number of particles with respect to the *z*-position, one can determine the best alignment (lowest N and smallest τ_D) and other parameters like diffusion coefficient and area parameters without the need for calibration [35]. However, this does not eliminate the possibility of photobleaching and stability issues because it still requires longer measurement times.

Another approach is via scanning FCS (SFCS). This method moves the detection volume through the membrane either parallel or perpendicular to it at high scanning rates [22, 36]. It can be achieved using the scanning unit of a laser scanning microscope (LSM). This reduces the residence time of the fluorophore in the detection volume, thus decreasing the probability of photobleaching. Consequently, long measuring times become feasible. In our discussion, SFCS will strictly refer to scanning the detection volume perpendicular to the membrane. How this is achieved and the process to analyze data from SFCS is discussed below.

1.2 Scanning FCS

SFCS is especially useful for free standing membranes in which the scanning path can be oriented perpendicular to the membrane plane (Fig. 2a). With this orientation, the focal volume passes through the membrane at only specific points in the scanning path. As mentioned above, this reduces the residence time of fluorophores in the detection volume, thus reducing photobleaching effects. The arrival times of the emitted photons can be recorded and grouped according to the scan rate of the LSM in a process called binning. For example, if one has a 300-s experiment on a system that has scan rate of 1.5 ms, grouping the data in 1.5-ms segments results in about 200,000 scans. The next step is to count the number of photons arriving at a particular position in the scan. A higher number of photon results in a higher intensity in a scan position. One can then create a pseudo-image with the scan number vs. the scan position (Fig. 2b). This pseudo-image shows the regions where the detection volume passes through the membrane as regions with high intensity values. Because of undulations of the membrane, this region may not be in the same position from scan to scan, but can be corrected by aligning them using appropriate software (Fig. 2c). By adding the intensity of the membrane contributions per scan, one can create a discrete fluorescence intensity trace (Fig. 2d). The intensity trace can then be autocorrelated as with point FCS (Fig. 2e) and described using two-dimensional diffusion of one component in a Gaussian elliptical detection volume [37]:

$$G(\tau) = \frac{1}{N}\left(1+\frac{\tau}{\tau_D}\right)^{-\frac{1}{2}}\left(1+\frac{\tau}{\tau_D S^2}\right)^{-\frac{1}{2}} \tag{5}$$

Here, we see an added parameter, the structure parameter S. $S = w_z/w_0$, where w_0 is the waist radius as discussed above and w_z is the axial (z-axis) extension of the detection volume. This represents the aspect ratio of the detection volume. This can be obtained through a calibration experiment of a well-characterized fluorescent dye (as with the aforementioned calibration experiment for w_0). The autocorrelation curve of the calibration measurement is then fitted with a simple 3D diffusion model (*see* Table 2) to obtain N, τ_D, and S.

Other advanced FCS techniques may be employed to show interactions of two differently labeled molecules (two-color fluorescence cross correlation Spectroscopy, two-color FCCS) or provide calibration-free measurements (two focus FCCS) [37–39].

A comprehensive discussion of FCS theory and application can be found in reference [40].

1.3 Model Membrane Systems

An important approach in studying membranes is to use membrane model systems to reduce sample-to-sample variability introduced when dealing with dynamic living systems. Commonly used model membrane systems are supported lipid bilayers (SLBs) and giant unilamellar vesicles (GUVs) (Fig. 3).

SLBs are attractive membrane systems for studying membrane dynamics as well as membrane protein interactions because they

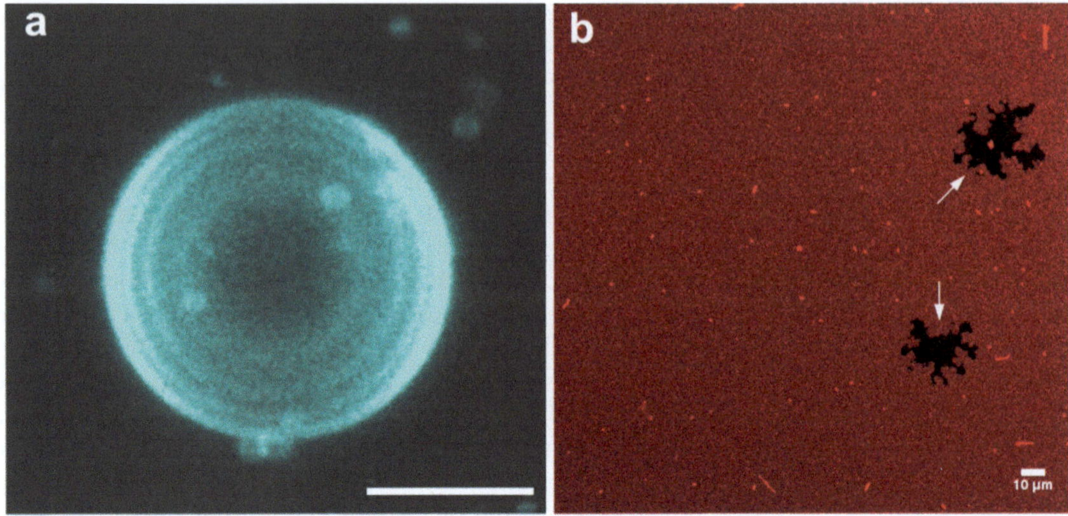

Fig. 3 Examples of model membranes that can be used with FCS. (**a**) Confocal *z*-projection of a giant unilamellar vesicle of about 20 μm in diameter with lipid composition of heart cardiolipin:egg phosphatidylcholine (1:2). The membrane is labeled with 0.01 mol% DiI (*blue*). (**b**) Confocal image of a supported bilayer with lipid composition of heart cardiolipin:egg phosphatidylcholine (1:4) labeled with 0.05 mol% DiD (*red*). The membrane appears as a continuous phase with the appearance of some membrane defects (*black*, indicated by *arrows*). Both scale bars represent 10 μm

can be analyzed with a combination of microscopy and surface analysis techniques [41]. Some common solid supports include mica [42], borosilicate glass [43], fused silica, and oxidized silicon [44]. A review on SLBs can be found in this reference [45]. One technique for SLB preparation that is protein-friendly is the adsorption and subsequent deposition of small unilamellar vesicles (SUVs) or proteo-SUVs on a solid support with the addition of calcium [46]. However, the solid support of the lipid bilayer may affect mobility of some molecules [41, 47].

GUVs are free-standing bilayers that are similar in size to eukaryotic cells and with virtually no curvature in the membrane area contained in the focal plane of a confocal microscope setup. The most common technique for GUV formation is via electroformation [48]. This setup uses a dried lipid film or proteolipid mixture on an electrode surface (usually Pt). However, drying the electrodes may be too harsh to some proteins. Furthermore, yield and size distribution is dependent on preparation conditions, which usually includes low ionic strength and osmolarity, and the characteristics of the electric field used (AC vs. DC current, voltage, frequency, etc.) [49–51]. Nevertheless, FCS studies involving GUVs have been successfully accomplished and include the characterization of lipid structures [25, 52] and protein–lipid interactions with SNARE proteins [53], the mechanosensitive channel MscL [54], Bcl-2 proteins [55], bacteriorhodopsin [56], and fibroblast growth factor 2 (FGF2) [57].

2 Materials

2.1 Supported Lipid Bilayers

1. Lipid mixtures: stock solutions prepared by dissolving the lipids of interest in chloroform. The lipids could be in the form of purified lyophilized lipids (for example, from Avant Polar Lipids, Alabaser, AL, USA) or extracted lipid mixtures (*see* **Note 1** for proper handling of lipids). Include about 0.01 % of a lipophilic fluorescent dye (*see* **Note 2** for information on lipophilic fluorescent dyes and **Note 3** for information on experiments involving membrane proteins).
2. Phosphate Buffered Saline (1× PBS Buffer): 2.7 mM KCl, 1.5 mM KH_2PO_4, 8 mM Na_2HPO_4, and 137 mM NaCl, pH 7.2. Weigh 0.2237 g KCl, 0.2041 g KH_2PO_4, 1.1357 g Na_2HPO_4, and 8.0065 g NaCl in a beaker. Fill to about 800 mL with water. Adjust the pH by adding NaOH. Add enough water to make 1 L solution. Store at room temperature.
3. SLB Buffer: 150 mM NaCl, 10 mM HEPES, pH 7.4. Weigh 1.916 g 4-(2-hydroxyethyl)-1-piperazineethanesulfonic acid (HEPES) and 8.7664 g NaCl in a beaker. Fill to about 800 mL with water. Adjust the pH by adding NaOH. Add enough water to make 1 L solution. Store at room temperature.
4. 1 M calcium chloride solution. Weigh 5.5492 g $CaCl_2$ and dissolve in enough water to make 50 mL of solution. Store at room temperature.
5. Mica sheets (NSC Mica Exports Ltd., New Delhi, India, www.nscmica.com or Electron Microscopy Sciences, Hatfield, PA, USA, http://www.emsdiasum.com/microscopy/products/preparation/mica.aspx).
6. Punch and die set (Precision Brand Products, Inc. Downers Grove, IL, USA, www.precisionbrand.com).
7. Optical glue, for example, Norland Optical Adhesive 88 (Norland Products, Inc., Cranbury, NJ, www.norlandprod.com).
8. Disposable glass cylinders, for example, 8 × 8 mm (height × diameter) cloning cylinders (Belco Glass, Inc., Vineland, NJ, USA).
9. Vacuum grease, for example, Glisseal (Borer Chemie, Zuchwil, Switzerland).

2.2 Giant Unilamellar Vesicle Components

1. Lipid mixtures (*see* Subheading 2.1).
2. 300 mM sucrose solution. Weigh 10.26 g of sucrose. Add enough water to make 100 mL solution. Store at 4 °C.
3. 2 mg/mL bovine serum albumin (BSA) blocking solution. Dissolve 100 mg of BSA in 50 mL of water. Store at 4 °C.
4. 300 mM glucose solution. Weigh 5.40 g of glucose. Add enough water to make 100 mL solution. Store at 4 °C. This can be

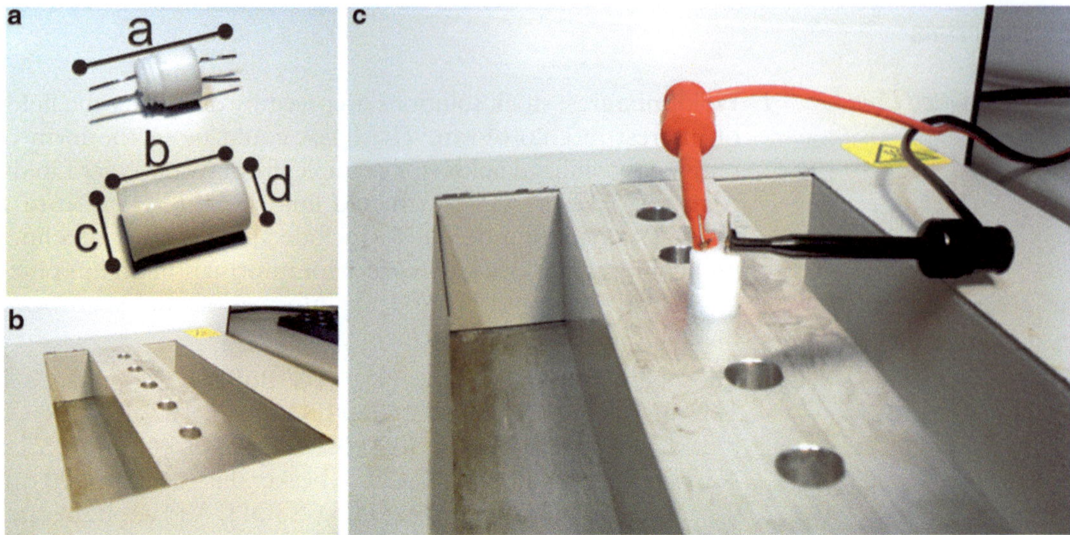

Fig. 4 Electroformation chamber and setup. (**a**) A homemade Pt-and-teflon chamber with the following dimensions is shown: *a* 30 mm, *b* 15 mm, *c* 8 mm, *d* 4 mm. The distance between the two electrodes is about 4 mm. (**b**) Homemade aluminum block is shown. (**c**) The connection of the chamber to the function generator and the whole setup is shown

another aqueous solution isoosmolar with the sucrose solution (*see* **Note 4**).

5. Homemade Teflon and Pt chambers (*see* Fig. 4a).
6. Homemade aluminum heating block (*see* Fig. 4b).
7. A function generator, for example, TTi TG315 (Turlby-Tandar Instruments Ltd., Huntingdon, Cambs, England).
8. Observation chambers, for example, Lab-Tek chambered cover glasses (Nalge Nunc Intl., Rochester, New York, USA, www.nuncbrand.com).

2.3 Confocal Microscope Setup with FCS Module

Several FCS setups are currently available in the market. Examples of such a system is the TCP SP8 (Leica Microsystems Ltd.) and the LSM 710 with ConfoCor3 (Carl Zeiss, Inc.).

The setup includes several lasers for excitation and several detector channels with fiber-coupled avalanche photodiodes (APDs). In general, an FCS setup consists of an inverted microscope in combination with high-numerical aperture objective. The incoming laser light is reflected by the dichroic mirror and focused by the microscope objective to a spot of approximately 0.3–0.5 μm in diameter. The fluorescence emission passes the dichroic and emission filters that removes residual Rayleigh and Raman scattered light. A pinhole in the image plane improves the axial resolution. For detection, APDs with single-photon sensitivity are usually used. The fluorescence trace can then be correlated with hardware or software correlator. For SFCS, one needs to record the arrival times of the photons to the APDs with adequate time resolution. If this is not possible with your commercial system, one alternative is to use the photon mode of the hardware correlator Flex 02-01D (Bridgewater, NJ, USA, http://correlator.com).

3 Methods

3.1 Preparation of Supported Lipid Bilayers (According to [46])

1. Evaporate the solvent of the lipid mixture under nitrogen flux and then subject to vacuum for at least 2 h to allow complete evaporation of the solvent.
2. Rehydrate to a final concentration of 10 mg/mL in PBS, resulting in a suspension made up of multi-lamellar vesicles. These can be separated into 10-μL aliquots and stored in −80 °C.
3. Dilute a 10-μL aliquot of the suspension with 140 μL of SLB Buffer
4. Vortex the suspension to mix evenly. Sonicate the suspension until the solution becomes clear resulting in the formation of SUVs.
5. While sonicating, prepare the mica support using a punch and die set to acquire your desired size and clean the mica surface by cleaving the upper layers with adherent tape (Fig. 5a) (*see* **Note 5**).

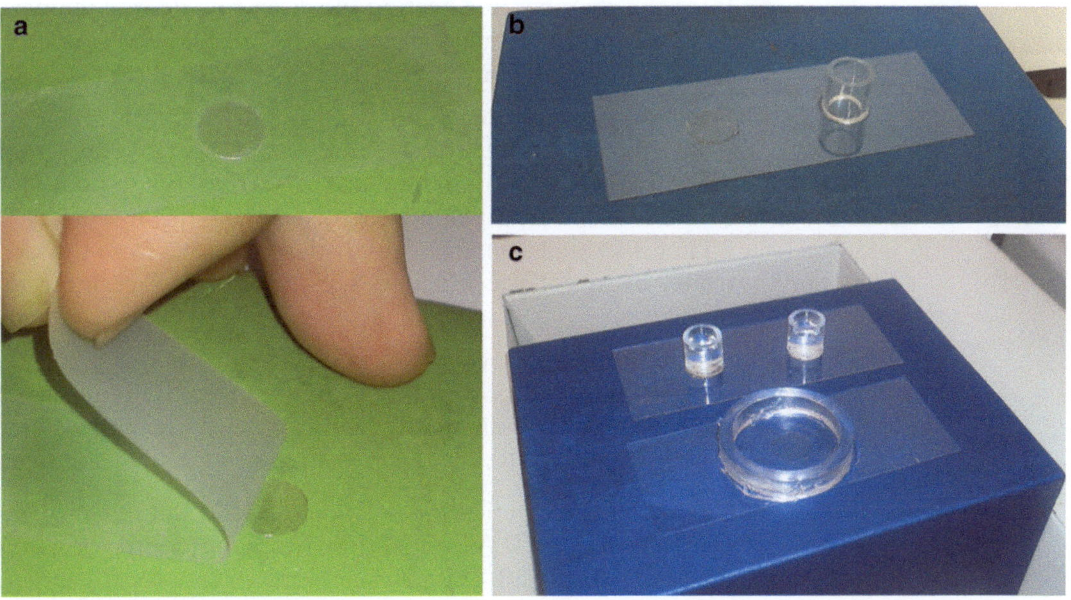

Fig. 5 Key steps and setup for SLB formation. (**a**) Mica disc is freshly cleaved prior to experiment. (**b**) It is then attached to a coverslip together with an 8 mm cylinder. (**c**) The chambers containing SLBs with buffer together with an alternative setup using wider chamber and vacuum seal as mentioned in **Note 7** are shown

6. Using the optical glue, attach the mica to a cover glass. Glue the glass cylinders to the outer rim of the mica to create a small chamber (Fig. 5b) (*see* **Note 6** for alternative reusable setup).
7. Pipet the clear SUV solution to the mica and add $CaCl_2$ to a final concentration of 3 mM.
8. Incubate the sample at 37 °C for 2 min to induce vesicle fusion.
9. Bring the sample to a temperature above the transition temperature of the lipid mixture (*see* **Note 7**) and incubate for 5 min.
10. Rinse several times with warm SLB Buffer to remove the non-fused vesicles (*see* **Note 8**).

3.2 GUV Formation

1. Clean the electroformation chamber with water, ethanol, and chloroform and let it dry.
2. Take the lipid mixtures and allow reaching room temperature before opening the vial. Dissolve the lipid mixtures in chloroform/methanol (2:1 v/v) to reach a concentration of 1 mg/mL.
3. Take 2.5–3 μL of the lipid stock and spread it on each of the Pt wires on the side that screws inside the electroformation chamber. Allow the solvent to evaporate completely.
4. Warm up adequate amounts of 300 mM sucrose to above the transition temperature of the lipid mixture. Fill the electroformation chamber with 350 μL of warm sucrose solution and close it by screwing the cap with the Pt wires.
5. Place the chamber in a heating block set to a desired temperature. Connect the cables of the function generator to each of the two Pt wires avoiding contact between them (Fig. 4c)
6. Start the function generator with a sinus wave at voltage of 1.4 V (RMS), for the case of an electrode distance of 4 mm, and a frequency of 10 Hz (*see* **Note 9**). Leave the setup for 2 h to allow electroformation. Then, decrease the frequency to 2 Hz for around 30 min to gently detach GUVs from the electrode. Turn off the function generator and disconnect the cables.
7. Fill the wells of the observation chamber with 700 μL of the BSA blocking solution and incubate for around 30 min. Rinse the wells of the observation chamber extensively with water.
8. Add 350 μL of 300 mM glucose solution.
9. Cut the end of a micropipette tip to widen the opening and gently take the GUV suspension from the electroformation chamber (usually 50–100 μL) (*see* **Note 10** for proper handling of produced GUVs).
10. Wait until the GUVs sink to the bottom of the chamber (about 5 min).

3.3 Setup Alignment and Calibration

To optimize conditions during data acquisition, the FCS setup needs to be aligned with a calibration measurement before use. The calibration also serves to compute for the parameters related to the shape of the detection volume (w_0 and S). A calibration measurement is done with a dye that is spectrally similar to the fluorophore that will be measured in the membrane. As an example, free Alexa-488 is a well-known dye in the green range spectrum. Other examples can be found in Table 2.

1. Switch on the lasers around 2 h before the measurements to allow stabilization.
2. Place 200 μL of 10 nM dye solution in an observation chamber similar to the one that will be used.
3. Set up the FCS configuration of the microscope that best fits the spectral properties of the fluorophore under study.
4. Select a laser power that gives maximum counts per molecule without significant photobleaching during the acquisition time. Depending on the microscope and sample, this is usually around 0.3–1 %.
5. Position the focal volume inside the solution at least 100 μm above the coverslip surface (*see* **Note 11**).
6. Set the size of the pinhole to 40 μm or 1 airy unit and align it with respect to the beam path. This involves selecting the X and Y positions of the pinhole that produce the highest fluorescence count rate.
7. In some setups, it may be necessary to adjust the position of the collimator to reach maximum count rate.
8. Adjust the correction collar of the objective to correct for glass thickness of the observation chamber by maximizing the count rate.
9. Perform an FCS experiment on the calibration dye solution. For example, perform three acquisitions of 20 s each.
10. Save the data for further analysis. If required for other programs, export the data in ASCII format.
11. Following the steps described below, fit the autocorrelation curve and estimate the diffusion time and the structure parameter (*see* **Note 12**).

3.4 Data Acquisition for Point FCS Measurements on SLBs

Online data acquisition is usually provided by commercial setups and can be used to monitor FCS data while measuring. With this, one can check instabilities in the fluorescence signal, photobleaching effects, as well as alterations in the shape of the autocorrelation curve.

1. Place your observation chamber in the microscope's sample holder
2. In the imaging mode, focus the sample on the SLB. Select the point in your SLB to position the focal volume and set the measurement conditions.
3. Select a laser power that gives maximum counts per molecule without significant photobleaching during the acquisition time. Depending on the microscope and sample, this is usually around 0.3–1 %. If necessary, start a test FCS measurement to check for the stability of the system to photobleaching and membrane undulations and movements, in order to optimize the measuring time and laser power.
4. In the FCS mode, scan in *z*-direction for the focal plane with maximum fluorescence intensity.
5. For statistical reasons, the measuring time has to be of the order of 10,000 times longer than the diffusion time under study (*see* **Note 13**). For example, for a probe that shows a diffusion time in the membrane of about 12 ms the measuring time should be 120 s. If the stability of the system allows it, perform three to five repetitions of the measurement.
6. Again in the imaging mode, select a point in the SLB where you want to conduct FCS measurements and make sure that you are in the focal plane with maximum fluorescence intensity.
7. Start an FCS measurement under the conditions set for the prior sample. Repeat the measurement several times in different positions of the membrane.
8. Save the obtained data. If necessary, export them to ASCII format.

3.5 Data Acquisition for SFCS on GUVs

1. In your microscope's imaging mode, set up the light path to use APDs as the detector. If this is not possible, perform the GUV selection and positioning of the focal volume in the normal imaging mode and change to APD detection for the FCS measurement.
2. Set the laser power to 0.3–1 %.
3. Focus the sample and look for a uniform GUV, for example, a GUV about 50 μm in size (*see* **Note 14** for selecting proper GUV for FCS analysis).
4. Once you have your desired GUV, place it on the center of your imaging field and zoom in. Find the focal plane of the equator of your GUV (where GUV appears largest) (*see* **Note 15**).
5. Align the scanning path of the microscope so that the GUV rim is at the center and perpendicular to the scanned line. We generally acquire 512 pixel per line with a zoom of 12 to have a pixel size of 0.03 μm.

6. Select the line scanning feature of your microscope at maximum speed (unidirectional scan, no averaging).
7. Start a time-lapse experiment of desired duration if saving the photon arrival times with the microscope software is possible. For a statistically relevant analysis, the total time should be 10,000 times longer than the predicted diffusion time of your fluorescent species (*see* **Note 13**).
8. If the microscope setup cannot save photon arrival times, one may use the software for the photon mode of the hardware correlator mentioned in Subheading 2.3 (*see* **Note 16** for more details).

3.6 Autocorrelation of SFCS Data

1. Using the data of the photon arrival times, bin the photon streams in bins of 100 ns to 5 μs depending on the scan rate and arrange it as a matrix/pseudo-image such that every row corresponds to one line scan (Fig. 2b).
2. To identify the membrane contributions to the fluorescence intensity signal, calculate the maximum of a running average per line scan.
3. To correct for membrane movement, shift all the line scans to align this maximum value (Fig. 2c).
4. Fit a Gaussian to the average of all the maximum value for all the line scans.
5. For each line scan, take only the values between -2.5σ and 2.5σ and add them to get one fluorescence intensity value. Doing this for all the line scans creates a discrete fluorescence intensity trace over time which can be autocorrelated using a multi-tau algorithm [58] (Fig. 2d).

3.7 Data Analysis and Fitting

1. Open the autocorrelation curve obtained with FCS-specific software, usually implemented in commercial systems, or on a program with mathematical fitting options, such as Origin or Matlab.
2. Plot the autocorrelation curves and discard those with distorted shapes. Qualitatively, the shape of the curve is related to the processes causing the fluorescence fluctuations. Among others, it depends on the type of particle motion. The curve decay provides information about the diffusion time of the molecules. In addition, the amplitude of the curve is a function of the area concentration of fluorescent particles in the focal plane (*see* **Note 17**).
3. For quantitative analysis, fit the autocorrelation curves with an appropriate model function (*see* Table 1). Usually, a nonlinear least-square fitting algorithm is used. A plot of the fitting residuals provides information about the quality of the fit.

Successful fitting of the data will provide an estimation of the diffusion time and number of particles in the focal area. Other parameters such as diffusion coefficient may then be derived.

4 Notes

1. Allow lipid stocks to equilibrate at room temperature before opening container vials to avoid water condensation. Phospholipids can get degraded in the presence of water or oxygen. Before closing the vials, flux an inert gas like N_2 or Argon into the vial and seal it with parafilm.
2. An area concentration of fluorescent dye between 1 and 1,000 molecules/μm^2 is usually desired for FCS on membranes [22]. For this reason, we suggest a concentration of 0.01–0.05 mol% of lipidic dye. The family of long-chain dialkylcarbocyanines, like DiD, DiI, and DiO (Invitrogen) span a wide range of wavelengths, and have been used to measure lateral organization of lipids in model membranes. Another possibility is the use of fluorescently labeled lipids, like rhodamine-phosphatidylethanolamine (Avanti Lipids) or BODIPY-cholesterol (Invitrogen). However, one should take into account that the fluorophore bound to the lipid can change the lipid behavior [59].
3. When membrane proteins are to be measured in FCS, one has to label them with fluorescent probe. If the systems allow it, proteins can be chemically labeled with reactive dyes specific for cysteines or amine groups. These dyes include the Alexa (Invitrogen) and Atto (Atto-TEC) family, which cover a broad range of wavelengths [60, 61]. Fusion of fluorescent proteins is also possible (like GFP), but care has to be taken since labeling may affect protein behavior and function. The following review contains further information on labeling proteins [62, 63].
4. Osmotic pressure can cause the vesicles to shrink or swell, or even burst. To avoid this, sugar solutions and buffers in and outside the GUVs should be isoosmolar. The concentration can range between 10 and 300 mM. The outside glucose solution may be substituted with a buffer with the same osmolarity (The PBS buffer in Subheading 2.1 can be used instead of 300 mM glucose solution.).
5. The mica support should be freshly cleaved prior to use to a very thin film without breaks or defects (Fig. 4a). Minimizing thickness of the mica is important as it influences the detection efficiency of the confocal microscope by reducing the photons arriving at the detector.
6. As an alternative to permanently attaching cylinders using optical glue, one can instead take reusable plastic or glass

cylinders which can be attached using vacuum grease or silicone. If one wishes to combine optical microscopy with other surface techniques like AFM, it may be better to use a wider chamber to fit the AFM cantilever. For example, a homemade plastic cylinder with inner diameter of 2 cm, outer diameter of 2.5 cm, and height of 0.5 cm would suffice (Fig. 4).

7. During heating, it is unavoidable to lose solvent. Make sure that the solution does not dry up. This will destroy the SLBs already formed in the surface of the mica. When working with temperatures above 45 °C, add more SLB buffer to compensate for evaporation and cover the chamber with another cover glass.

8. Washing steps remove calcium ions and unfused vesicles from the surface of the SLB. The unfused vesicles result in bigger fluorescent particles that may affect the FCS measurement [41]. We suggest that you wash the SLB thoroughly but also carefully to prevent contact of mica with air which may destroy the SLBs.

9. The electroformation conditions are dependent on several factors as described in [48–51]. AC currents are advantageous because they produce electro-osmotic effects which help in lipid swelling and liposome formation from the electrodes [50]. A recent study shows the relationship between AC-field strength (V/mm) and frequency [64].

10. Always pipette and mix GUV suspensions extremely gently. Cut-off pipette tips to avoid shear that can destroy GUVs.

11. Distortions in the detection volume affect FCS results. These distortions may arise among others from improper axial positioning, differences in refractive index along the beam path, and coverslip thickness [22, 32]. Ensuring that calibration measurements are done away from the coverslip surface is therefore very important.

12. By fitting the autocorrelation curve of the calibration measurement with the 3D-diffusion model with Gaussian detection, one can estimate the diffusion time, τ_D of the dye and the structure parameter S of the detection volume. Using the diffusion coefficient of the dye obtained from literature and Eq. 4, the waist radius, w_0, can be calculated. With this, one can then calculate for the diffusion coefficient of one's sample by using Eq. 4. Different microscope setups will have different detection volume parameters (w_0 and S), as such there is no exact true value for these parameters. It is then good to keep a record of these calibration measurements to ensure the consistency of measurements over several months and detect any problems in the microscope setup.

13. Accurate FCS data are obtained when appropriate time resolution for the species diffusion is taken into account [22, 33].

Therefore, it is advisable to measure a period that is at least 10,000 times the predicted diffusion time of the species being observed.

14. With the electroformation method, GUVs with sizes between 20 and 300 μm are obtained depending on the lipid composition. Larger vesicles are more suitable for FCS measurements because they resemble better to a plane and cause less photobleaching problems. When using proteoliposomes, it may be difficult to get large enough vesicles. Be careful to only select truly unilamellar, tensed, homogeneous, and immobile vesicles.
15. The usual scanning mode of a typical LSM scans in the horizontal (x–y) plane. One has to find the region of the GUV that will exhibit zero curvature on the vertical plane so that the LSM can scan perfectly perpendicular on a flat membrane. This usually coincides with the equator of the GUV (the focal plane where the GUV appears largest). Performing the scan elsewhere will result in a non-perpendicular scanning, and may distort the autocorrelation curves during analysis.
16. If using the hardware correlator, one must open the software for the photon mode and set up the analysis time, data storage, etc. Using the imaging mode of your microscope, perform continuous scanning on the area of interest as mentioned in the procedures. Start acquisition on the hardware correlator. Stop the experiment (acquisition and continuous scanning) once the desired time has elapsed.
17. Qualitative observations of the autocorrelation function can give an idea of the diffusion characteristics of your species of interest. These observations include the decay time and amplitude of the autocorrelation curve. The decay time estimates the type of diffusion. A steep decay is indicative of transport phenomenon, while slower decay corresponds to random Brownian diffusion or even anomalous diffusion. The amplitude shows the amount of particles in the detection volume; larger amplitude corresponds to smaller number of particles in the detection volume.

References

1. Kholodenko BN, Hancock JF, Kolch W (2010) Signalling ballet in space and time. Nat Rev Mol Cell Biol 11:414–426
2. van Meer G, Voelker DR, Feigenson GW (2008) Membrane lipids: where they are and how they behave. Nat Rev Mol Cell Biol 9:112–124
3. Helenius A, Simons K (1975) Solubilization of membranes by detergents. Biochim Biophys Acta 415:29–79
4. Ellis JA, Jackman MR, Luzio JP (1992) The post-synthetic sorting of endogenous membrane proteins examined by the simultaneous purification of apical and basolateral plasma membrane fractions from Caco-2 cells. Biochem J 283:553–560
5. Bünger S, Roblick UJ, Habermann JK (2010) Comparison of five commercial extraction kits for subsequent membrane protein profiling. Cytotechnology 61:153–159

6. Von Jagow G, Schägger H (eds) (1994) A practical guide to membrane protein purification. Academic, San Diego
7. Santoni V, Molloy M, Rabilloud T (2000) Membrane proteins and proteomics: un amour impossible? Electrophoresis 21:1054–1070
8. Edwards A, Arrowsmith C, Christendat D et al (2000) Protein production: feeding the crystallographers and NMR spectroscopists. Nat Struct Biol 7:970–972
9. Gräslund S, Nordlund P, Weigelt J et al (2008) Protein production and purification. Nat Methods 5:135–146
10. Hühmer AFR, Aced GI, Perkins MD et al (1997) Separation and analysis of peptides and proteins. Anal Chem 69:29R–57R
11. Yamamoto K, Soong R, Ramamoorthy A (2009) Comprehensive analysis of lipid dynamics variation with lipid composition and hydration of bicelles using nuclear magnetic resonance (NMR) spectroscopy. Langmuir 25:7010–7018
12. Ricchelli F, Gobbo S, Moreno G et al (1999) Changes of the fluidity of mitochondrial membranes induced by the permeability transition. Biochemistry 38:9295–9300
13. Muller DJ (2008) AFM: a nanotool in membrane biology. Biochemistry 47:7986–7998
14. Goksu EI, Vanegas JM, Blanchette CD et al (2009) AFM for structure and dynamics of biomembranes. Biochim Biophys Acta 1788:254–266
15. Cambi A, Lidke D (2012) Nanoscale membrane organization: where biochemistry meets advanced microscopy. ACS Chem Biol 7:139–149
16. Baumgart T, Hammond AT, Sengupta P et al (2007) Large-scale fluid/fluid phase separation of proteins and lipids in giant plasma membrane vesicles. Proc Natl Acad Sci USA 104: 3165–3170
17. Simons K, Ikonen E (1997) Functional rafts in cell membranes. Nature 387:569–572
18. Jacobson K, Mouritsen OG, Anderson RGW (2007) Lipid rafts: at a crossroad between cell biology and physics. Nat Cell Biol 9:7–14
19. Schwille P (2001) Fluorescence correlation spectroscopy and its potential for intracellular applications. Cell Biochem Biophys 34: 383–408
20. Hess ST, Huang S, Heikal AA et al (2002) Biological and chemical applications of fluorescecnce correlation spectroscopy: a review. Biochemistry 41:697–705
21. Garcia-Saez AJ, Buschhorn SB, Keller H et al (2011) Oligomerization and pore formation by equinatoxin II inhibit endocytosis and lead to plasma membrane reorganization. J Biol Chem 286:37768–37777
22. Ries J, Schwille P (2008) New concepts for fluorescence correlation spectroscopy on membranes. Phys Chem Chem Phys 10:3487
23. Malengo G, Andolfo A, Sidenius N et al (2008) Fluorescence correlation spectroscopy and photon counting histogram on membrane proteins: functional dynamics of the glycosylphosphatidylinositol-anchored urokinase plasminogen activator receptor. J Biomed Opt 13:031215
24. Xu L, Pallikkuth S, Hou Z et al (2011) Dysferlin forms a dimer mediated by the C2 domains and the transmembrane domain in vitro and in living cells. PLoS One 6:e27884
25. Chiantia S, Schwille P, Klymchenko AS et al (2011) Asymmetric GUVs prepared by MβCD-mediated lipid exchange: an FCS study. Biophys J 100:L1–L3
26. Ganguly S, Chattopadhyay A (2010) Cholesterol depletion mimics the effect of cytoskeletal destabilization on membrane dynamics of the serotonin1a receptor: a zFCS study. Biophys J 99:1397–1407
27. Guo L, Gai F (2010) Heterogeneous diffusion of a membrane-bound pHLIP peptide. Biophys J 98:2914–2922
28. Rigler R, Mets Ü, Widengren J et al (1993) Fluorescence correlation spectroscopy with high count rate and low background: analysis of translational diffusion. Eur Biophys J 22: 169–175
29. Magde D, Elson E, Webb W (1972) Thermodynamic fluctuations in a reacting system—measurement by fluorescence correlation spectroscopy. Phys Rev Lett 29:705–708
30. Haustein E, Schwille P (2007) Fluorescence correlation spectroscopy: novel variations of an established technique. Annu Rev Biophys Biomol Struct 36:151–169
31. Petrov EP, Schwille P (2008) State of the art and novel trends in fluorescence correlation spectroscopy. In: Resch-Genger U (ed) Standardization and quality assurance in fluorescence measurements II, vol 6, Springer series on fluorescence. Springer, Berlin, pp 145–197
32. Enderlein J, Gregor I, Patra D et al (2005) Performance of fluorescence correlation spectroscopy for measuring diffusion and concentration. Chemphyschem 6:2324–2336
33. Tcherniak A, Reznik C, Link S et al (2009) Fluorescence correlation spectroscopy: criteria for analysis in complex systems. Anal Chem 81:746–754
34. García-Sáez AJ, Schwille P (2008) Fluorescence correlation spectroscopy for the study of

membrane dynamics and protein/lipid interactions. Methods 46:116–122

35. Humpolíčková J, Gielen E, Benda A et al (2006) Probing diffusion laws within cellular membranes by Z-scan fluorescence correlation spectroscopy. Biophys J 91:L23–L25
36. Ries J, Chiantia S, Schwille P (2009) Accurate determination of membrane dynamics with line-scan FCS. Biophys J 96:1999–2008
37. Ries J, Schwille P (2006) Studying slow membrane dynamics with continuous wave scanning fluorescence correlation spectroscopy. Biophys J 91:1915–1924
38. Bacia K, Schwille P (2007) Practical guidelines for dual-color fluorescence cross-correlation spectroscopy. Nat Protoc 2:2842–2856
39. Dertinger T, Loman A, Ewers B et al (2008) The optics and performance of dual-focus fluorescence correlation spectroscopy. Opt Express 16:14353–14368
40. Ries J, Weidemann T, Schwille P (2012) Fluorescence correlation spectroscopy. In: Egelman E (ed) Comprehensive biophysics. Academic, San Diego, pp 210–245
41. Chiantia S, Ries J, Kahya N et al (2006) Combined AFM and two-focus SFCS study of raft-exhibiting model membranes. Chemphyschem 7:2409–2418
42. Egawa H, Furusawa K (1999) Liposome adhesion on mica surface studied by atomic force microscopy. Langmuir 15:1660–1666
43. Cremer PS, Boxer SG (1999) Formation and spreading of lipid bilayers on planar glass supports. J Phys Chem B 103:2554–2559
44. Tamm LK, McConnell HM (1985) Supported phospholipid bilayers. Biophys J 47:105–113
45. Castellana ET, Cremer PS (2006) Solid supported lipid bilayers: from biophysical studies to sensor design. Surf Sci Rep 61:429–444
46. Garcia-Saez AJ, Chiantia S, Schwille P (2007) Effect of line tension on the lateral organization of lipid membranes. J Biol Chem 282:33537–33544
47. Przybylo M, Sykora J, Humpolickova J et al (2006) Lipid diffusion in giant unilamellar vesicles is more than 2 times faster than in supported phospholipid bilayers under identical conditions. Langmuir 22:9096–9099
48. Angelova MI, Dimitrov DS (1986) Liposome electroformation. Faraday Discuss 81:303
49. Angelova MI, Dimitrov DS (1987) Swelling of charged lipids and formation of liposomes on electrode surfaces. Mol Cryst Liq Cryst Inc Nonlin 152:89–104
50. Dimitrov DS, Angelova MI (1988) Lipid swelling and liposome formation mediated by electric fields. Bioelectrochem Bioenerg 19:323–336
51. Angelova MI, Soléau S, Méléard P et al (1992) Preparation of giant vesicles by external AC electric fields. Kinetics and applications. Progr Colloid Polym Sci 89:127–131
52. Chiantia S, Ries J, Schwille P (2009) Fluorescence correlation spectroscopy in membrane structure elucidation. Biochim Biophys Acta 1788:225–233
53. Bacia K (2004) SNAREs prefer liquid-disordered over "raft" (liquid-ordered) domains when reconstituted into giant unilamellar vesicles. J Biol Chem 279:37951–37955
54. Doeven MK, Folgering JHA, Krasnikov V et al (2005) Distribution, lateral mobility and function of membrane proteins incorporated into giant unilamellar vesicles. Biophys J 88:1134–1142
55. García-Sáez AJ, Ries J, Orzáez M et al (2009) Membrane promotes tBID interaction with BCLXL. Nat Struct Mol Biol 16:1178–1185
56. Kahya N (2002) Spatial organization of bacteriorhodopsin in model membranes. Light-induced mobility changes. J Biol Chem 277:39304–39311
57. Steringer JP, Bleicken S, Andreas H et al (2012) Phosphatidylinositol 4,5-bisphosphate (PI(4,5)P2)-dependent oligomerization of fibroblast growth factor 2 (FGF2) triggers the formation of a lipidic membrane pore implicated in unconventional secretion. J Biol Chem 287:27659–27669
58. Magatti D, Ferri F (2001) Fast multi-tau real-time software correlator for dynamic light scattering. Appl Optics 40:4011–4021
59. Chiantia S, Kahya N, Schwille P (2007) Raft domain reorganization driven by short- and long-chain ceramide: a combined AFM and FCS study. Langmuir 23:7659–7665
60. Dertinger T, Pacheco V, von der Hocht I et al (2007) Two-focus fluorescence correlation spectroscopy: a new tool for accurate and absolute diffusion measurements. Chemphyschem 8:433–443
61. Ries J, Ruckstuhl T, Verdes D et al (2008) Supercritical angle fluorescence correlation spectroscopy. Biophys J 94:221–229
62. Marks KM, Nolan GP (2006) Chemical labeling strategies for cell biology. Nat Methods 3:591–596
63. Prescher JA, Bertozzi CR (2005) Chemistry in living systems. Nat Chem Biol 1:13–21

64. Politano TJ, Froude VE, Jing B et al (2010) AC-electric field dependent electroformation of giant lipid vesicles. Colloids Surf B Biointerfaces 79:75–82

65. Petrášek Z, Schwille P (2008) Precise measurement of diffusion coefficients using scanning fluorescence correlation spectroscopy. Biophys J 94:1437–1448

66. Culbertson CT, Jacobson SC, Ramsey JM (2002) Diffusion coefficient measurements in microfluidic devices. Talanta 56: 265–272

Chapter 14

Analyses of Protein–Protein Interactions by In Vivo Photocrosslinking in Budding Yeast

Takuya Shiota, Shuh-ichi Nishikawa, and Toshiya Endo

Abstract

Recent development of methods for genetic incorporation of unnatural amino acids into proteins in live cells enables us to analyze protein interactions by site-specific photocrosslinking. Here we describe a method to incorporate *p*-benzoyl-L-phenylalanine (*p*Bpa), a photoreactive unnatural amino acid, into defined positions of a target protein in living yeast cells. Photocrosslinking using the *p*Bpa-incorporated proteins has been proven to be a powerful method for analyzing protein–protein interactions at the spatial resolution of amino-acid residues. Since photocrosslinking can be performed for *p*Bpa-incorporated proteins that are properly assembled into a protein complex in living cells, this method will allow us to reveal protein–protein interactions of the target proteins at work.

Key words Protein interactions, Photocrosslink, Unnatural amino acids, Suppressor tRNA, Yeast

1 Introduction

Site-specific photocrosslinking has been proven to be a powerful tool for analyzing protein–protein interactions at the spatial resolution of amino-acid residues. This method was first developed using in vitro transcription/translation systems involving tRNAs chemically acylated with wide variety of naturally unavailable amino acids [1]. Unnatural amino acids with photoreactive benzophenone side chains such as *p*-benzoyl-L-phenylalanine (*p*Bpa) can be incorporated into defined positions of target proteins. Upon excitation at 350–365 nm, the benzophenone group reacts with nearby carbon–hydrogen (C–H) bonds, yielding a covalent cross-link (Fig. 1a) [2]. However, the stoichiometric nature of the chemical acylation reaction severely limits the amount of proteins generated. This method is thus limited to in vitro analyses and cannot be used to analyze protein–protein interactions in vivo.

In order to overcome this limitation, methods to produce protein variants with site-specifically incorporated unnatural amino acids in living *E. coli*, yeast or CHO cells have been developed by

Doron Rapaport and Johannes M. Herrmann (eds.), *Membrane Biogenesis: Methods and Protocols*, Methods in Molecular Biology, vol. 1033, DOI 10.1007/978-1-62703-487-6_14,

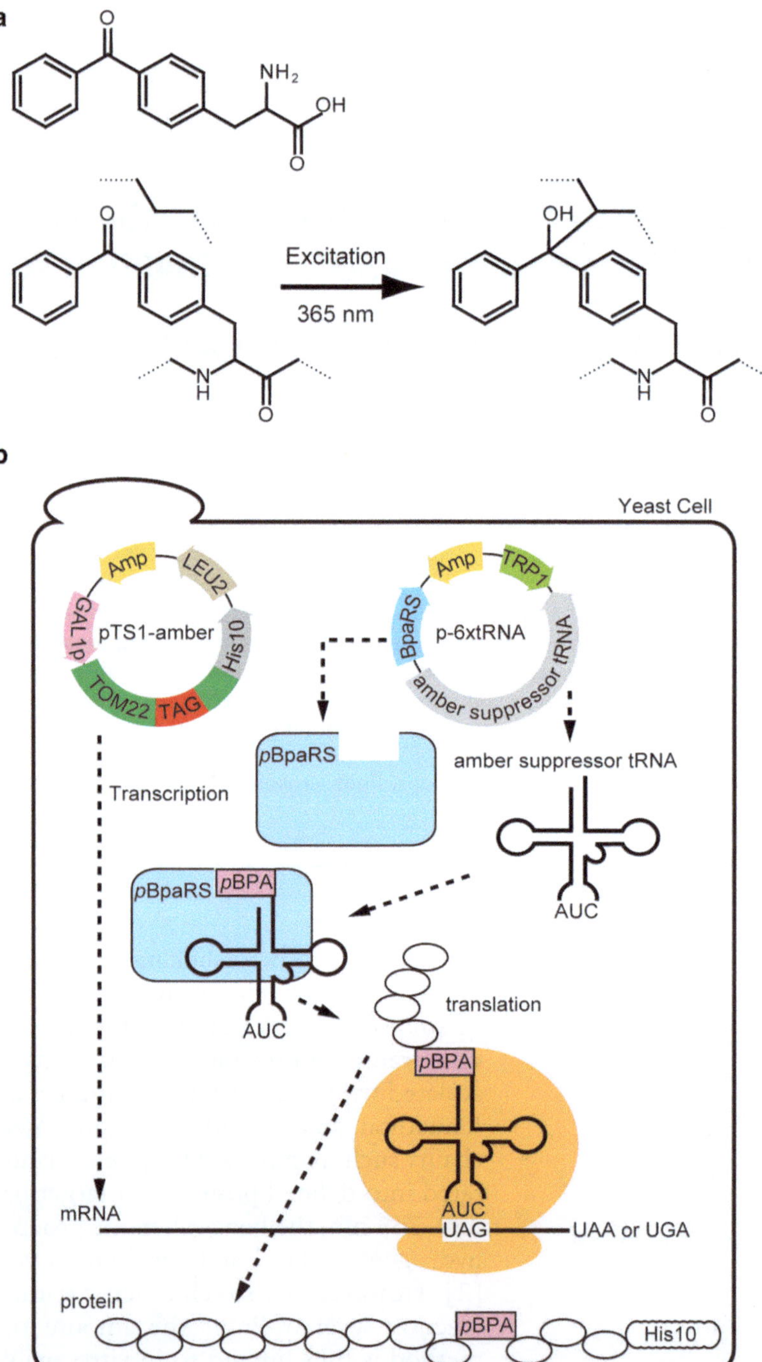

Fig. 1 Site-specific incorporation of *pBpa* into proteins expressed in yeast. (**a**) Chemical structure of *p*Bpa and illustration of photocrosslink between *p*Bpa and a nearby C–H bond are shown. (**b**) Outline of site-specific incorporation of *p*Bpa into Tom22-His10 in yeast cells. (**b**, Modified from Shiota et al. (2012), with permission from Yodosha [8])

the Schultz's group and Yokoyama's group [3–5]. These methods employ orthogonal pairs of tyrosyl $tRNA_{CUA}$, which recognizes the *amber* codon, and modified tyrosyl-tRNA synthetase (TyrRS), which aminoacylates the tyrosyl $tRNA_{CUA}$ with a defined unnatural amino acid. In the case of in vivo photocrosslinking in yeast, *E. coli* tyrosyl $tRNA_{CUA}$ and *E. coli pBpaRS*, which aminoacylates tyrosyl $tRNA_{CUA}$ with *p*Bpa, were expressed in yeast cells with a target-protein gene containing an *amber* mutation at a defined position. When such yeast cells were grown in the presence of *p*BPA, *p*BPA was incorporated into the *amber* position of the target protein (Fig. 1b). Since the *p*BPA-containing target protein can be transported to its proper subcellular location, protein–protein interactions can be analyzed by photocrosslinking under physiological conditions. This method can be easily scaled up, which enables us to identify cross-linked partner proteins by mass-spectrometry analyses.

We applied this in vivo photocrosslinking method to analyze protein–protein interactions between subunits of the mitochondrial outer membrane translocator complex, the TOM40 complex, in the yeast *Saccharomyces cerevisiae.* By changing the positions of introduction of BPA throughout the sequence, we could construct a protein interaction map of a multifunctional subunit, Tom22, at work [6]. This chapter describes the method for in vivo photocrosslinking using living yeast cells.

2 Materials

2.1 Media

1. 20 % glucose (1 L): Dissolve 200 g of D-glucose in 750 mL distilled water. Adjust to 1 L, autoclave, and store at room temperature.
2. 20 % galactose (1 L): Dissolve 200 g of D-galactose in 750 mL distilled water. Adjust to 1 L, autoclave, and store at room temperature.
3. 100× (–Trp –Leu) supplement (250 mL): Dissolve 400 mg of adenine sulfate, 400 mg of uracil, 400 mg of L-histidine-HCl, and 600 mg of L-lysine-HCl in 200 mL of 10 mM HCl. Filter sterilize and store at room temperature (*see* **Note 1**).
4. 100 mM Bpa: Dissolve 13.5 mg of DL-2-amino-3-(*p-benzoylphenyl*)pentanoic acid (Bachem F-2800) in 50 μL of 1 N NaOH (*see* **Note 2**).
5. SD (–Trp –Leu) (1 L): Dissolve 6.7 g of yeast nitrogen base without amino acids (BD Difco) in 900 mL distilled water. Autoclave and add 100 mL of 20 % glucose and 10 mL of 100× (–Trp –Leu) supplement after cooling the solution to ~50 °C.

6. SD (–Trp –Leu) agar (1 L): Dissolve 6.7 g of yeast nitrogen base without amino acids (BD Difco) and 20 g of agar in 900 mL distilled water. Autoclave and add 100 mL of 20 % glucose and 10 mL of 100× (–Trp –Leu) supplement, cool medium to ~65 °C and pour plates.
7. SGal (–Trp –Leu) (1 L): Dissolve 6.7 g of yeast nitrogen base without amino acids (BD Difco) in 900 mL distilled water. Autoclave and add 100 mL of 20 % galactose and 10 mL of 100× (–Trp –Leu) supplement after cooling the solution to ~50 °C.
8. SGal (–Trp –Leu + Bpa) (50 mL): Mix 50 mL of SGal (–Trp –Leu) and 50 μL of 100 mM BPA in a sterile 200 mL Erlenmeyer flask (*see* **Note 3**).

2.2 Purification of Photocrosslinked Products

Prepare all solutions using ultrapure water (e.g., Milli Q water).

1. 1 M Tris–HCl, pH 7.5 (1 L): Dissolve 121.1 g of Tris in 750 mL water and adjust pH to 7.5 with HCl. Make up to 1 L, autoclave, and store at room temperature.
2. 1 M Tris–HCl, pH 8.0 (1 L): Dissolve 121.1 g of Tris in 750 mL water and adjust pH to 8.0 with HCl. Make up to 1 L, autoclave, and store at room temperature.
3. 5 N NaOH (200 mL): Dissolve 40 g of NaOH in 150 mL water and make up to 200 mL. Store at room temperature.
4. 0.5 M EDTA, pH 8.0 (500 mL): Dissolve 93 g of Na_2 EDTA $2H_2O$ in 400 mL water and adjust pH to 8.0 with 5 N NaOH. Make up to 500 mL. Store at room temperature.
5. 5 M NaCl (500 mL): Dissolve 146.2 g of NaCl in 350 mL water and make up to 500 mL. Autoclave and store at room temperature.
6. 4 M imidazole-HCl, pH 8.0 (200 mL): Dissolve 54.4 g of imidazole in 150 mL water and adjust pH to 8.0 with HCl. Make up to 200 mL and store at 4 °C in a bottle wrapped with aluminum foil.
7. 20 % Triton X-100 (200 mL): Dissolve 40 mL of Triton X-100 in 150 mL water and make up to 200 mL. Store at room temperature.
8. TE buffer (100 mL): Mix 10 mL 1 M Tris–HCl pH 8.0, 2 mL 0.5M EDTA pH 8.0, and 88 mL water. Store at room temperature.
9. β-mercaptoethanol.
10. 100 % (w/v) trichloroacetic acid.
11. Acetone.
12. Ni-NTA agarose: Equilibrate with Triton buffer before use.

13. SDS buffer (100 mL): Dissolve 1 g of SDS in 80 mL water. Add 5 mL of 1 M Tris–HCl pH 8.0 and 3 mL of 5 M NaCl to the SDS solution and make up to 100 mL. Store at room temperature.
14. Triton buffer (200 mL): Mix 10 mL of 1 M Tris–HCl pH 8.0, 6 mL of 5 M NaCl, 5 mL 20 % of Triton X-100, 1 mL of 4 M imidazole-HCl pH 8.0, and 178 mL water. Store at room temperature.
15. Elution buffer (100 mL): Mix 5 mL of 1 M Tris–HCl pH 8.0, 3 mL of 5 M NaCl, 2.5 mL of 20 % Triton X-100, 10 mL of 4 M imidazole-HCl pH 8.0, and 79.5 mL water. Store at room temperature.
16. 1.5 % (w/v) sodium deoxycholate (10 mL): Dissolve 150 mg sodium deoxycholate in 10 mL water. Store at room temperature.
17. SDS-PAGE sampling buffer: 125 mM Tris–HCl, pH 6.8, 2 mM EDTA, 2 %(w/v) SDS, 1 % (w/v) sucrose, 0.03 % (w/v) bromophenol blue, 2 %(v/v) β-mercaptoethanol.
18. 5 mL microcentrifuge tubes (Ina optica ST-500 or Agros Technologies T2076).
19. Eppendorf tubes.

2.3 Strain and Plasmids

1. The yeast strain W303-1A (*MATa ade2-1 his3-11,15 ura3-1 leu2-3,112 trp1-1 can1-100*) was used to express site-specifically *p*Bpa-incorporated proteins.
2. p-6xtRNA: An yeast multicopy plasmid with the *TRP1* marker harboring six tandem copies of the *E. coli amber* suppressor tRNA genes and a gene expressing the *E. coli p*BpaRS in yeast (A gift from Peter G. Schultz, The Scripps Research Institute).
3. The plasmid pTS1 was used to express Tom22-His$_{10}$ from the *GAL1* promoter in yeast cells (*see* **Note 4**). This plasmid was constructed by introducing the *GAL1* promoter-*TOM22-HIS*10-*CMK1* terminator cassette into the polylinker site of pRS315 (a single copy plasmid with the *LEU2* marker).

2.4 Apparatus for Photocrosslinking

1. High intensity UV lamp: Type B-100AP (UVP).
2. 90 mm dishes.

3 Methods

3.1 Construction of Yeast Strains for In Vivo Photocrosslinking Experiments

1. Amber mutations can be introduced into the target plasmid (e.g., pTS1 in our case) by PCR-based site-directed mutagenesis such as the QuikChange site-directed mutagenesis kit (Stratagene) (*see* **Note 5**).

2. The mutagenized plasmid and p-6xtRNA were introduced into W303-1A by yeast transformation procedures [7]. Trp + Leu + transformants were selected on SD (–Trp –Leu) plates.

3.2 Yeast Culture Conditions

1. Inoculate 5 mL of SD (–Trp –Leu) with a transformant colony. Grow cells to an early stationary phase at 30 °C.
2. Centrifuge the culture at 1,000 × *g* for 3 min at room temperature. Aspirate off the supernatant.
3. Prepare a sterile 200 mL Erlenmeyer flask containing 50 mL of SGal (–Trp –Leu + BPA).
4. Suspend yeast cells with 1 mL of SGal (–Trp –Leu + Bpa), which was taken from the flask, and inoculate into the flask containing SGal (–Trp –Leu + Bpa).
5. Cover the flask with aluminum foil.
6. Grow cells at 30 °C to mid logarithmic (2–4 × 10^7 cells/ml) phase (~16 h) with appropriate aeration (*see* **Note 6**).

3.3 Photocross-linking

1. Place the UV lamp on the bench with the UV bulb facing down (Fig. 2a). Turn on the light and wait until the output light power becomes constant (~10 min) (*see* **Note 7**).
2. Transfer half (25 mL) of the culture to a 90 mm dish. Transfer the rest (25 mL) of the culture to a 50-mL Falcon tube, cover the tube with aluminum foil, and keep at room temperature.
3. Place the 90 mm dish containing the culture on the bench. Remove the lid from the dish. Place the UV lamp right above the dish so that the distance between the surface of the bench and the surface of the UV bulb becomes 5 cm (*see* **Note 8**) (Fig. 2b). Expose the culture to UV light for 10 min at room temperature.
4. Transfer the UV-irradiated culture to a 50-mL Falcon tube.
5. Centrifuge both the irradiated and unirradiated cultures at 1,000 × *g* for 3 min at room temperature. Aspirate off the supernatant.
6. Suspend the cells with TE buffer at 1.2 × 10^8 cells/mL.
7. Dispense the cell suspension to 1-mL aliquots in Eppendorf tubes. Snap freeze in liquid nitrogen and store at –30 °C until use.

3.4 Purification of Photocrosslinked Products

1. Thaw 1 mL of the cell suspension by immersing the tube in water.
2. Vortex the tube to make sure that cells are dispersed uniformly.
3. Add 67 μL of 5 N NaOH and 70 μL of β-mercaptoethanol to the cell suspension. Mix gently by inverting the tube several times.

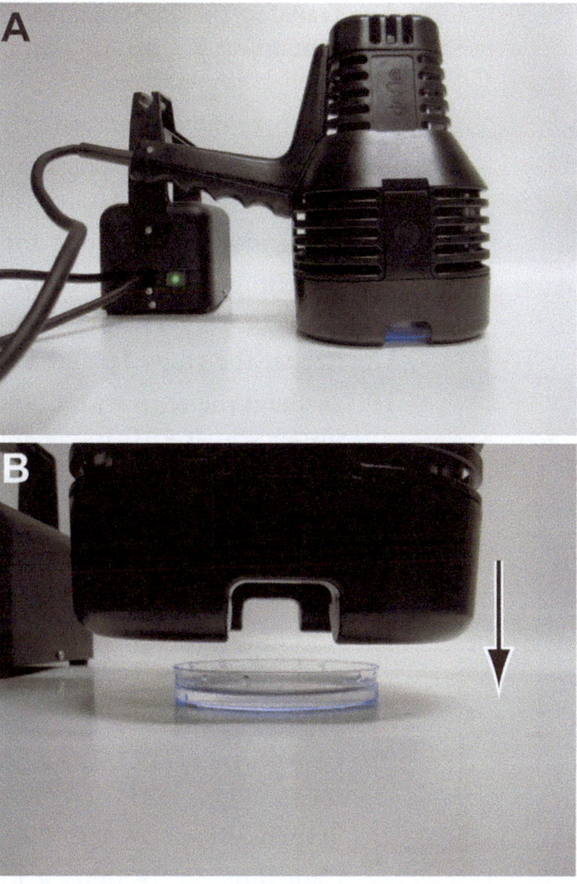

Fig. 2 Procedure for UV photocrosslinking. (**a**) Type B-100AP UV lamp (UVP) used for photocrosslinking. (**b**) UV irradiation of yeast culture. The UV lamp are being placed (moving downward on the direction of the *arrow*) above the dish containing yeast culture. (Reproduced from Shiota et al. (2012), with permission from Yodosha [8])

4. Keep the cell suspension on ice for 4 min.
5. Add 125 μL of 100 % (w/v) trichloroacetic acid to the cell suspension. Mix the suspension by vortexing.
6. Keep the suspension on ice for 10 min.
7. Centrifuge the suspension at 14,000 × *g* for 10 min at 4 °C. Remove the supernatant.
8. Suspend the pellet with 600 μL of ice cold acetone.
9. Centrifuge the suspension at 14,000 × *g* for 10 min at 4 °C. Remove the supernatant.
10. Suspend the pellet with 600 μL of SDS buffer (*see* **Note 9**).
11. Heat the suspension at 98 °C for 5 min.
12. Centrifuge the suspension at 14,000 × *g* for 10 min at 4 °C.

13. Transfer the supernatant to a 5-mL microcentrifuge tube containing 2.4 mL of Triton buffer and 100 μL (bed volume) of Ni-NTA agarose.
14. Incubate the suspension at 4 °C for 1 h with gentle agitation.
15. Centrifuge the suspension at 500×*g* for 10 min at 4 °C. Remove the supernatant without disturbing the resin pellet.
16. Suspend the resin with 1 mL of Triton buffer.
17. Centrifuge the suspension at 500×*g* for 10 min at 4 °C. Remove the supernatant without disturbing the resin pellet.
18. Repeat the **steps 16** and **17** three times.
19. Suspend the resin with 600 μL of Elution buffer. Incubate the suspension at 4 °C for 15 min with gentle agitation.
20. Centrifuge the suspension at 500×*g* for 10 min at 4 °C.
21. Transfer the supernatant to a new Eppendorf tube (*see* **Note 10**).
22. Add 13 μL of 1.5 % (w/v) sodium deoxycholate to the solution. Mix by vortexing.
23. Add 65 μL of 100 % (w/v) trichloroacetic acid to the cell suspension. Mix by vortexing.
24. Keep the suspension on ice for 10 min.
25. Centrifuge at 14,000×*g* for 10 min at 4 °C. Remove the supernatant.
26. Suspend the pellet with 600 μL of ice cold acetone.
27. Centrifuge at 14,000×*g* for 10 min at 4 °C. Remove the supernatant.
28. Suspend the pellet with 30 μL of SDS-PAGE sampling buffer. Heat the sample at 98 °C for 5 min.
29. Analyze the sample by SDS-PAGE and immunoblotting. An interaction map of Tom22 can be created using a series of the *amber* mutant of the *TOM22-HIS*$_{10}$ gene (Fig. 3).

Fig. 3 (continued) the eluted materials. (**b**) Detection of Tom22-Tom20 cross-linked products. Photocrosslinked materials were purified from protein extracts prepared from the yeast cells expressing Tom22-His$_{10}$ with *p*Bpa incorporated at residue 48 (Tom22(48*p*Bpa)-His$_{10}$) as above and subjected to SDS-PAGE and immunoblotting with anti-Tom22 (αTom22) or anti-Tom20 (αTom20) antibodies. 22-20 indicates position of the cross-linked product between Tom22 and Tom20. (**c**) *p*Bpa was introduced into various positions in the cytosolic domain of Tom22 (indicated with *dots*) in vivo and subjected to photocrosslinking analyses as in (**a**). The amounts of the cross-linked products with Tom20 were quantified and plotted against residue number; the largest amount for residue 48 was set to 100 %. The *red* letters in the sequence show acidic residues. (**d**) Protein-interaction mapping of Tom22. The residues in *red* indicate positions used for site-specific incorporation of *p*Bpa. "20" highlighted in *orange*, "40" highlighted in *pink* and "50" highlighted in *blue* show positions of the residues cross-linked to Tom20, Tom40, and Tim50, respectively. (**b**, **c**, Reproduced from Shiota et al. (2011) [6]; **d**, Reproduced from Shiota et al. (2012), with permission from Yodosha [8])

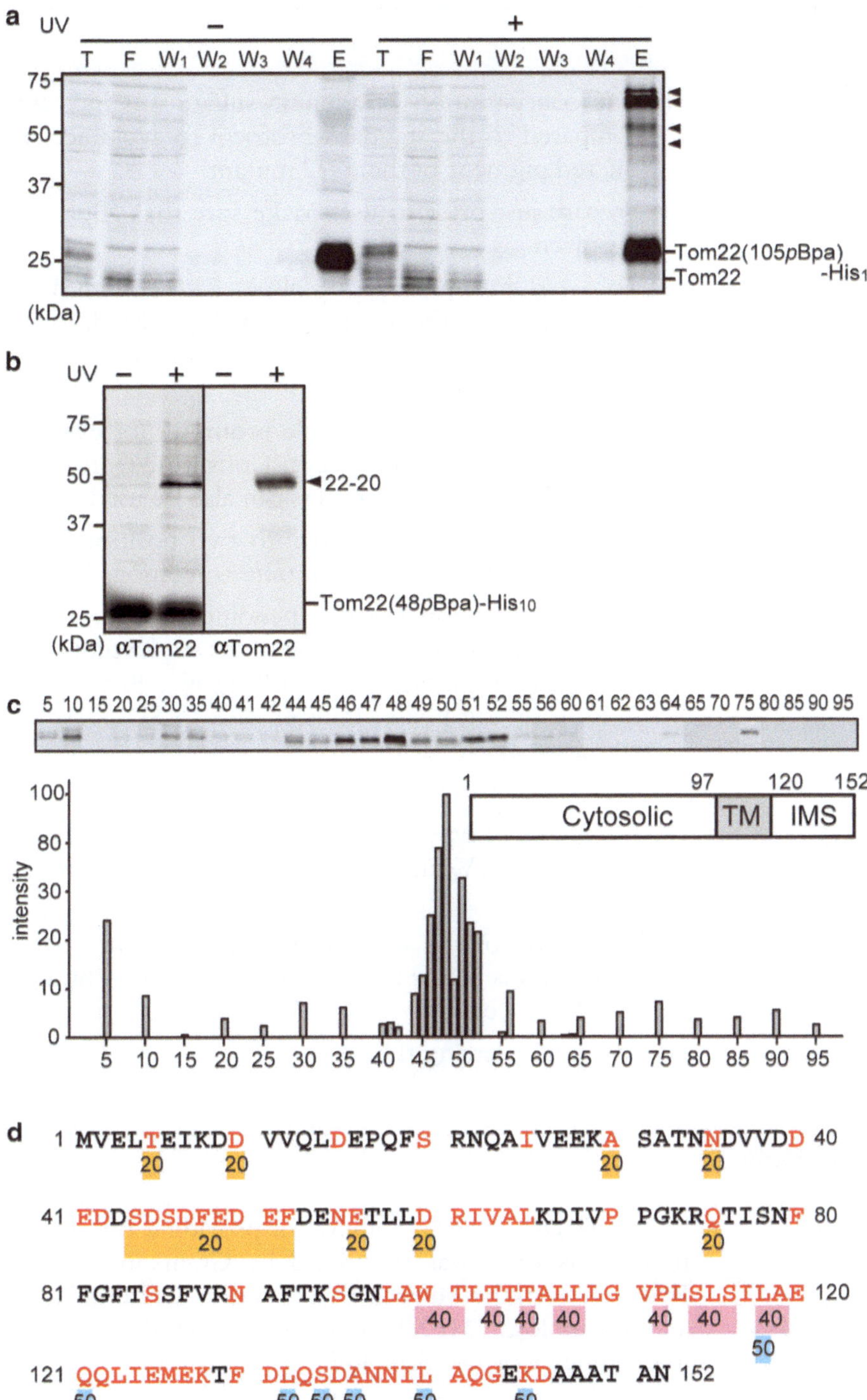

Fig. 3 Site-specific photocrosslinking of Tom22 expressed in yeast. (**a**) Purification of photocrosslinked products. The yeast cells expressing *p*Bpa-incorporated Tom22-His$_{10}$ at residue 105 (Tom22(105*p*Bpa)-His$_{10}$) were irradiated with UV (+UV) or kept in dark (−UV). Protein extracts were prepared and subjected to affinity purification with Ni-NTA agarose. Each fraction was analyzed by SDS-PAGE and immunoblotting with anti-Tom22 antibodies. *Arrowheads* indicate the cross-linked products. T, 2.5 % of the protein extracts subjected to affinity purification; F, 2.5 % of the flow through fractions; W1-4, 100 % of the wash fractions; E, 100 % of

4 Notes

1. The concentration of adenine sulfate was increased twofold compared to the standard protocol to avoid accumulation of the red pigment of the *ade2* mutant.
2. Prepare just prior to use. Make sure that *p*Bpa is completely dissolved.
3. To avoid precipitation of *p*Bpa, add 100 mM Bpa to the medium while the medium is being stirred. If precipitated materials are visible, heat the medium to ~65 °C to dissolve *p*Bpa completely.
4. We used the *GAL1* inducible promoter for expression of the *TOM22 amber* mutant genes. Constitutive promoters such as the *GPD* or *ADH* promoters can also be used.
5. *p*Bpa can be incorporated into two positions of the target protein by introducing two *amber* mutations [6].
6. Culture condition varies depending on strains and plasmids used. In the case of expression of the *amber* mutant genes from constitutive promoters, we usually grow cells in SD (−Trp −Leu) to an early log phase, transfer cells to SD (−Trp −Leu + Bpa) and grow at 30 °C for 3–5 h.
7. Wear protective goggles.
8. In the case of type Type B-100AP UV lamp, this is achieved by placing the UV lamp directly on the bench with the bulb side down.
9. Check pH of the solution by a pH test strip. If pH of the solution was below 5, neutralize the solution by adding 10 μL of 1 M Tris base.
10. Avoid transferring the resin.

Acknowledgments

We thank Dr. Peter G. Schultz for providing the p-6xtRNA plasmids. This work was supported by Grants-in-Aid for Scientific Research from the Ministry of Education, Culture, Sports, Science and Technology (MEXT) and a grant from the Japan Science and Technology Corporation (JST).

References

1. Ellman J, Mendel D, Anthony-Cahill S et al (1991) Biosynthetic method for introducing unnatural amino acids site-specifically into proteins. Methods Enzymol 202: 301–336
2. Dormán G, Prestwich GD (1994) Benzophenone photophores in biochemistry. Biochemistry 33:5661–5673
3. Farrell IS, Toroney R, Hazen JL et al (2005) Photo-cross-linking interacting proteins with a

genetically encoded benzophenone. Nat Methods 2:377–384

4. Chin JW, Cropp TA, Anderson JC et al (2003) An expanded eukaryotic genetic code. Science 301:964–967
5. Hino N, Okazaki Y, Kobayashi T et al (2005) Protein photo-cross-linking in mammalian cells by site-specific incorporation of a photoreactive amino acid. Nat Methods 2:201–216
6. Shiota T, Mabuchi H, Tanaka-Yamano S et al (2011) In vivo protein-interaction mapping of a mitochondrial translocator protein Tom22 at work. Proc Natl Acad Sci USA 108: 15179–15183
7. Adams A, Gottschling DE, Kaiser CA et al (1998) Methods in yeast genetics, a Cold Spring Harbor Laboratory course manual. Cold Spring Harbor Laboratory Press, Cold Spring Harbor
8. Shiota T, Nishikawa S, Endo T (2012) Protein interaction mapping using the *in vivo* site-specific photo crosslinking technique. Exp Med 30:1799–1805 (in Japanese)

Chapter 15

Sedimentation Velocity Analytical Ultracentrifugation in Hydrogenated and Deuterated Solvents for the Characterization of Membrane Proteins

Aline Le Roy, Hugues Nury, Benjamin Wiseman, Jonathan Sarwan, Jean-Michel Jault, and Christine Ebel

Abstract

This chapter is a step-by-step protocol for setting up, realizing, and analyzing sedimentation velocity experiments in hydrogenated and deuterated solvents, in the context of the characterization of membrane protein, in terms of homogeneity, association state, and amount of bound detergent, based on a real case study of the membrane protein BmrA solubilized in *n*-Dodecyl-β-D-Maltopyranoside) detergent.

Key words Sedimentation velocity, Analytical ultracentrifugation, Membrane proteins, Homogeneity, Association state, Bound detergent, BmrA, Detergent, Heavy water, D_2O, SEDFIT

1 Introduction

1.1 Sedimentation Velocity for Detergent Solubilized Membrane Proteins

The membrane proteins perform a wide range of essential cellular functions. Pores, channels, pumps, and transporters control the transport of ions and metabolites between the cell and the extracellular environment or between cellular compartments. The characterization of the interactions and structures of membrane proteins is often difficult based on their low natural abundance, their low over expression, and their difficult extraction from their natural lipid environment. The number of available structures is increasing rapidly, but few membrane protein structures are still known as compared with soluble proteins [1].

Detergents are most commonly used to solubilize, stabilize, and manipulate membrane proteins for the characterization of their function and structure. The stability and composition of the protein–detergent complexes depend on the type and concentration of the detergent (*see*, e.g., [2–4]). Above the critical micelle concentration (CMC), the detergent is bound in quantities that are highly variable and can be very large and thus modify considerably

Doron Rapaport and Johannes M. Herrmann (eds.), *Membrane Biogenesis: Methods and Protocols*, Methods in Molecular Biology, vol. 1033, DOI 10.1007/978-1-62703-487-6_15, © Springer Science+Business Media, LLC 2013

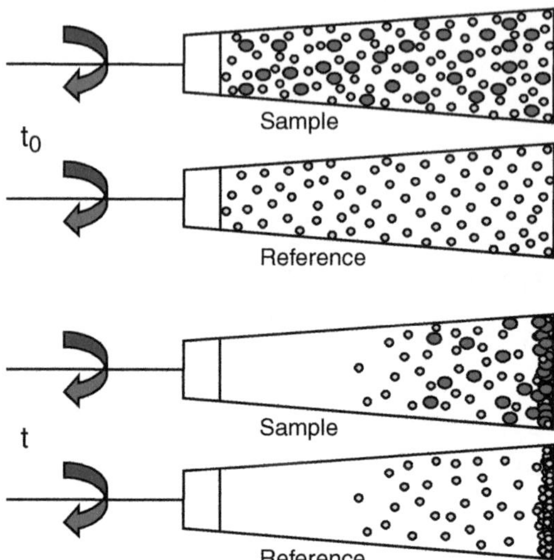

Fig. 1 Schematic representation of the sedimentation in SV-AUC for a detergent solubilized membrane protein. The centerpiece comprises two channels, filled with the sample and reference solvent. *Red balls* represent the protein detergent complex and *light blue balls* the detergent micelles. At time 0 (*top*), the solutions in each of the two channels are homogeneous. SV experiments measure the evolution of the particles concentrations in solution when submitted to centrifugal force with time. Upon centrifugation (*bottom*), the sedimentation of the two types of particles differs: the detergent–protein complexes sediment faster than the detergent micelles. If there were aggregates, they would sediment faster than the complexes. We used solvent exchange columns as the last step of sample preparation prior to AUC and double sector boundary capillary type cells (described in Fig. 2) to mask the sedimentation of glycerol and detergent micelles (Color figure online)

the size and mass of the complexes [5]. It will affect the mass, density, and shape of the protein–detergent complex, which are parameters that in turn affect the value of the sedimentation coefficient measured in sedimentation velocity (SV) analytical ultracentrifugation (AUC) experiments.

Sedimentation velocity is a powerful technique to determine the molar mass, M, and hydrodynamic (Stokes) radius, R_H, of macromolecules in solution [6–9]. SV studies macromolecules in solution that are subjected to a large centrifugal field. SV combines the separation of the macromolecules (Fig. 1) and the analysis of their transportation in view of a rigorous thermodynamics. AUC is complementary to the methods based on size determination (e.g., size exclusion chromatography, dynamic light scattering) that probe R_H. Sedimentation velocity (SV) was proved to be a very efficient method to characterize detergent-solubilized membrane proteins: particularly the homogeneity, the amount of bound detergent, and the oligomeric structure [10–13].

The transport in the ultracentrifuge is determined by the sedimentation coefficient, s, and by the diffusion coefficient, D, which are directly related, for noninteracting species, to M_b/R_H and to R_H, respectively. $M_b = M(1-\rho\bar{v})$, the buoyant molar mass of the particle, is directly related to the molar mass of the particle, M, and to $(1-\rho\bar{v})$, the buoyant term. The latter depends on the solvent density, ρ, and on the particle partial specific volume, $\bar{v}$ (i.e., the inverse of the particle density). Detergent micelles and protein–detergent complex have variable values for $\bar{v}$, since $\bar{v}$ depends on the type of detergent and of complex composition. The comparison of sedimentation profiles obtained with absorbance and with interference, gives a way to evaluate the amount of bound detergent [7, 11–13]. Measuring sedimentation in solvents of different densities is a complementary way to evaluate the particle composition [14, 15]. There are three different SV software packages that have been developed and are freely available: SEDFIT/SEDPHAT [16], which will be used here, ULTRASCAN [17] and SEDANAL [18].

This chapter presents a detailed SV protocol for the study of membrane protein solubilized with detergent. The protocol describes sample preparation for investigating the sedimentation velocity of a membrane protein in two solvents of different densities but same composition, apart from the use of heavy water, D_2O, in place of usual water H_2O, which should not change the properties of the sample, and help to determine the particle density and composition [14, 15, 19]. The two samples are prepared using solvent exchange columns as a last step prior to AUC, which will allow to have exactly the same well-defined solvent in the reference solvent and sample compartments of the two-channel centerpiece. We checked the efficiency of the solvent exchange by a test where the loaded sample was the deuterated buffer in the column equilibrated in hydrogenated buffer, in the protocol described in Subheading 3.2, **step 3**: the densities of the recovered fraction and of the elution buffer were virtually identical. Sedimentation velocity is done using synthetic boundary centerpieces, designed by the supplier (Beckman) to measure diffusion coefficients (*see*, e.g., [20]), and used here to provide exactly the same column height in the two compartments of the two-channel centerpiece (Fig. 2). Other slightly different centerpieces are available for the same purpose (Aviv Technology [21]). Such a design is to provide interference SV data of high quality, in addition to absorbance data, even in the case of a complex solvent where solvent components may sediment [22, 23]. It will allow to evaluate the amount of free detergent micelles in excess to that present in the solvent. Protocols without detergent in the reference solvent allow the determination of the total micelle concentration [11–13].

The protocol presented here is based on a real case study of the membrane protein, BmrA, solubilized with *n*-Dodecyl-*β*-D-Maltopyranoside (DDM) in a solvent containing 10 % glycerol. BmrA from *Bacillus subtilis* belongs to the ABC (ATP-binding cassette) superfamily of membrane transporters and is involved in

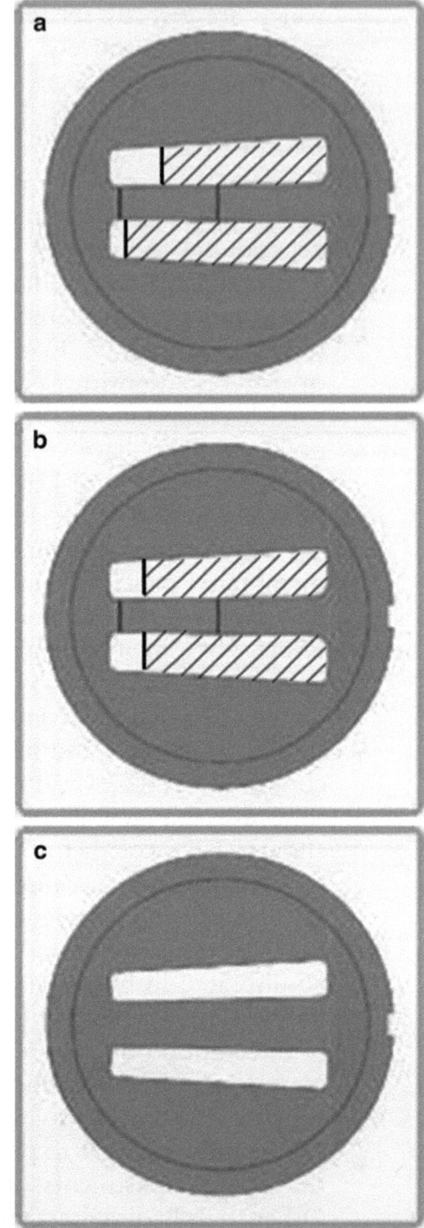

Fig. 2 AUC centerpieces. (**a, b**) Double sector boundary capillary type cell of optical path length 12 mm are here used for the detergent solubilized protein samples. Centerpieces with other design may also be used (see text). (**a**) The reference compartment is filled with a larger volume than the sample compartment (440 μL vs. 420 μL). (**b**) A capillary channel between the two channels allows the perfect matching of their heights upon low speed centrifugation (typically 30 min at 7,000 rpm), a small volume (≈20 μL) of the solvent fluxing to the sample channel. (**c**) To investigate SV of the detergent micelles, the more usual 12 mm optical path length double sector centerpieces. The sample and reference channels were filled with the same volume (400 μL) of buffers with and without detergent, respectively

multidrug resistance [24]. It was purified in an active form [25] in a study combining size exclusion chromatography, [^{14}C]DDM- and phospholipid-binding determination. This study provided an estimate of the hydrodynamic radius of 5.6 ± 4 nm, of DDM binding of 1.6 ± 0.6 g/g, and of phospholipids binding of 0.07 g/g. SV and equilibrium AUC performed using optical detection at 280 nm allowed to describe the protein as a dimer in solution.

1.2 Theoretical Background

1.2.1 Sedimentation Velocity: General Case

Sedimentation is a transport method akin to diffusion and sedimentation [6–9]. Sedimentation velocity measures in a rotor spinning at high angular velocity, ω, in the centrifuge, the evolution of the weight concentration, c, with time, t, and radial position, r. For each homogeneous ideal solute, given the sector-shaped cells used in AUC, the transport is described by the Lamm equation:

$$(\partial c / \partial t) = -(1/r)\partial / \partial r\left[r\left(cs\omega^2 r - D\partial c / \partial r\right)\right] / \partial r, \quad (1)$$

where s and D are the sedimentation and diffusion coefficients of the macromolecule. s is defined as the ratio of the macromolecule velocity (cm/s) to the centrifugal field ($\omega^2 r$ in cm/s^2). s is expressed in Svedberg unit S (1 S = 10^{-13} s). For an ideal solution, s and D (cm^2/s), from the Svedberg and Stokes-Einstein equations, are related to the buoyant mass M_b and the hydrodynamic radius R_H (also referred to as the Stokes radius R_S). M_b (thus s) and D also depend on the solvent density ρ and viscosity η.

$$s = M_b D / RT \quad (2)$$

$$D = kT / 6\pi\eta R_H \quad (3)$$

From the combination of the two equations:

$$s = M_b / N_A 6\pi\eta R_H \quad (4)$$

$R = N_A k$ being gas constant, N_A Avogadro's number, and T the absolute temperature. The buoyant molar mass M_b depends on the molar mass, M, and the partial specific volume, $\bar{v}$, of the particle:

$$M_b = M(1 - \rho\bar{v}). \quad (5)$$

The ratio of R_H to the minimum theoretical hydrodynamic radius R_{min} of non-hydrated volume, V, of the particle defines the frictional ratio f/f_{min}:

$$R_H = f/f_{min} R_{min}, \quad (6)$$

V and R_{min} being defined by M and $\bar{v}$:

$$V = (4/3)\pi R_{min}^3 = M\bar{v} / N_A. \quad (7)$$

Non-ideality effects in concentrated samples influence s and, in less extent, D. The sedimentation and diffusion coefficients at infinite dilution, s_0, and D_0, may be derived from the linear approximations (*see*, e.g., [26]). Such effects are often observed with samples of membrane proteins after a concentration step, because detergent micelles are often concentrated with proteins.

The sedimentation coefficient is expressed as $s_{20,w}$ after correction for solvent density and viscosity in relation to the density and viscosity of water at 20 °C ($\rho_{20,w} = 0.99832$ g/mL; $\eta_{20,w} = 1.002$ mPa/s):

$$s_{20,w} = s\left[(1-\rho_{20,w}\bar{v})/(1-\rho\bar{v})\right](\eta/\eta_{20,w}) \quad (8)$$

1.2.2 Sedimentation in a Deuterated Solvent

The subscripts H (generally omitted) and D refer to the hydrogenated and deuterated solvents, respectively. The molecular mass of the particle increases to M_D in the D_2O solvent, by the substitution of labile hydrogen atoms for deuterium ones. For the solvent as for the particle, isotopic substitution changes the masses but not the volume properties of the molecules: R_H is unchanged and $M_D\bar{v}_D = \bar{v}M$, $\bar{v}$ standing for $\bar{v}_H$ and M for M_H. A general formulation for M_b is derived, which is valid in hydrogenated and deuterated solvents. In H_2O, $M_D/M = 1$ (*see* **Note 1**).

$$M_b = M(M_D/M - \rho\bar{v}). \quad (9)$$

1.2.3 SV of Membrane Proteins

Expressing the protein–detergent complex (subscript pd) as composed of protein and detergent (subscripts p and d, respectively), with B_d the amount (g/g) of bound detergent, the buoyant molar mass, and minimum radius for the complex can be written:

$$M_{b\,pd} = M_p\left[\left((M_D/M)_p - \rho\bar{v}_p\right) + B_d\left((M_D/M)_d - \rho\bar{v}_d\right)\right], \quad (10)$$

$$R_{min\,pd} = (3/4\pi N_A)^{1/3}[M_p\bar{v}_p + B_d M_p\bar{v}_d]^{1/3}. \quad (11)$$

Then, combining Eqs. 6, 10, and 11 into Eq. 4:

$$s = M_p[((M_D/M)_p - \rho\bar{v}_p) + B_d((M_D/M)_d - \rho\bar{v}_d)] \\ /[N_A 6\pi\eta\, f/f_{min}(3/4\pi N_A)^{1/3}\,[M_p\bar{v}_p + B_d M_p\bar{v}_d]^{1/3})] \quad (12)$$

s_{20w} is derived from Eq. 12:

$$s_{20w} = s[[(1-\rho_{20w}\bar{v}_p) + B_d(1-\rho_{20w}\bar{v}_d)]/[((M_D/M)_p - \rho\bar{v}_p) \\ + B_d((M_D/M)_d - \rho\bar{v}_d)]][\eta/\eta_{20w}]. \quad (13)$$

*1.2.4 The **c(s)** Analysis of Sedimentation Velocity Data*

We analyze the SV data with the $c(s)$ analysis of the software SEDFIT [27]. Detailed instructions on how to use the software can be found on its website [16]. The analysis considers the solution contains a continuous distribution of a large number of types of particles characterized by s, and quantified by their signal. The $c(s)$ method deconvolutes the effects of diffusion broadening for obtaining resolution distributions. This is approximately but efficiently done by assuming that all species have the same partial specific volume (this is actually not the case because detergent micelle and complexes have different $\bar{v}$), the same shape (i.e., value of f/f_{min}), which gives a relationship between s and D (Eqs. 2–7). s and D are related through the inputs given for $\bar{v}$, f/f_{min} (which can be fitted), ρ and η. The $c(s)$ method allows to obtain a high resolution distribution of sedimentation coefficients. The $c(s)$ method is in its basic form sometimes inappropriate, e.g., when the detergent micelles float (*see* **Note 2**). SEDFIT algebraically accounts for the systematic noise of the experimental data. The different contributions in the $c(s)$ distribution are characterized by their s-values and signals in a very robust way. The superimposition of the $c(s)$ obtained for samples at different concentrations helps to decide if a model of noninteracting species can be considered. In the case of association–dissociation equilibrium, while a symmetrical boundary may be obtained in the $c(s)$ analysis, the mean s-value should increase with the concentration (*see*, e.g., [4, 28–30]). Analysis of interacting systems in SV is possible but requires more sophisticated analysis. Interpretation of the s-values presented here considers noninteracting species, as was assessed for BmrA [25].

1.2.5 The Noninteracting Species Analysis

When there is a small number (one or two) of noninteracting particles, a second type of analysis may be appropriate: The noninteracting species analysis proposed, e.g., in SEDFIT [16], for which independent values of s and D—thus M_b and R_H—are fitted. The s-value is obtained with precision, while the D-value may be overestimated (if there is unconsidered heterogeneity) or underestimated (for concentrated nonideal samples).

1.2.6 Analysis of Absorbance and Interference Signals

In addition, the $c(s)$ and noninteracting species analysis characterize the signals, in absorbance, A, or fringe shift, ΔJ, for each sedimenting species. A and ΔJ are related to concentration, c, in mg/mL, optical path length, l (cm), extinction coefficient $E_{0.1\%}$ (Absorbance for 1 mg/mL and $l = 1$ cm) and increment of index of refraction, $\partial n/\partial c$ (mL/g), of the particle, and laser wavelength ($\lambda = 6.75 \times 10^{-5}$ cm or 6.55×10^{-5} cm for Beckman ultracentrifuges):

$$A = E_{0.1\%}\, l c. \tag{14}$$

$$\Delta J = (\partial n/\partial c\, /\, \lambda) l (c\, /\, 1{,}000). \tag{15}$$

An estimate for B_d of the protein–detergent complex can be obtained from the combination of A and ΔJ. Since $B_b c_p$ is the concentration of bound detergent:

$$A = \left(E_{0.1\%p} + B_d E_{0.1\%d}\right) l c_p. \tag{16}$$

$$\Delta J = ((\partial n/\partial c_p + B_b \partial n/\partial c_d)\, /\, \lambda) l \left(c_p\, /\, 1{,}000\right). \tag{17}$$

The global analysis of absorbance and interference data in the software SEDPHAT [31] was described for the case of membrane proteins in [11] to provide directly the concentration of detergent and protein in the $c(s)$ profiles.

B_d is an important characteristic of the protein–detergent complex. But the values of the extinction coefficients and of the refractive index increments for p and d needed for determining B_d from A and ΔJ are sometimes known with uncertainties. Detergent very often absorbs slightly in the UV-range, due to chemical degradation, or to the fact their micelles solvate absorbing impurities. The refractive index increment varies depending on the type of detergents (see the paragraph below). Its value may be unknown and affected by the presence of bound lipids. Bound ligand may change the extinction coefficient of the protein. Of course the uncertainties in the extinction coefficients and refractive increments affect the precision on B_d. For this reason, we propose in the next paragraph a procedure which analyzes sedimentation velocity data obtained in different solvent densities. Note that size exclusion coupled to online refractive index, light scattering, and absorbancy detectors, a complementary technique to AUC for characterizing membrane protein homogeneity and association state [13, 15, 32], requires the same input values, the uncertainties of which affect in the same way the precision on B_d, and also the derived molar masses.

1.2.7 Analysis of SV Experiments of Membrane Proteins in One Solvent

The sedimentation coefficient s is related to M, B_d, and R_H (Eqs. 4 and 12). Experimental values of R_H may be available. Then s and R_H may be combined to calculate M_b, the value of which may be compared to theoretical values for different association states and B_d, in order to determine or reject a given association state. Alternatively, considering different association states and B_d, R_H may be calculated with hypotheses on the shape factor and, with M_b, used to calculate theoretical values of s for the same purpose.

1.2.8 Combined Analysis of SV Experiments in Hydrogenated and Deuterated Solvents

From Eq. 9, equilibrium sedimentation experiments—which provide M_b, when performed in D_2O and H_2O solvents (of different densities), allow the determination of the partial specific volume and mass of the sedimenting particle [33, 34]. The analysis assumes its composition is the same in D_2O and H_2O solvents. From Eqs. 4 and 9 and assuming in addition that shape does not change, SV allows the same. The $c(s)$ analysis allowing to discriminate the different species in solution makes now SV efficient for this purpose [10, 19, 35]. The errors associated to this determination however depend on the relative density of the particle and solvent and are not determined trivially [19]. For this reason, we have proposed a graphical procedure [14], which was used to decipher the controversial association state of two different membrane proteins [14, 15].

In Eq. 12, s is expressed as a function of the molecular characteristics of the protein–detergent complex. The unknowns are the association state, f/f_{min}, B_d (or B_{d+l}), and very often $\bar{v}_d$ ($\bar{v}_{d+l}$) if there are lipids or any bound compound. The text below will be written with the subscript d only for clarity but is applicable to d + l (even hydration may be considered), the unknown content in lipid in membrane protein being one motivation of this approach. The protocol uses a graphical representation to account for the experimental possibilities of detergent/lipid binding and respond to the question: Are the sedimentation coefficients measured in H_2O/D_2O compatible with a given association state?

Equation 12 should be satisfied in different solvents. s is determined experimentally in each solvent. But M_p can be only a linear combination of the mass of the polypeptide chain(s) present in the sample (for glycosylated protein, *see* **Note 3**). f/f_{min} lies within a narrow range except for an unfolded or very anisotropic particle (see below: numerical values). $(M_D/M)_p$ and $(M_D/M)_d$, $\bar{v}_p$, ρ and η, can be estimated from chemical composition or measured (see below: numerical values) Data analysis is made according to Eq. 12, considering different hypothesis of M_p (i.e., the association states) for the protein part and of the shape factor f/f_{min}. If the oligomeric state (M_p) and shape factor (f/f_{min}) are appropriately chosen, a unique pair of (B_d; $\bar{v}_d$) values must account for sedimentation in H_2O and D_2O buffers. We used the software *Maple* to draw the implicit function $B_d = f(\bar{v}_d)$, based on different assumptions as to the magnitude of f/f_{min} and aggregational state. For each inputs of M_p and f/f_{min}, a figure is plotted on which four curves $B_d = f(\bar{v}_d)$ are superimposed, based on the limiting values of the experimental sedimentation coefficients in the hydrogenated and deuterated solvents, thus a total of four s-values: $(s_H + \Delta s_H)$; $(s_H - \Delta s_H)$; $(s_D + \Delta s_D)$; $(s_D - \Delta s_D)$. The areas delimited by the four curves define all possible mathematical solutions for $B_d = f(\bar{v}_d)$ that are compatible with the experimental data. Among them, we have to exclude some irrelevant ones ($\bar{v}_d$ is between the value for the pure detergent and that expected for a reasonable content of lipid; B_d is related to the transmembrane area of membrane proteins [5]). This procedure gives a more precise analysis for membrane protein complexes with a large value of $1/\bar{v}$ value, i.e., close to water (for reasons underlined in [19]). It was very efficient in the study of membrane proteins binding lipids and detergents with $\bar{v}$ in the range 0.92–1.06 [14, 15]. The protocol is presented below in the less favorable case of BrmA complexed with DDM.

1.2.9 Numerical Values

Molar mass for protein are determined from amino-acid sequence. An estimation of the total mass of the glycosylated polypeptide can be obtained by mass spectroscopy (*see*, e.g., [15]). $(M_D/M)_p$ can be calculated from amino-acid sequence too. We calculate for BmrA, for the typical value of 80 % of exchange for the labile hydrogens, in 100 % D_2O, a value of 1.012, which is very usual and can be considered as a typical value. For detergents the chemical structure should be considered, with 100 % of exchange for the labile hydrogens in 100 % D_2O.

The partial specific volume is calculated from the amino-acid sequence and tabulated data, for example, in SEDNTERP [36] or in the calculator menu of the software SEDFIT [16]. A typical value is 0.74 mL/g. We consider $\bar{v} = 0.63\text{ml} / \text{g}$ for the glycosylated moieties of glycoprotein according to [37], while the precise value depends on the sugar type as seen, e.g., in the tabulated data given in SEDNTERP [36]. Values for the detergents depend on their chemical structure [5, 38, 39]: for DDM: $\bar{v}$ is 0.82 mL/g [10] (0.81–0.837 mL/g from [5]); for $C_{12}E_8$: 0.95 mL/g [10]; for LAPAO that floats in aqueous solvents: 1.002 mL/g [14], fluorinated surfactants are more dense, with $\bar{v} \approx 0.6\,\text{mL/g}$ [40]; for some amphiphilic polymers, we measured $\bar{v} \approx 0.8\,\text{mL} / \text{g}$ [19, 41]. For lipids, the value for $\bar{v}$ of 0.981 mL/g reported for egg yolk phosphatidylserine [38] is often used, while $\bar{v}$ depends on the type of lipids, *see*, e.g., [38]; $\bar{v} = 1.06\,\text{mL/g}$ was calculated for the lipids of the purple membrane [42]. $\bar{v}$ can be measured by density measurements on anhydrous material (*see*, e.g., [10]), by SV experiments in D_2O and H_2O (*see*, e.g., [19]), or calculated from chemical structure [43, 44].

The extinction coefficient is calculated for the proteins from amino-acid composition. The softwares given above for $\bar{v}$ can be used for this purpose. Regarding the refractive index increment, for the protein component, $\partial n/\partial c$ varies few with amino-acid composition, a value of 0.187 mL/g is generally considered for membrane proteins [32]. For proteins of small size or of atypical composition, for which atypical values of $\partial n/\partial c$ are possible [45], $\partial n/\partial c$ can be calculated from amino-acid composition, at the wavelength of the laser (655 or 675 nm depending on the Beckman XLI models) in the calculator menu of the software SEDFIT [16]. For polysaccharides, we consider 0.15 mL/g, the value for dextran from [46]. For detergents, $\partial n/\partial c$ varies in a quite large range: from 0.07 to 0.18 mL/g from the detergent manufacturer Anatrace [47], from refs. [32, 48–50], and from our measurements [10, 14, 15]. For fluorinated surfactants, *see* [40, 51] and for amphiphilic polymers [19, 41, 52]. For lipids in water, we consider 0.1 or 0.12 mL/g (the value given for dimyristoyl phosphatidylglycerol by [53]). $\partial n/\partial c$ can be measured by sedimentation velocity (*see*, e.g., [10]) or light scattering [50].

The value for f/f_{min} is moderately sensitive to shape, except for unfolded non-compact proteins [54]. A typical value for globular compact particles is 1.25, 1.5 for a moderately anisotropic shape, 1.8 for a significative anisotropic shape (*see*, e.g., [55]). Glycosylated proteins are characterized by larger values, such as 1.5–1.8, for a moderate asymmetrical shape (*see*, e.g., [15]).

2 Materials

1. The protein solution in volume and concentration required for SV measurements in H_2O and D_2O buffers, and with two protein concentrations (*see* **Notes 4** and **5**). Our study used two frozen aliquots of 410 μL each of BmrA at 2.5 mg/mL as determined by Bradford assay. BmrA was previously purified in DDM as previously described [25]. The solvent was Buffer H: 50 mM Tris-HCl pH 8, 150 mM NaCl, 10 % glycerol, 0.02 % DDM.
2. 50 mL (a minimum of 15 mL) of the hydrogenated and deuterated solvents (Buffer H and Buffer D). The solvent should be well defined and preferentially contain salt above 100 mM (*see* **Note 6**). It will be used for the solvent exchange, SV, density, and viscosity measurements.
3. 2 mL (a minimum of 0.5 mL) of hydrogenated and deuterated solvents but without detergent.
4. PD MiniTrap G-25 column (GE Healthcare) to exchange the solvent toward Buffer H and Buffer D, and the GE Healthcare instructions 28-9225-29 AB, describing the used gravity protocol.
5. A density-meter (DMA5000 Anton Paar).
6. A viscosity-meter (AMVn Anton Paar) with associated software Anton Paar VisioLab.
7. An analytical ultracentrifuge (Optima XLI Beckman) with associated software Beckman Proteom Lab v 6.0. A rotor (8-hole AnTi-50, Beckman) (*see* **Note** 7).
8. AUC cell assemblies equipped with sapphire windows and 2-channel Epon Synthetic Boundary capillary-type (Beckman) or Titanium, 12 mm optical path length (Nanolytics) (*see* **Note** 7) with notice for cell montage (Beckman, provided with the rotor).
9. Counterbalance adapted for Beckman analytical ultracentrifuge, provided by Beckman. When using Titanium 12 mm optical path length centerpieces provided by Nanolytics, we use a heavier counterbalance provided by Nanolytics. Note that radial calibration is done once a week using Beckman counterbalance.

10. The software SEDNTERP created by D. Hayes, T. Laue, J. Philo, and available free [36] for calculating the parameters relevant to SV analysis (*see* **Note 8**).
11. The software SEDFIT created by P. Schuck and available free [16], for the analysis of AUC experiments (*see* **Note 9**).
12. A UV spectrophotometer in order to check the absorbance of the stock samples (e.g., Biophotometer Eppendorf).
13. The amino-acid sequence as a text file.
14. Homemade Excel sheets for all simple calculations.
15. We use the Maple software for the graphical analysis of SV for the membrane protein in Buffer H and Buffer D.

3 Methods

3.1 Prior to the Experiment

1. Use the software SEDFIT for calculating, from the amino-acid composition of the protein, $M = 64{,}519$ Da, $\bar{v} = 0.751 \text{ml}/\text{g}$, $\partial n/\partial c = 0.185$ mL/g, and the extinction coefficient at 280 nm, $E_{0.1\%,\,280} = 0.602$ mg/mL cm. We calculate for BmrA, for 80 % of exchange for the exchangeable H, in 100 % D_2O, $(M_D/M)_p = 1.012$. Compile what is known about complex composition and hydrodynamic properties of the membrane protein complex. For BmrA in DDM, we know from size exclusion chromatography, [^{14}C]DDM-binding and phospholipid quantification, that $R_H = 5.6 \pm 4$ nm, DDM and phospholipid binding are 1.6 ± 0.6 g/g and 0.07 g/g, respectively [25].
2. Find in literature or measure (*see* Subheading 1.2.9) $\bar{v}$, $E_{0.1\%,\,280}$ and $\partial n/\partial c$ for the detergent, and optionally the aggregation number for an estimate of the molar mass of the micelle. For DDM: $\bar{v}$ is 0.82 mL/g, $E_{0.1\%,\,280} = 0$ mg/mL cm, $\partial n/\partial c = 0.143$ mL/g, $M_{\text{micelle DDM}} = 67$ kDa [10]. DDM has seven exchangeable hydrogens, from which we calculate $(M_D/M)_d = 1.014$.
3. Decide the samples to test, i.e., the dilutions to be made, the AUC centerpieces to be used, given their path length (l) and the sample volume (V) (*see* **Note 5**). In the present case, in each buffer, from 410 µL of frozen protein at 2.5 mg/mL, we expect after the solvent exchange step (*see* below Subheading 3.2, **steps 3** and **4**) 1.2 mL at 0.85 mg/mL. We decide to measure in SV one undiluted sample with $l = 1.2$ cm, $V = 420$ µL providing $A_{\text{init}} = 0.602 \times 1.2 \times 0.85 = 0.6$ and one sample diluted three times, prepared by mixing 140 µL of the undiluted sample and 280 µL of buffer, providing $A_{\text{init}} = 0.2$ with $l = 1.2$ cm. We decided to duplicate the measurements for the two dilutions, in case of cell linkage or if the communication between the two

channels of the boundary cell was blocked—which happens hopefully very rarely. The four samples will be loaded in cells equipped with sapphire windows and 2-channel Epon Synthetic Boundary capillary-type centerpiece, and positioned in two sets of opposed holes in the 8-hole rotor. In addition, we will measure in a classical 12 mm two-channel centerpiece, the sedimentation of detergent micelle, the sample and reference channels being filled with the same volume (400 μL) of buffers with and without detergent, respectively. This cell will be placed opposite to the calibration cell in the 8-hole rotor.

4. Decide the temperature that will be used for the SV experiments: here, 20 °C, because our sample is stable at that temperature (*see* **Note 10**).

5. Prepare 50 mL of the hydrogenated and deuterated solvents (Buffer H and Buffer D). The solvent should be well defined and preferentially contain salt above 100 mM (*see* **Note 6**). We prepared them in our study from stock solutions: 1 M Tris–HCl pH 8.0; 4 M NaCl; 10 % DDM; 100 % glycerol; Ultrapure MilliQ water (for Buffer H) or 100 % D_2O (for Buffer D). Buffer H is 50 mM Tris–HCl pH 8.0, 150 mM NaCl, 0.02 % DDM, and 10 % glycerol. For its preparation, add 2.5 mL of 1 M Tris–HCl pH 8.0, 1.875 mL of 4 M NaCl, 5 mL of 100 % glycerol, mix and make up to 50 mL with water. Filter with a 0.2 μm disposable filter. Important: before adding the detergent take a 1–2 mL aliquot for investigating the behavior of DDM alone in AUC. After taking the aliquot add 0.1 mL of 10 % DDM, mix and store at 4 °C. Buffer D is 50 mM Tris–HCl pD 8.0 (pH 7.59), 150 mM NaCl, 0.02 % DDM, and 10 % glycerol. To minimize the dilution of the D_2O directly dissolve 303 mg Tris base and 438 mg NaCl in about 30 mL D_2O, add 5 mL 100 % glycerol and mix. Adjust pH to 7.59 with HCl (Note: pH 7.59 = pD 8.0). Mix and make up to 50 mL with D_2O. Filter with a 0.2 μm disposable filter. Take an aliquot of 1–2 mL for investigating the sedimentation of DDM in the deuterated buffer, then add 0.1 mL of 10 % DDM, mix and store at 4 °C.

6. Measure the density and the viscosity of the solvents at 20 °C: Buffer H: $\rho = 1.039$ g/mL, $\eta = 1.558$ mPa/s (1 mPa/s = 1 cP (cp)) and Buffer D: $\rho = 1.132$ g/mL, $\eta = 1.796$ mPa/s. SEDNTERP [36] give estimates of the solvent density and viscosity from tabulated data in some buffers.

7. Simulate the sedimentation of our system. First: Estimate *s* for the DDM micelle and considering a globular complex, with 1 g/g bound DDM to the protein with different association states. The mass of the complex is twice that of BmrA: thus ≈130 kDa for BmrA as a monomer (65 kDa), ≈260 kDa for

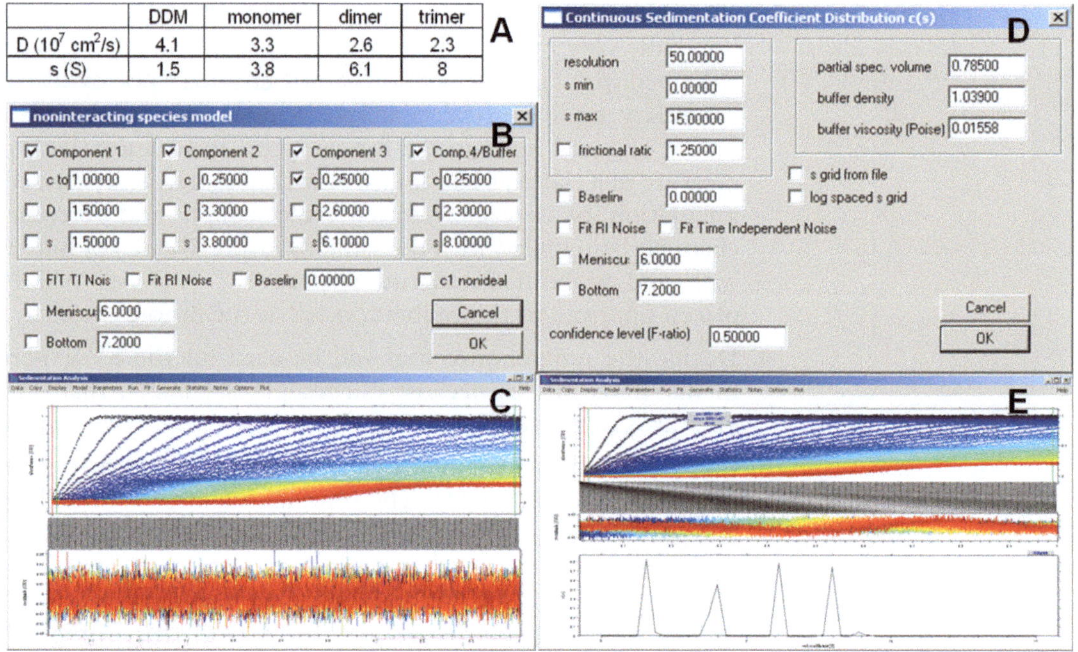

Fig. 3 SV simulations. (**a**) Diffusion and sedimentation used for the simulation. (**b**) SEDFIT window for generating the SV profiles. (**c**) Generated profiles. (**d**) SEDFIT window for the parameters of the *c*(*s*) analysis. (**e**) Result of the *c*(*s*) analysis

the expected dimer, ≈390 kDa for a trimer; we consider for the complex a mean $\bar{v}$ value of 0.785 mL/g, the mean of the protein and the detergent values. For DDM micelles, we consider $M = 60$ kDa, $\bar{v} = 0.82$ mL / g. Eqs. 3–7 are used to calculate *s*- and *D*-values, the other inputs being solvent density and viscosity of the hydrogenated buffer; temperature and frictional ratio, $f/f_{min} = 1.25$, i.e., a globular compact shape. We use homemade excel sheets. We can also use the calculation facilities in the software SEDFIT, as detailed in [54]. The calculated values are reported in Fig. 3a.

8. Second step of the simulation: Use SEDFIT to simulate the sedimentation in our standard rotor speed for SV experiments of 42,000 rpm (revolution per minute), i.e. 130,000 × *g* with the rotor used. Open SEDFIT; *generate/single/fit M and s instead for s and D*: no/*dr* = $1e^{-3}$: OK/*rotor speed*: change to 42,000 rpm/*simulate…*: yes/*acceleration…*: OK/*Time interval of scan (sec)*: write 600 (i.e., 10 min)/*number of simulated scans*: write 60 (i.e., 10 h)/*std of noise* = $1e^{-2}$: OK. Fill the parameter box Fig. 3b: component box should be marked to be integrated in the simulation; *c* is the total signal for component 1 and the fraction of the total signal for the other components. *D* and *s* are filled according to Fig. 3a. Set meniscus = 6.0, then: OK; create a temporary folder and save the generated

data; a window opens showing in the top panel the superimposition of simulated sedimentation profiles: the vertical and horizontal axis are the concentration (stated as absorbance) and radial position, respectively. Drag the vertical green line at right to $r \approx 7$ cm/*Enter*. Change the scale: *display/data range*. Copy screen is shown in Fig. 3c. The SV profiles, at 42,000 rpm, show four boundaries distinguishable by eyes. This is a particularly favorable situation, while however the *c*(*s*) analysis would deconvolute broad boundaries corresponding to non-separated sedimentation boundaries for different species, the sedimentation duration (overnight) is appropriate, as well as the rotor speed. Note the maximum rotor speed is imposed by the centerpieces (*see* **Note 7**).

9. Third step of the simulation, actually optional, but here used as an introduction to the *c*(*s*) analysis: the *c*(*s*) analysis of the software SEDFIT is applied to the simulated data. In the menu line of SEDFIT, *Model/Continuous c(s) distribution*. A window opens with the parameters of the *c*(*s*) analysis (Fig. 3d). We will consider 50 species (in this rapid analysis) sedimenting between 0 and 15 S, denominated as *Resolution*, *s min* and *s max*. The *partial spec. volume*, *buffer density*, and *viscosity* (in Poise) are changed to 0.785, 1.039, and 0.01558, respectively. *Frictional ratio* is set at 1.25. *Meniscus* and *bottom* should be at 6.0 and 7.2 cm according to our simulation. The *confidence level (F-ratio)* is set at 0.5 (*see* **Note 11**); corresponding to the fact we will not use regularization procedure for the data here. Once the parameters are changed, close the window (*OK*), and use *Run* in SEDFIT menu (*see* **Note 12**). Two message windows appear (type *OK* and *OK*) before the result of the *c*(*s*) analysis shown in Fig. 3e. Four sedimenting species are identified, as expected.

3.2 Experiment

1. Unfroze the protein solution and dilute it in Buffer H to have a final volume of 1 mL.
2. On the first day, do the SV experiment in H_2O buffer.
3. Exchange the solvent, for half of the protein solution, using PD MiniTrap G-25 column, according to the protocol of the purchaser (reference in Subheading 2, **item 4**) in cold room. Equilibrate the column with Buffer H (8 mL), deposit 0.5 mL of the protein solution, and discard the flow-through. Then add 1 mL of Buffer H and recover the eluate containing the protein.
4. Add 0.2 mL of Buffer H to the recovered protein to have a final volume of 1.2 mL. Prepare the three-times diluted samples by mixing 150 μL of the protein sample with 300 μL of Buffer H and store the undiluted and diluted samples in Buffer H at the appropriate temperature (here 4 °C).

5. Equilibrate the centrifuge at 20 °C.
6. Prepare the AUC cell assemblies. Details concerning cell assemblies and practical steps for cell montage and rotor preparation are described as the video and technical notice that can be downloaded [56, 57]. We prepared one cell equipped with Ti-centerpiece with $l = 1.2$ cm (for following the sedimentation of the detergent micelles in the buffer) and four cells equipped with Epon Synthetic Boundary capillary-type centerpieces with $l = 1.2$ cm (for undiluted and diluted samples in duplicate), and comprising sapphire windows, allowing measurements with absorbance and interference optics. Fill the sample compartments with 420 μL of sample, and the solvent compartment with 440 μL of Buffer H for the boundary centerpiece. For the usual two-channel centerpiece, fill the two compartments with 400 μL.
7. Once the cells are closed, check that cells that will be placed in opposite positions in the rotor have the same weights: the four 12 mm Epon synthetic boundary cells, and the 12 mm Titanium cell and the counterbalance (required for instrumentation calibration). Place the cells in the rotor. Put the rotor into the analytical ultracentrifuge, and initiate vacuum and temperature equilibration at 20 °C. Start the software Beckman Proteom Lab. Give a title for each cell and a—common—localization for the generated data. Start the AUC at 3,000 rpm, i.e. 665 × *g*. For each of the cells, check that nice interference patterns are obtained, and if not, adapt the laser delay and the laser duration: typically 0.1° (in previous version of the software Proteom Lab, the brightness and/or contrast have also to be adjusted). Acquire for each cell a wavelength scan between 240 and 600 nm at a radial position of 6.5 cm (the mean radial position of the compartments) (*see* **Note 13**). Check that the absorbance corresponds to expectation. In the specific experiment described here as an example, it was not the case. We measure an absorbance of 1.8 for the undiluted sample and 0.6 for the diluted one. These values are larger than expected. We thus decided to measure SV at 280 and 405 nm (the analysis at 405 nm will not be detailed here). Start for all cells a radial scan at 280 and 405 nm, using interference for control (*see* **Note 14**).
8. Start an SV experiment to equilibrate the meniscus of the boundary cells, at 7,000 rpm, i.e. 3,600 × *g*. Check after about 30 min on the sedimentation profiles that the levels of the two channels equilibrate in the synthetic boundary cells: the position of the two meniscuses—the interfaces between air and the solution, followed in absorbance as a negative spike in the reference solvent compartment and a positive spike in the sample compartment—should reach the same position. Stop the

data acquisition and centrifugation. After breaking the vacuum, remove the rotor and rotate it to provoke the displacement of an air bubble within the sample compartment in each cell. Replace the rotor in the analytical centrifuge, and initiate vacuum and temperature equilibration at 20 °C. Wait for temperature equilibration before the SV experiment, but at least 30 min, irrespective of the temperature reading (*see* **Note 15**).

9. Increase the angular velocity of AUC to 42,000 rpm. Once the rotor speed is reached, check the interference pattern. If noisy, adapt the interference parameters (laser delay and laser duration: typically 0.1°) and start SV measurements overnight (*see* **Note 16**).
10. The next morning (day 2), stop the data acquisition and centrifugation. After breaking the vacuum, remove the rotor, set off AUC, and remove the cells from the rotor. Rotate the cells to provoke the displacement of an air bubble within the sample compartment if the samples are to be recovered for analytical purpose. Dismount and clean the cell assemblies.
11. Zip the raw data and duplicate them for saving and analysis.
12. On the same day, do the solvent exchange and SV experiment in D_2O buffer (Subheading 3.2, **steps 3–11**).

3.3 Data Analysis

1. Open the software SEDFIT and select (*see* **Note 17**), starting with the more concentrated sample at 280 nm, a set of 20–30 SV profiles corresponding to the whole sedimentation process (*see* one example in Fig. 4a, top). We select all profiles for 6 h of sedimentation, plus one over two profiles for the 3 next hours (a total of 30 profiles. Fix the meniscus (air-sample interface) and bottom position (red and blue lines) and the radial limits for the fit (green lines), avoiding the pellet region at the bottom of the cell: maximum limit for the fit is typically between 7 and 7.1 cm).
2. Analyze the loaded set of SV profile in the model of the $c(s)$ analysis. We first consider (as in Subheading 3.1, **step 9**, see also the related notes) 50 particles, in the range 0–15 S without regulation procedure, with $f/f_{min} = 1.25$ corresponding to compact particles, and fitting the systematic noises and baseline. Partial specific volume is the mean value between the protein and detergent, solvent density and viscosity are those measured or estimated in Subheading 3.1, **step 6**. f/f_{min} and the meniscus are not fitted, we use the command *run*. From the results shown, e.g., in Fig. 4a, we check that the s-range is appropriate, with no indication of species above 15 S and a reasonable quality of fit as indicated by the superposition of the experimental curves and those resulting from the analysis. A second analysis is done with the same parameters but fitting f/f_{min} and the meniscus. We thus use the command *Fit*.

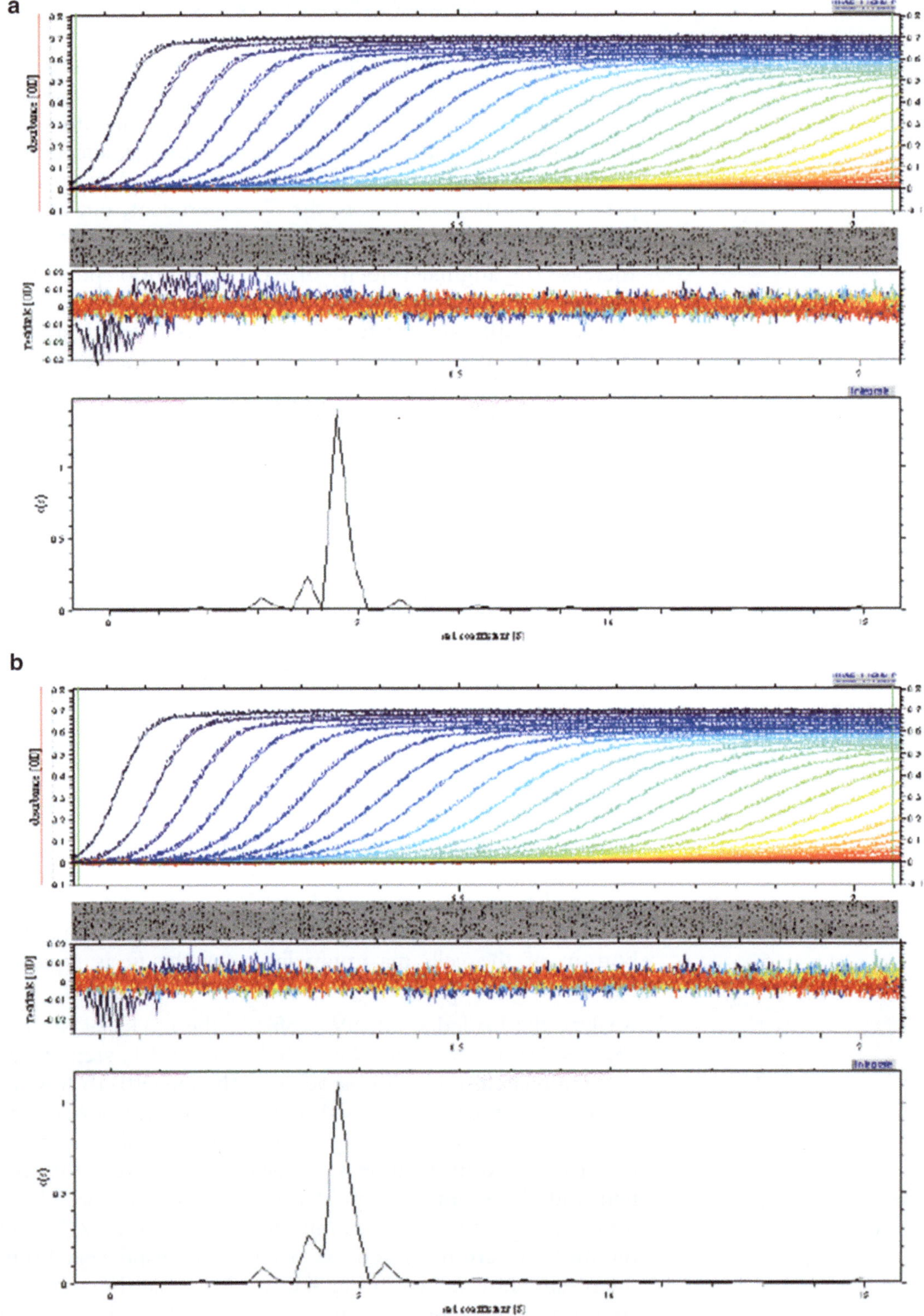

Fig. 4 Fitting f/f_{min} and meniscus in the $c(s)$ analysis. Copy screens for the first and second steps we propose for the $c(s)$ analysis of SEDFIT for BmrA diluted three times in Buffer H. (**a**) With $f/f_{min} = 1.25$. (**b**) Meniscus and f/f_{min} fitted. Note the improvement of the quality of the fit, while f/f_{min} and meniscus are only slightly changed upon fit, e.g., fitted $f/f_{min} = 1.23$

The quality of the fit (Fig. 4b) increased slightly as attested by the better superposition of the fitted and experimental SV profiles (top panels) and the observation of the residuals (middle panels), the significant decrease of the root mean square deviation (rmsd). The details of the bottom panels are however only slightly different. The fitted value of f/f_{min} may compensate the poor precision on the inputs for the partial specific volume, density, and viscosity. It has not been considered as a relevant value for characterizing the complex.

3. Then, the final $c(s)$ analysis is made considering 200 particles and a regularization (F-ratio of 0.68) leading to a more regular $c(s)$ distribution and avoiding possible irrelevant details. We typically do not fit f/f_{min} and the meniscus at that stage. We thus use the command *Run*. Check that the result of the analysis is reasonably good. Save (by copy screen) the parameters of the fits (as in Fig. 3d) and the resulting figure (here shown in Fig. 5c) in your electronic lab book. Copy the $c(s)$ data in a spreadsheet software (e.g., Excel). Integrate the main peaks to obtain mean values for s, signals, and percentages.
4. Apply **steps 2** and **3** for all cells at 280 nm and with interference optics.
5. The resulting screens at that stage are shown for the four investigated samples (undiluted/diluted in Buffer H/D) on the left panels of Fig. 5 for data acquired at 280 nm and on the right panels for data acquired with interference optics.
6. Analyze also the buffers in the same way. The related analyses are shown in Fig. 6.
7. Compare the results by superposing the $c(s)$ (Fig. 7).
8. Summarize the data from the $c(s)$ analysis (Table 1).
9. Comment the sedimentation of the detergent micelles from the data for the buffer with detergent in the sample channel, and the buffer without detergent in the reference channel. Fig. 6 shows that detergent is not detected at 280 nm as expected for DDM. The sedimentation of the micelles is clearly detected using interference. The fit for the detergent is reasonably good even if the detergent micelle concentration is low. In addition, the solvent contains glycerol while the volume of the reference solvent and sample (solvent *plus* detergent) in the two channels of the centerpiece is not strictly the same. The experimental s-values for DDM micelles are 1.56 S in Buffer H and 0.70 S in Buffer D, which, considering the parameters of DDM, makes s_{20w} of 3.0 (Eq. 8) and 2.7 S (Eq. 13, with $B_d = 0$), respectively. This corresponds grossly to the literature value of 3.1 S [10]. The detergent samples are here only controls, since the protocol for the protein sample is defined to mask the contribution of detergent micelles.

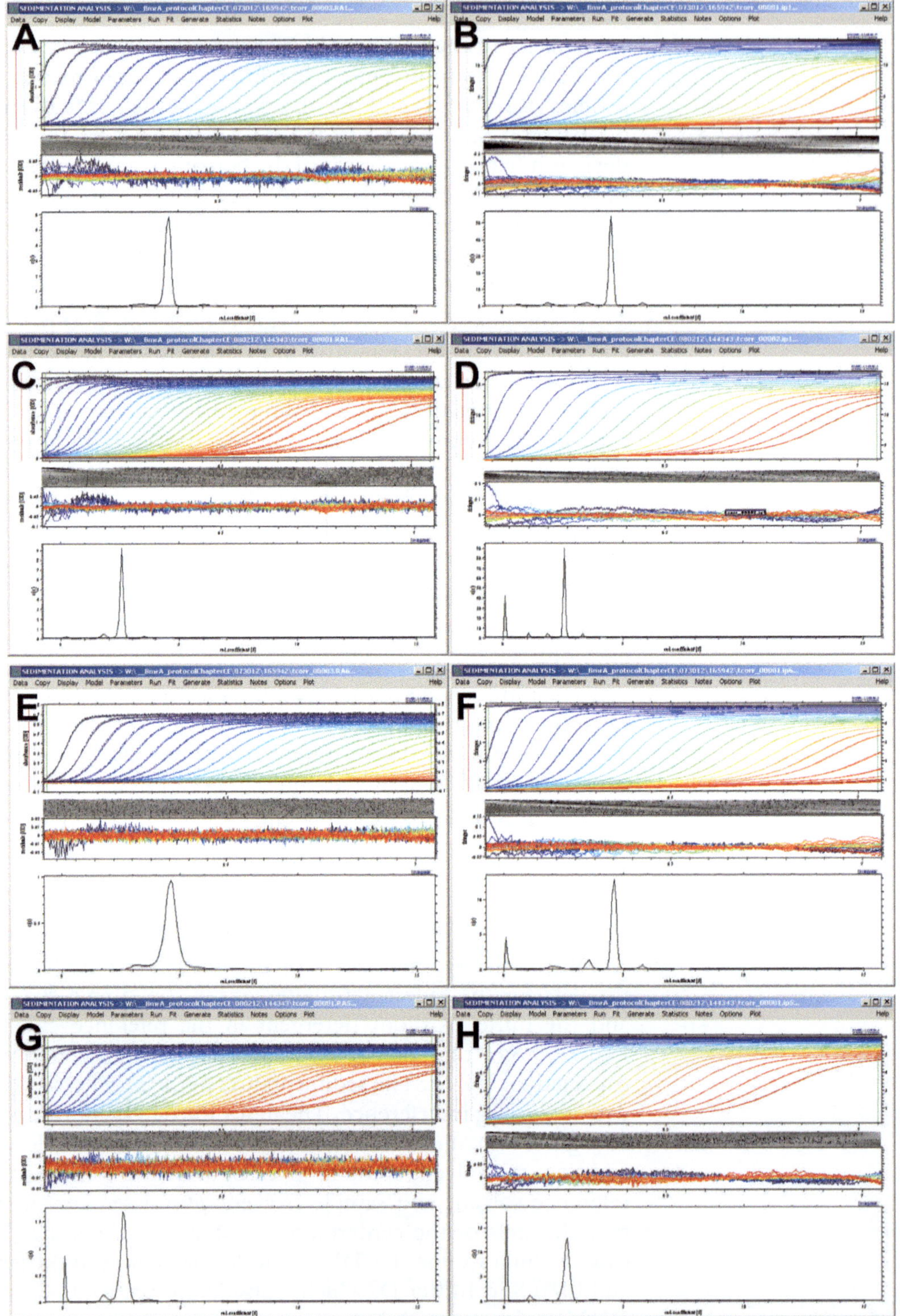

Fig. 5 Sedimentation velocity *c*(*s*) analysis of all BmrA samples. Print screens were obtained from the analysis with the software SEDFIT. Panels (**a**, **c**, **e**, **g**): A_{280} data. Panels (**b**, **d**, **f**, **h**): Interference data. Panels (**a**, **b**) and (**c**, **d**): undiluted BmrA in Buffer H and Buffer D, respectively. Panels (**e**, **f**) and (**g**, **h**): diluted three times BmrA in Buffer H and Buffer D, respectively

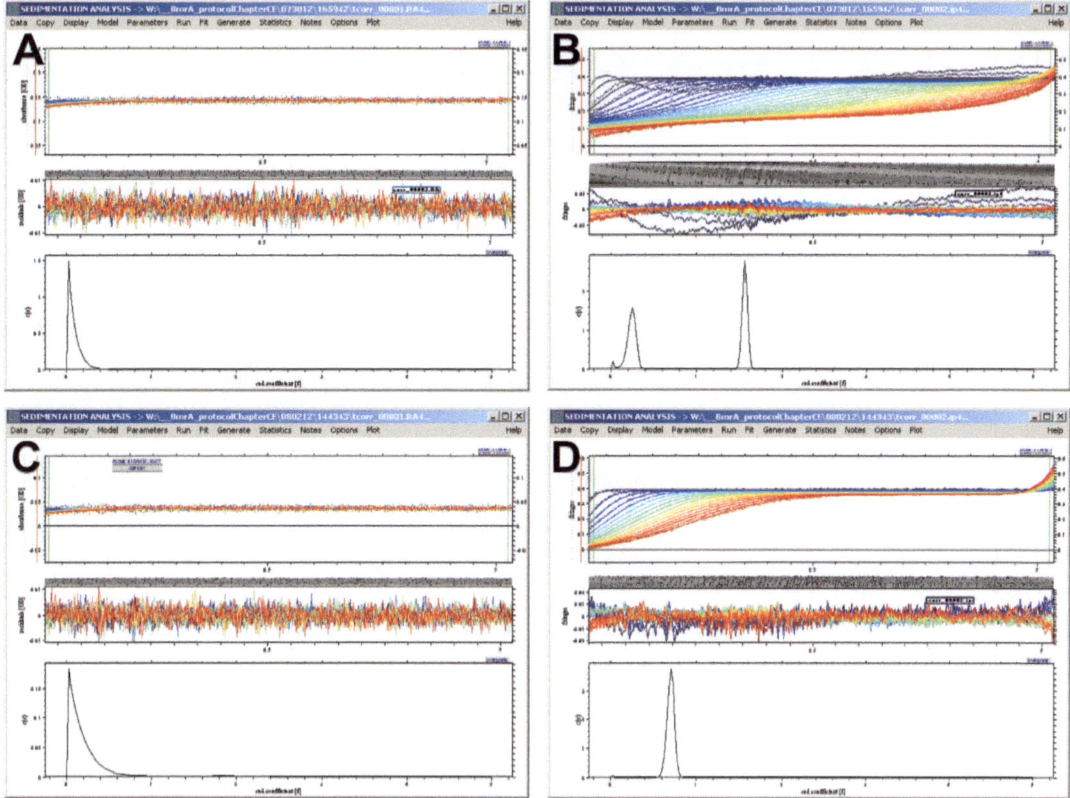

Fig. 6 Sedimentation velocity *c*(*s*) analysis of detergent micelles. Sample channels contain Buffer H or D (with detergent) and reference channel the related buffer but without detergent. Panels (**a**, **b**): Buffer H, data acquired at 280 nm and with interference, respectively. Panels (**c**, **d**): Buffer D, data acquired at 280 nm and with interference, respectively

10. Examine the results of the *c*(*s*) analysis for the BmrA samples (Fig. 5 and Table 1). The peak at *s*-values close to zero for the protein samples in Fig. 5d, f, g, h results most probably from a poor deconvolution of the baseline and are not considered further. The sedimentation of the detergent micelle (paragraph above) is not detected in BmrA samples, meaning that there is no observable excess of detergent in these samples when compared to the solvent. All BmrA samples have one main contribution (>80 %) in addition minor species sedimenting slower or faster. The main species is detected at the same *s*-value at 280 nm and with interference optics. The sedimentation is slower in Buffer D when compared to Buffer H, as expected from the larger density and viscosity of Buffer D. The proportions of the larger species and/or the mean *s*-values are expected to increase in the case of equilibrium of association. Analyze the changes of the percentage of the different species with concentration. They do not change. Analyze the slight changes in the *s*-values for the main peak

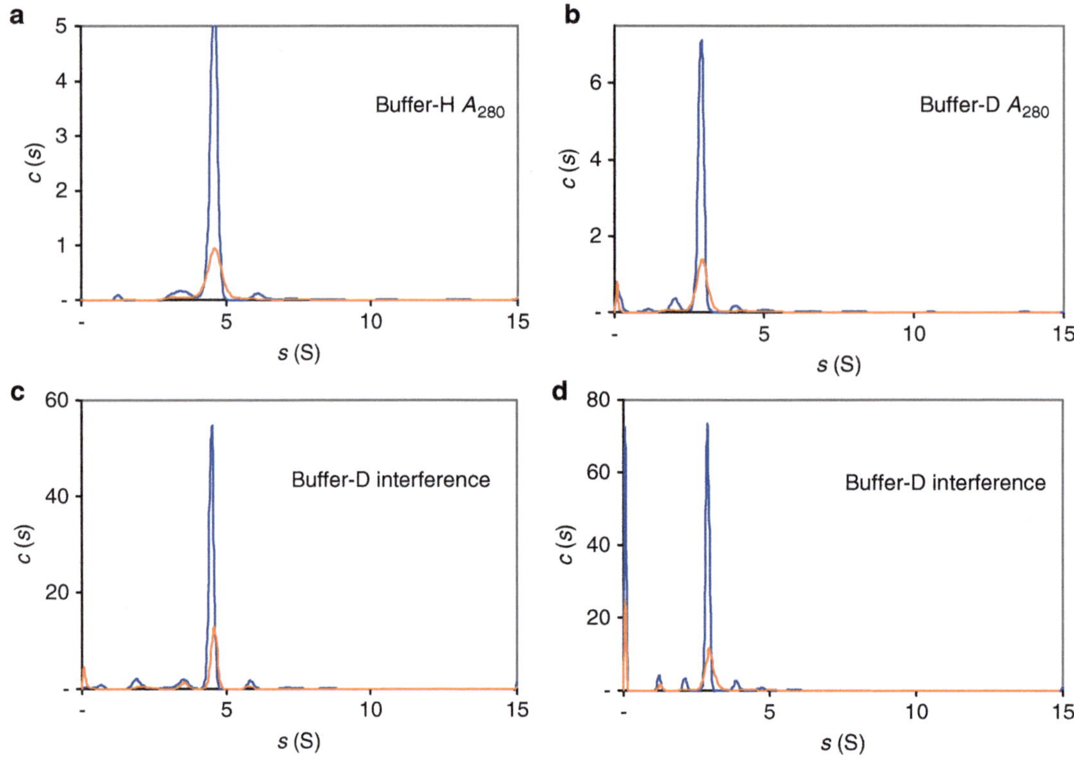

Fig. 7 Superposition of *c*(*s*) distribution. Superposition of *c*(s) distribution for undiluted BmrA (*blue*), BmrA diluted three times, in absorbance at 280 nm (Panels (**a** and **b**)) and in interference optics (Panels (**c** and **d**)) in hydrogenated (Panels (**a** and **c**)) and deuterated (Panels (**b** and **d**)) solvents

Table 1
***c*(*s*) Analysis of BmrA in Buffer H and Buffer D**

	Smaller species				Main peak					Aggregates	
	A_{280}		ΔJ		A_{280}			ΔJ		A_{280}	ΔJ
	s (S)	% A_{280}	*s* (S)	% ΔJ	*s* (S)	A_{280}	% A_{280}	*s* (S)	% ΔJ	% A_{280}	% ΔJ
Undiluted Buffer H	3.7	7	3.4	7	4.56	1.80	87	4.49	86	6	7
Dilution 3 Buffer H	3.2	6	3.5	10	4.57	0.61	86	4.58	82	8	8
Undiluted Buffer D	1.8	7	1.8	5	2.54	1.78	87	2.54	87	6	9
Dilution 3 Buffer D	1.7	6	1.1	7	2.58	0.59	84	2.60	85	10	8

Data are obtained from the integration of the *c*(*s*) peaks. A_{280} is given for the experimental optical path of 1.2 cm

Table 2
Characterization of BmrA in Buffer H derived from the *c*(*s*) analysis

Parameter	Value from Buffer H	Value from Buffer D
s (S)	4.6 ± 0.1	2.6 ± 0.1
s_{20w} (S)	8.4 ± 0.2	8.1 ± 0.2
$\Delta J/A_{280}$	5.6	6.5
B_d (g/g)$^{(1)}$	0.3	0.6
R_H (nm)	5.6 ± 0.4	
M_b (kDa)	45.5 ± 4.2	
M_{theo} (kDa)	129.0	
B_d (g/g)$^{(2)}$	0.9 ± 0.2	
f/f_{min}	1.32	

The parameters are derived from the measurement made for the diluted sample. $s_{20,w}$ is calculated according to Eq. 13, using solvent densities and viscosities given in Subheading 3.1, **step 6** of the text ($B_d = 0$ because the complex is here considered), $\bar{v} = 0.785\text{ml}/\text{g}$ (corresponding to nearly 1 g/g bound DDM), $(M_D/M) = 1.000$ and 1.013 in Buffer H and D, respectively. $B_d^{(1)}$ is calculated from $\Delta J/A_{280}$ considering the values of $E_{0.1\%,280}$ and $(\partial n/\partial c)$ given in Subheading 3.1, **steps 1** and **2** (Eqs. 16 and 17). R_H is from [25]. M_b is from s_H and R_H (Eq. 4). M_{theo} is the molecular mass of the protein considered as a dimer from amino-acid composition. $B_d^{(2)}$ is derived from M_b (from s_{exp} and R_H) and M_{theo} (Eq. 10 with $M_D/M = 1$). f/f_{min} is calculated from R_H and $B_d^{(2)}$ (satisfying s_{exp} and M_{theo}), using Eqs. 11 and 6

with concentration. A slight increase would reflect an equilibrium of association. We observe for the main peak of BmrA a slight decrease of s with increasing concentration. This nonideal behavior is related to inevitable excluded volume effect [26]. The linear extrapolation of $1/s$ to infinite dilution would give s_0. But we have here only two concentrations. We estimate at infinite dilution: $s_H = 4.6 \pm 0.1$ S. $s_D = 2.6 \pm 0.1$ S. Report these values in a table (Table 2), where derived results will be reported. The invariance of s (apart from non-ideality) with concentration indicates that the samples may be considered as composed of noninteracting species: a peak in the $c(s)$ represents a species.

11. Attempt to analyze the data in SEDFIT with the noninteracting species model for determining independent values of s and D, thus M and R_H is not appropriate here because of the sample heterogeneity.
12. Analyze further the s- and signal-values from the $c(s)$ analysis. The s-value in Buffer H is in the order of magnitude of that calculated for a putative dimer with 1 g/g of bound DDM (Fig. 3a). The s_{20w} roughly calculated with the hypothesis of

1 g/g of bound DDM are similar and close to the value described for other ABC multidrug transporters [58].

13. Estimate bound DDM to BmrA from A_{280} and interference signals of the main peak in the $c(s)$ distribution. We use the extinction coefficients and increments of refractive indices given in Subheading 3.1, **steps 1** and **2** and Eqs. 16 and 17, using a homemade excel sheet, to calculate the protein concentration and the bound detergent B_d. We consider only the diluted samples, since the absorbance in the ultracentrifuge was very large, thus potentially imprecise, for the undiluted samples. The values $\Delta J/A_{280}$ actually change, however, only slightly for the undiluted samples: the values are 5.6 and 6.5 in Buffer H and in Buffer D, respectively. Ratio $\Delta J/A_{280}$ for the diluted samples is reported in Table 2 and is similar. $\Delta J/A_{280}$s are also similar in the two solvents. We derive in Buffer H, $B_{det} = 0.3$ g/g and in Buffer D $B_d = 0.6$ g/g (details are given in the caption of Table 2). These values are significantly lower than that—1.6 ± 0.6 g/g—estimated in Ravaud et al. by a radiolabelled detergent binding [25]. We suspect that the absorbance at 280 nm does not reflect the true protein concentration. Indeed the actual absorbance was larger by a factor of about three than expected from the concentration estimated from Bradford.

14. Combine the s_H-value with $R_s = 5.6 \pm 0.4$ nm estimated from size exclusion chromatography by Ravaud et al. [25] to derive—using Eq. 4—the buoyant molar mass of the complex, 51.5 kDa, then—using Eq. 10—B_d considering the most probable dimeric association state. We derive $B_d = 0.9$ g/g bound detergent lower than the previously determined value. Since the particle is defined, the frictional ratio can be derived from R_s—using Eq. 10. The analysis reported in Table 2 gives $f/f_{min} = 1.32$ corresponding to a compact, globular or slightly elongated, species.

15. Combine graphically the data in hydrogenated and deuterated buffers to determine the association state and the amount of bound detergent and lipid from the data in these solvents that differs in terms of density. We use the software Maple to draw implicit functions of Eq. 12 (*see* **Note 18**). The numerical values used as input are, for the protein: M_p, i.e., a multiple of the molar mass of the polypeptide, $\bar{v}_p$ and $(M_D/M)_p$, for the "detergent" representing detergent plus lipid, $(M_D/M)_d$, for the solvents, ρ_H, ρ_D, η_H, and η_D (estimated prior the experiment, *see* Subheading 3.1) and the minimum and maximum experimental values of s_H and s_D determined from the $c(s)$ analysis above. We have generated figures for different association states, monomer, dimer, and tetramer of BmrA, and different frictional ratios of 1.25, 1.5, and 1.8, respectively. Some are

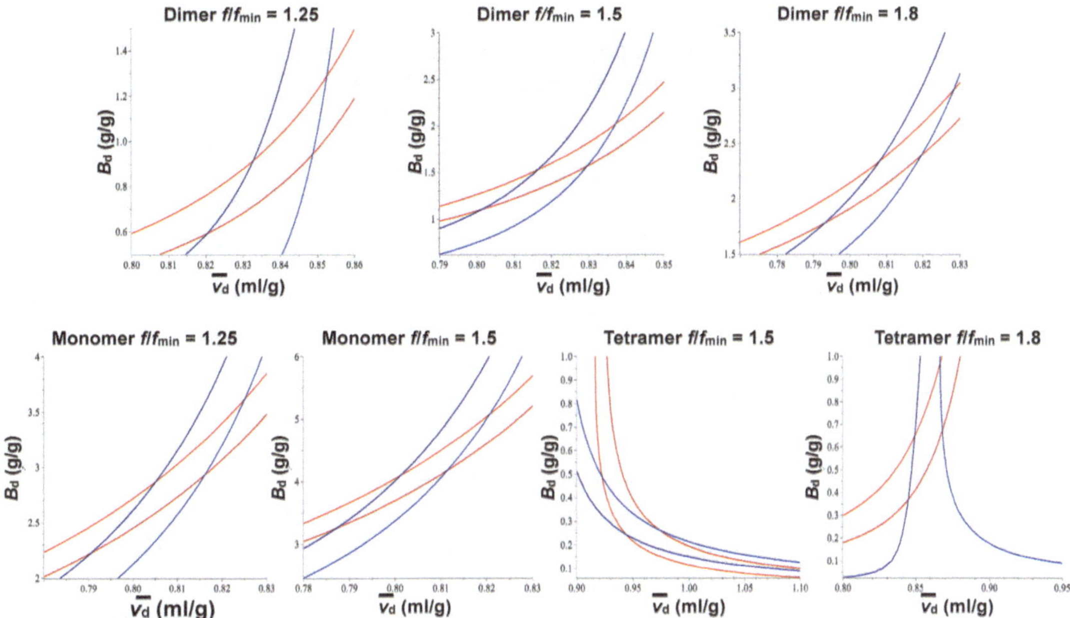

Fig. 8 Combined analysis of SV in Buffer H and D. The analysis is made in terms of a complex composed of two pseudo-components: the protein and bound (detergent + lipid + any bound compound). On each panel corresponding to a different hypothesis of f/f_{min}, the four curves give the ensemble of solutions (B_d; $\bar{v}_d$), with B_d the amount in g/g of protein and $\bar{v}_d$ the partial specific volume for bound detergent plus lipid. Lines are drawn for the limiting experimental values of the sedimentation coefficients determined from the *c*(*s*) analysis: $s_H - \Delta s_H$, $s_H + \Delta s_H$, $s_D - \Delta s_D$, and $s_D + \Delta s_D$. Mathematical solutions (B_d; $\bar{v}_d$) are defined by the area encircled by the four curves. Clearly irrelevant mathematical solutions ($B_d > 3$ g/g; $\bar{v}_d < 0.815\text{ml}/\text{g}$; $\bar{v}_d > 0.84\text{ml}/\text{g}$) are also excluded

shown in Fig. 8. The area corresponding to the intersection of the four plots on each figure defines all the mathematical solutions ($\bar{v}_d$; B_d) compatible with the experimental data. We eliminate non-reasonable values of $\bar{v}_d$—which should be in the range 0.815 mL/g (a low value for DDM) to 0.84 mL/g (90 % DDM and 10 % lipids)—and B_d—which should be in the order of magnitude of 1 g/g: we thus reject $B_d > 3$ g/g. The minimum and maximum $\bar{v}_d$ and B_d values defining this area and not leading to relevant solutions are reported in Table 3. We conclude: The s_H/s_D data are not compatible with a monomer. The s_H/s_D data are only compatible with a tetramer. The s_H/s_D data are compatible with a compact or slightly elongated dimer, and also a dimer with a very anisotropic shape, which, because BmrA is unglycosylated, is unlikely. We note that the dimer with f/f_{min} intermediate between 1.25 and 1.5, corresponds to a solution centered on $\bar{v}_d = 0.83\text{ ml}/\text{g}$; $B_d = 1.2$ g/g, close to the $\bar{v}_d$ value of DDM, and to the complex composition expected from complementary measurements.

Table 3
Results of the graphical analysis of the sedimentation velocity of BmrA in hydrogenated and deuterated solvents

N	M_p (Da)	f/f_{min}	$\bar{v}_d$ (mL/g)	B_d (g/g)
2	129,056	1.25	0.82–0.84	0.6–1.0
2	129,056	1.5	0.815–0.837	1.5–2.0
2	129,056	1.8	0.815–0.828	2.3–3.0

N is the association state of BmrA. M_p and f/f_{min} are the hypothesis. $\bar{v}_d$ and B_d are the results. Subscript *d* states for detergent plus lipid plus any bound compound

3.4 Conclusion

We recommend to investigate in hydrogenated and deuterated solvents the sedimentation (s_H/s_D method) of membrane proteins. They should not be undergoing auto-association. This condition can be assessed by preliminary sedimentation velocity experiments performed in the hydrogenated solvent at different protein concentrations. The s_H/s_D method requires, however, for its analysis the accurate determination of the viscosities of the solvents. The s_H/s_D method is not difficult to set up and is not time consuming. The graphical representation of the results allows an easy evaluation of the meanings of the results. What are the association states that are or not compatible with the data? What are the characteristics of the bound detergent, lipids…, i.e., all the bound components, being considered as a whole? The s_H/s_D method appears as a complementary strategy for the hydrodynamic characterization of membrane proteins. The analysis of detergent binding based on the absorbance of the protein appears in the case of BmrA not appropriate, probably because the extinction coefficient is not well estimated for this protein. One cause could be some bound ligand. The s_H/s_D method is more appropriate when the partial specific volume of the protein and of the detergent differs significantly. It is particularly interesting when lipids are bound in poorly characterized amounts. The s_H/s_D analysis comforts the description of solubilized in DDM BmrA as a globular or moderately asymmetric compact dimer binding DDM at a level of 0.9 +/− 0.3 g/g.

4 Notes

1. The equations are also valid for isotopically labelled molecules. In that case, in H_2O, $M_D/M \neq 1$, since M_D/M takes into account the excess of mass related to isotopical substitution. The contribution for H-D exchange of labile hydrogens of the particle with that of the solvent has to be considered to

calculate M_D/M in H_2O and D_2O. For intermediate solvent composition, the linearly derived M_D/M-value is considered.

2. When the partial specific volume of the detergent differs strongly from that of the protein complexes, it may be appropriate to use the analysis option *continuous c(s) with one discrete species*, allowing to consider the sedimentation of the detergent micelle. This is particularly appropriate when the detergent micelle floats, e.g., LAPAO and is detected together with membrane protein complexes that sediment.
3. If considering a two-component system, as we do in the combined analysis of s_H/s_D, for a glycosylated membrane protein, subscript pg has to be considered instead of p in Eqs. 10–13. If there are lipids, d stands for d + l. The partial specific volume has to be considered for pg and d + l are mean weighted values, as $E_{0.1\%}$ and $\partial n/\partial c$. Of course the amount of bound detergent plus lipid obtained directly from the two-component analysis is per gram of glycosylated protein, a value that has to be multiplied by (M_{pg}/M_p) to derive the value per gram of non-glycosylated protein. A second and very often used formalism (used here in Eqs. 16 and 17) considers the complex particle composed of protein binding different components (subscript *i*), such as detergent, sugar, lipids, and ligand. In the second formalism, $M_{b\ pd}$ and $R_{min\ pd}$ are written as below, and expression of s and s_{20w} are derived accordingly. Be careful that in the second formalism, M_p, $(M_D/M)_p$, $\bar{v}_p$ (and also $E_{0.1\%}$ and $\partial n/\partial c$ in Eqs. 16 and 17), relate to the protein.

$$M_{bpd} = M_p\left[((M_D / M)_p - \rho\bar{v}_p) + \sum B_i((M_D / M)_i - \rho\bar{v}_i)\right].$$

$$R_{minpd} = (3 / 4\pi N_A)^{1/3}\left[M_p\bar{v}_p + \sum B_i M_p\bar{v}_i\right]^{1/3}.$$

Note also that in the two formalisms, hydration may be considered, as bound water. The amount of bound water in g/g is typically 0.3 g/g. Bound water will only contribute to $M_{b\ pd}$ when the solvent density differs significantly from that of pure water.

4. The measurement at different concentrations is made for evaluating nonideal effects and for attesting that the protein is not undergoing association equilibrium. For a rapid equilibrium between monomer and dimer, one peak at *s*-value intermediate between monomer and dimer is expected (*see*, e.g., [28, 59]). For more complex interacting systems, *see*, e.g., [29, 30]. We typically use a dilution series between 1, 2, 4 and 1, 4, 16. The range of concentrations of the present protocol would not be large enough to conclude about protein–protein equilibrium. From other SV experiments on BmrA (not shown), we know it

does not associate or dissociate with concentration. Measurements at two protein concentrations are here made as duplicates.

5. An absorbance between 0.1 and 1.2 in the ultracentrifuge is optimal. Absorbance can be measured between 230 and 650 nm (typically 280 nm). The absorbance and volume of the stock sample determines the choice of the optical path length and the extent of the dilution series. The centerpieces with optical path lengths of 1.2, 0.3, and 0.15 cm require typically sample volumes of 420 μL, 100 μL, and 50 μL, respectively.

6. Lowering solvent salt below 100 mM may lead to significant hydrodynamic non-ideality for charged proteins—which is the usual case. The non-ideality is related to the differential sedimentation of the counter-ions and the macromolecule and to electrostatic repulsion between proteins [26].

7. Rotor 8-hole AnTi 50 is used for AUC experiments performed at angular velocity up to 50,000 rpm, while 4-hole AnTi 60 rotor tolerates up to 60,000 rpm. Epon centerpieces are not to be used above 42,000 rpm, while Al and Ti centerpieces can be used up to 60,000 rpm. Spin analytical [21] provides centerpieces specially designed for meniscus matching. A default in scan time recording in the ProteomeLabTM XL-A/XL-I Graphical User Interface Version 6.0 (Beckman) was recently revealed [60], which is corrected by using the software SEDFIT [16], version 14.1.

8. The software SEDNTERP [36] allows evaluating, from amino-acid composition, M, $\bar{v}$ and extinction coefficients of proteins, and, from the solvent composition, ρ and η. For unusual solvents, ρ and η can be measured experimentally: we use the density-meter DMA 5000 and a viscosity-meter AMVn (Anton Paar).

9. The software SEDFIT uses numerical solution of the Lamm equation for analyzing SV experiments in different ways. Detailed help, including step-by-step tutorials, is available online [16]. SEDFIT incorporates the possibility of accounting for the systematic noise of the experimental data, a procedure we routinely apply. Note the software SEDPHAT, which will not be described here and is available free on SEDFIT web page [16], may be used for analyzing globally different sets of data, e.g., SV data obtained at different concentrations.

10. SV measurements can be done between 4 and 20 °C on our AUC. The temperature of 20 °C is less time consuming for temperature equilibration of the rotor prior to SV experiments. In addition, density and viscosity measurements are easier at 20 °C.

11. Use F-Ratio = 0.5 for the analysis without regularization procedure. Use F-Ratio of 0.68 or 0.95 for analysis with regularization.

12. Use *Run* when the parameters *frictional ratio*, *meniscus*, or *bottom* do not have to be fitted (they are unchecked in Figure 3D). Use *fit* if one, two or all of them have to be fitted, after having ticked them in the *c*(*s*) parameter window (Figure 3D). Varying these parameters implies the software has to simulate new simulation profiles. On the other side, the software does not have to simulate new simulation profiles for the calculation of *baseline*, *Fit RI noise*, and *Fit Time independent noise*, which will be calculated, if ticked in Figure 3D, in the *Run* command.
13. In addition to angular velocity (3,000 rpm) and temperature (20 °C), parameters to be defined in the wavelength scan for each cell are the number of measurements to be averaged [2] and the wavelength increment (1 nm).
14. Radial scan at 3,000 rpm allows checking the proper filling of the cells, since the meniscus, i.e., the interface between air and the solutions (sample and solvent), are easily detected. It will measure A_{280} as a function of r at time "zero." A flat profile is expected, since the concentration is the same at all radial positions (at 3,000 rpm, sedimentation is not significant unless for very large assemblies). The quality of the signal is also checked at that step (noise should be below 0.01 absorbance unit). The parameters to be defined for each of the cell are: the wavelength (280 nm), the range of radial position (5.8–7.2 cm), the number of A_{280} measurements to be averaged for each radial position [2], the radial step (0.003 cm), the mode of acquisition (continuous).
15. Temperature equilibration is very important because sedimentation depends on solvent viscosity which changes significantly with temperature.
16. Global parameters for SV experiments: 42,000 rpm, 20 °C. The duration for scanning the absorbance of one cell at one wavelength is typically 1.5–2 min. In the method, we define: the interval between scans (1 min., much below 5 min in such a way that the successive measurements are done as soon as possible); the total number of scans (typically 200 for exceeding overnight and stopping manually the AUC in the morning—other protocols are possible); an overlay of four scans (for each of the cells, the superposition of the last four scans is displayed upon centrifugation). The parameters for each of the cells are those given in **Note 14**.
17. Because we register here SV profiles at two different wavelengths acquired sequentially, profiles with either odd (or even) numbers have to be loaded for the *c*(*s*) analysis, which considers a set of SV profiles acquired at one wavelength or with interference.

18. The calculation are based on the capability of the software Maple to draw implicit functions, i.e., functions that do not directly relate x and y but can be written $f(x,y) = 0$. Here the unknown x and y are $\bar{v}_d$ and B_d (subscript d is for detergent + lipid + any bound compound). f is derived from Eq. 12 by grouping the two members of the equation at left. The function f is expressed with the variables f/f_{min}, η, ρ, $(M_D/M)_p$, $(M_D/M)_d$, and s. These parameters will be varied to draw different plots on the same figure. Common parameters for the figure are $\bar{v}_p$, N_A, and M_p. We have written three command lines in Maple. The first line writes the implicit function:

```
> f := (ffo,eta,rho,MDMHp,MDMHd,s)-> s-10^15*
(M*(MDMHp-rho*vbar)+M*y*(MDMHd-rho*x))/
(6*Pi*Na*eta*ffo*(3/(4*Pi*Na)*(M*vbar+M*
y*x))^(1/3));
```

ffo stays for f/f_{min}, **eta** for η, **rho** for ρ, **MDMHp** for $(M_D/M)_p$, **MDMHd** for $(M_D/M)_d$, and **s** for s. **M** stays for M_p, **vbar** for $\bar{v}_p$, **Pi** is π, and **Na** is N_A. The solutions of the implicit function are **y** and **x**, representing B_d and $\bar{v}_d$. The second line defines the common parameters for a figure with different plots. The first and third numbers ($\bar{v}_p$ and M_p, in mL/g and Da) have to be adapted according to the study.

```
> (vbar,Na,M) := (0.751,6.02*10^(23),129056);
```

The third line contains the variables of the implicit functions, the scales of the figure, the color and style of the plots, and the labels of the figure. The variables of f are the input parameters (f/f_{min}, η, ρ, $(M_D/M)_p$, $(M_D/M)_d$, s). The units for η and ρ are centipoise and g/mL. We draw the functions f corresponding to a given frictional ratio and the input parameters of the minimum and maximum values of the experimental sedimentation coefficients estimated in Buffer H, and the minimum and maximum values of the experimental sedimentation coefficient estimated in buffer. Each of these four functions is calculated twice, because they will be drawn with two different color and style (line and point). **x** is for $\bar{v}_d$ and **y** for B_d. The minimum and maximum values of their scale on the figure to be drawn have to be chosen.

```
>implicitplot([f(1.25,1.558,1.039,1.0,1.0,4.5),
f(1.25,1.558,1.039,1.0,1.0,4.7),f(1.25,1.796,1.132,
1.012,1.014,2.5),f(1.25,1.796,1.132,1.012,1.014,2.7
)],x=0.8..0.86,y=0.5..1.5,color=[orange,red,blue,n
avy],     style=[line,line,line,line],labels=["Vbar
det","Bound det"],labeldirections=[HORIZONTAL,VERTI
CAL],font=[helvetica,roman,36],thickness=5);
```

The functions are drawn by the command "execute the entire worksheet." The vertical and horizontal scales of the figure are possibly modified, and the command re-executed, to obtain a figure in which the area defined by the intersection of the four plots is clearly shown. This area defined all the mathematical solutions ($\bar{\nu}_d$; B_d) compatible with the inputs. The minimum and maximum $\bar{\nu}_d$ and y for B_d-values defining this area are noted.

Other inputs are then investigated. For a same experiment, the input to be changed are, in the second command line, **M** (M_p), and, in the third command line, **f/fo** (f/f_{min})—to be changed 8 times—and the scales **x** ($\bar{\nu}_d$) and **y**—(B_d)—to be defined large before being adapted. We generally consider, for each relevant association state, frictional ratios of 1.25, 1.5, and 1.8.

Acknowledgments

This work used the AUC platform of the Grenoble Instruct centre (ISBG; UMS 3518 CNRS-CEA-UJF-EMBL) with support from FRISBI (ANR-10-INSB-05-02) and GRAL (ANR-10-LABX-49-01) within the Grenoble Partnership for Structural Biology (PSB).

References

1. http://blanco.biomol.uci.edu/Membrane_Proteins_xtal.html.
2. Musatov A, Robinson NC (2002) Cholate-induced dimerization of detergent- or phospholipid-solubilized bovine cytochrome C oxidase. Biochemistry 41:4371–4376
3. Fisher LE, Engelman DM, Sturgis JN (2003) Effect of detergents on the association of the glycophorin a transmembrane helix. Biophys J 85:3097–3105
4. Josse D, Ebel C, Stroebel D, Fontaine A, Borges F, Echalier A, Baud D, Renault F, Le Maire M, Chabrieres E, Masson P (2002) Oligomeric states of the detergent-solubilized human serum paraoxonase (PON1). J Biol Chem 277:33386–33397
5. le Maire M, Champeil P, Møller JV (2000) Interaction of membrane proteins and lipids with solubilizing detergents. Biochim Biophys Acta 1508:86–111
6. Lebowitz J, Lewis MS, Schuck P (2002) Modern analytical ultracentrifugation in protein science: a tutorial review. Protein Sci 11:2067–2079
7. Ebel C (2004) Analytical ultracentrifugation for the study of biological macromolecules. Prog Colloid Polym Sci 127:73–82
8. Howlett GJ, Minton AP, Rivas G (2006) Analytical ultracentrifugation for the study of protein association and assembly. Curr Opin Chem Biol 10:430–436
9. Ebel C (2007) Analytical ultracentrifugation. State of the art and perspectives. In: Uversky VN, Permyakov EA (eds) Protein structures: methods in protein structure and stability analysis. Nova, New York, pp 229–260
10. Salvay AG, Ebel C (2006) Analytical ultracentrifuge for the characterization of detergent in solution. Prog Colloid Polym Sci 131: 74–82
11. Salvay AG, Santamaria M, le Maire M, Ebel C (2007) Analytical ultracentrifugation sedimentation velocity for the characterization of detergent–solubilized membrane proteins Ca++-ATPase and ExbB. J Biol Phys 33: 399–419
12. le Maire M, Arnou B, Olesen C, Georgin D, Ebel C, Moller JV (2008) Gel chromatography and analytical ultracentrifugation to determine the extent of detergent binding and aggregation, and stokes radius of membrane proteins using sarcoplasmic reticulum Ca2+-ATPase as an example. Nat Protoc 3:1782–1795
13. Ebel C (2011) Sedimentation velocity to characterize surfactants and solubilized membrane proteins. Methods 54:56–66
14. Nury H, Manon F, Arnou B, le Maire M, Pebay-Peyroula E, Ebel C (2008) Mitochondrial bovine ADP/ATP carrier in detergent is predominantly monomeric but also forms multimeric species. Biochemistry 47:12319–12331

15. Dach I, Olesen C, Signor L, Nissen P, le Maire M, Møller JV, Ebel C (2012) Active detergent solubilized H+, K+-ATPase is a monomer. J Biol Chem 287:41963–41978
16. Peter Schuck's Software Home Page. http://www.analyticalultracentrifugation.com, p Download Sedfit and Sedphat and description
17. Borries Demeler's Ultrascan Software Page. http://www.ultrascan.uthscsa.edu/
18. Walter Stafford's Programm Sedanal. http://rasmb.bbri.org/sedanal.html.
19. Gohon Y, Pavlov G, Timmins P, Tribet C, Popot J-L, Ebel C (2004) Partial specific volume and solvent interactions of amphipol A8-35. Anal Biochem 334:318–334
20. Pavlov G, Finet S, Tatarenko K, Korneeva E, Ebel C (2003) Conformation of heparin studied with macromolecular hydrodynamic methods and X-ray scattering. Eur Biophys J 32:437–449
21. Web Site of Spin Analytical. http://www.spin-analytical.com/index.php
22. Schuck P (2004) A model for sedimentation in inhomogeneous media. II. Compressibility of aqueous and organic solvents. Biophys Chem 108:201–214
23. Schuck P (2004) A model for sedimentation in inhomogeneous media. I. Dynamic density gradients from sedimenting co-solutes. Biophys Chem 108:187–200
24. Steinfels E, Orelle C, Fantino JR, Dalmas O, Rigaud JL, Denizot F, Di Pietro A, Jault JM (2004) Characterization of YvcC (BmrA), a multidrug ABC transporter constitutively expressed in Bacillus subtilis. Biochemistry 43:7491–7502
25. Ravaud S, Do Cao MA, Jidenko M, Ebel C, Le Maire M, Jault JM, Di Pietro A, Haser R, Aghajari N (2006) The ABC transporter BmrA from Bacillus subtilis is a functional dimer when in a detergent-solubilized state. Biochem J 395:345–353
26. Solovyova A, Schuck P, Costenaro L, Ebel C (2001) Non-ideality by sedimentation velocity of halophilic malate dehydrogenase in complex solvents. Biophys J 81:1868–1880
27. Schuck P (2000) Size-distribution analysis of macromolecules by sedimentation velocity ultracentrifugation and lamm equation modeling. Biophys J 78:1606–1619
28. Buisson M, Valette E, Hernandez JF, Baudin F, Ebel C, Morand P, Seigneurin JM, Arlaud GJ, Ruigrok RW (2001) Functional determinants of the Epstein-Barr virus protease. J Mol Biol 311:217–228
29. Schuck P (2010) Sedimentation patterns of rapidly reversible protein interactions. Biophys J 98:2005–2013
30. Cole JL, Correia JJ, Stafford WF (2011) The use of analytical sedimentation velocity to extract thermodynamic linkage. Biophys Chem 159:120–128
31. Balbo A, Minor KH, Velikovsky CA, Mariuzza RA, Peterson CB, Schuck P (2005) Studying multiprotein complexes by multisignal sedimentation velocity analytical ultracentrifugation. Proc Natl Acad Sci USA 102:81–86
32. Hayashi Y, Matsui H, Takagi T (1989) Membrane protein molecular weight determined by low-angle laser light-scattering photometry coupled with high-performance gel chromatography. Methods Enzymol 172:514–528
33. Edelstein SJ, Schachman HK (1967) The simultaneous determination of partial specific volumes and molecular weights with microgram quantities. J Biol Chem 242:306–311
34. Edelstein SJ, Schachman HK (1973) Measurement of partial specific volume by sedimentation equilibrium in H2O-D2O solutions. Methods Enzymol 27:82–98
35. Brown PH, Balbo A, Zhao H, Ebel C, Schuck P (2011) Density contrast sedimentation velocity for the determination of protein partial-specific volumes. PLoS One 6:e26221
36. John Philo's Software Home Page. http://www.jphilo.mailway.com/, p Download Sednterp
37. Philo J, Talvenheimo J, Wen J, Rosenfeld R, Welcher A, Arakawa T (1994) Interactions of neurotrophin-3 (NT-3), brain-derived neurotrophic factor (BDNF), and the NT-3.BDNF heterodimer with the extracellular domains of the TrkB and TrkC receptors. J Biol Chem 269:27840–27846
38. Tanford C, Reynolds JA (1976) Characterization of membrane proteins in detergent solutions. Biochim Biophys Acta 457:133–170
39. Tiefenbach K-J, Durchschlag H, Jaenicke R (1999) Spectoscopic and hydrodynamic investigations of nonionic and zwitterionic detergents. Prog Colloid Polym Sci 113:135–141
40. Breyton C, Gabel F, Abla M, Pierre Y, Lebaupain F, Durand G, Popot J-L, Ebel C, Pucci B (2009) Micellar and biochemical properties of (hemi)fluorinated surfactants are controlled by the size of the polar head. Biophys J 97:1077–1086
41. Sharma KS, Durand G, Gabel F, Bazzacco P, Le Bon C, Billon-Denis E, Catoire LJ, Popot JL, Ebel C, Pucci B (2012) Non-ionic amphiphilic homopolymers: synthesis, solution properties, and biochemical validation. Langmuir 28:4625–4639
42. Gohon Y, Dahmane T, Ruigrok RW, Schuck P, Charvolin D, Rappaport F, Timmins P, Engelman DM, Tribet C, Popot J-L, Ebel C (2008) Bacteriorhodopsin/amphipol complexes: structural and functional properties. Biophys J 94:3523–3537
43. Durchschlag H, Zipper P (1994) Calculation of the partial volume of organic compounds and polymers. Prog Colloid Polym Sci 94:20–39

44. Durchschlag H, Zipper P (1997) Calculation of partial specific volumes and other volumetric properties of small molecules and polymers. J Appl Crystallogr 30:803–807
45. Zhao H, Brown PH, Schuck P (2011) On the distribution of protein refractive index increments. Biophys J 100:2309–2317
46. Beri RG, Walker J, Reese ET, Rollings JE (1993) Characterization of chitosans via coupled size-exclusion chromatography and multiple-angle laser light-scattering technique. Carbohydr Res 238:11–26
47. http://www.affymetrix.com
48. Csucs G, Ramsden JJ (1998) Solubilization of planar bilayers with detergent. Biochim Biophys Acta 1369:304–308
49. Arnaud N, Georges J (2001) On the analytical use of the Soret-enhanced thermal lens signal in aqueous solutions. Anal Chim Acta 445:239–244
50. Strop P, Brunger AT (2005) Refractive index-based determination of detergent concentration and its application to the study of membrane proteins. Protein Sci 14:2207–2211
51. Abla M, Durand G, Breyton C, Raynal S, Ebel C, Pucci B (2012) A diglucosylated fluorinated surfactant to handle integral membrane proteins in aqueous solution. J Fluor Chem 134:63–71
52. Bazzacco P, Sharma KS, Durand G, Giusti F, Ebel C, Popot J-L, Pucci B (2009) Trapping and stabilization of integral membrane proteins by hydrophobically grafted glucose-based telomers. Biomacromolecules 10:3317–3326
53. Riske KA, Politi MJ, Reed WF, Lamy-Freund MT (1997) Temperature and ionic strength dependent light scattering of DMPG dispersions. Chem Phys Lipids 89:31–44
54. Salvay AG, Communie G, Ebel C (2012) Sedimentation velocity analytical ultracentrifugation for intrinsically disordered proteins. In: Intrinsically disordered protein analysis, vol 2, Methods and experimental tools. Springer, New York, pp 91–104
55. Tabarani G, Reina JJ, Ebel C, Vives C, Lortat-Jacob H, Rojo J, Fieschi F (2006) Mannose hyperbranched dendritic polymers interact with clustered organization of DC-SIGN and inhibit gp120 binding. FEBS Lett 580: 2402–2408
56. Balbo A, Zhao H, Brown PH, Schuck P (2009) Assembly, loading, and alignment of an analytical ultracentrifuge sample cell. J Vis Exp (33):e1530
57. https://www.beckmancoulter.com/wsrportal/techdocs?docname=LXLAI-IM-10.
58. Boncoeur E, Durmort C, Bernay B, Ebel C, Di Guilmi A-M, Croizé J, Vernet T, Jault JM (2012) PatA and PatB form a functional heterodimeric ABC multidrug efflux transporter responsible for the resistance of Streptococcus pneumoniae to fluoroquinolones. Biochemistry 51:7755–65
59. Schuck P (2003) On the analysis of protein self-association by sedimentation velocity analytical ultracentrifugation. Anal Biochem 320:104–124
60. Zhao H, Ghirlando R, Piszczek G, Curth U, Brautigam CA, Schuck P (2013) Recorded Scan Times Can Limit the Accuracy of Sedimentation Coefficients in Analytical Ultracentrifugation. Anal Biochem. 437:104–108

Chapter 16

Membrane Partitioning and Translocation Studied by Isothermal Titration Calorimetry

Carolyn Vargas, Johannes Klingler, and Sandro Keller

Abstract

The ability to bind to and translocate across lipid bilayers is of paramount importance for the extracellular administration of intracellularly active compounds in cell biology, medicinal chemistry, and drug development. A combination of the so-called uptake and release experiments performed by high-sensitivity isothermal titration calorimetry provides a powerful and universally applicable tool for measuring membrane binding and translocation of various compound classes in a label-free manner in solution. The protocol presented here is designed for a quantitative analysis of microcalorimetric uptake and release titrations. In contrast with simpler approaches described previously, it is applicable also to electrically charged solutes, such as peptides and proteins, experimentally and clinically relevant surfactants, drugs, metal ions, and other ionic compounds.

Key words Ionic solutes, Lipid bilayer, Membrane binding, Membrane permeability, Microcalorimetry, Uptake and release

1 Introduction

Many peptides, proteins, small molecules, ions, and other biologically relevant compounds interact with biological membranes by partitioning into their lipid matrix or by binding to peripheral or integral membrane proteins. Furthermore, effector molecules often need to translocate across one or several membranes to fulfil their functions inside a cell or a subcellular compartment. These membrane interactions play decisive roles in both basic biological as well as applied pharmaceutical research, where biomolecules and drugs are critically assessed for membrane binding and for their ability to traverse lipid bilayers by passive diffusion to reach their intracellular targets and achieve the desired effects.

In most cases, membrane interactions do not follow a stoichiometric binding model but rather a surface partition equilibrium in which the solute (i.e., peptide, protein, small molecule, or ion) partitions between the aqueous phase (usually buffer) and

Doron Rapaport and Johannes M. Herrmann (eds.), *Membrane Biogenesis: Methods and Protocols*, Methods in Molecular Biology, vol. 1033, DOI 10.1007/978-1-62703-487-6_16, © Springer Science+Business Media, LLC 2013

the lipid bilayer phase [1]. A broad range of methods are used for studying such membrane partition equilibria, including absorbance [2], circular dichroism [3], fluorescence [4], and other kinds of optical spectroscopies, nuclear magnetic resonance (NMR) spectroscopy [5, 6], and surface plasmon resonance (SPR) spectroscopy [7]. In 1999, Heerklotz et al. [8] established a combination of the so-called uptake and release titrations by high-sensitivity isothermal titration calorimetry (ITC) as a particularly powerful and universally applicable tool for simultaneously measuring the membrane affinity of and the membrane permeability to nonionic compounds. Advantages of ITC include excellent accuracy and precision and independence from isotopic or spectroscopic labels as well as from immobilization [9].

However, the applicability of the above ITC protocol is limited to nonionic, that is, electrically uncharged compounds. For electrically charged solutes, such as metal ions, amino or nucleic acids, and most biopolymers, the membrane partition equilibrium is modulated by electrostatic effects, that is, attraction or repulsion between surface-bound and free solute [1]. In other words, the partition equilibrium is established between the bilayer and the aqueous phase adjacent to the bilayer, where the concentration of ions can differ substantially from that in the bulk aqueous phase because of electrostatic attraction to or repulsion from the charged membrane. In many cases, the latter can be accounted for by a simple electrostatic model based on Gouy–Chapman theory [10–12]. By taking into account electrostatic effects at the membrane surface with the aid of this model, we could extend the ITC uptake and release strategy [13] to a broad range of biologically important compounds such as proteins and peptides, ionic surfactants, and drugs. With the aid of this assay, it could be shown, for example, that a cell-penetrating peptide dubbed penetratin cannot cross lipid bilayers by passive diffusion [14]. Fluorescence correlation spectroscopy and a model-free dialysis assay confirmed this finding and thus validated the ITC approach [15]. Moreover, ITC uptake and release titrations analyzed in terms of Gouy–Chapman theory were used to demonstrate that a doxycycline derivative developed for photoactivated gene expression can indeed reach its nuclear target by membrane permeation [16], that the cellular internalization of cyclic nucleotides [17] and photoactivatable capsaicin derivatives [18] is not due to membrane diffusion, and that phototriggers for protons can be modified in a controlled manner such that, upon irradiation, they acidify lipid membranes on either only one or both sides [19]. It could also be shown that the velocity of transmembrane diffusion is the decisive parameter determining a detergent's suitability for thermodynamically controlled membrane solubilization and reconstitution [20, 21], which has far-reaching implications for the use of detergents in the handling and biophysical investigation of membrane proteins [22].

Detailed protocols have been published for planning, performing, and analyzing uptake and release experiments on nonionic solutes by ITC [9] and on both ionic and nonionic solutes by fluorescence spectroscopy [23]. By contrast, to the best of our knowledge, there exists no in-depth description of the rather sophisticated methodology required for a quantitative analysis of ITC uptake and release titrations involving ionic compounds. While the protocol presented in this chapter also includes a brief step-by-step description of the basic experimental procedure, the main focus is on the analysis of uptake and release data obtained from high-sensitivity ITC measurements using ionic solutes (e.g., peptides, proteins, and charged small molecules). To demonstrate this approach, we take the cell-penetrating peptide penetratin [14, 15] as a well-characterized example of an ionic solute that interacts with lipid bilayers. For data analysis, we present a simple and user-friendly spreadsheet program based on Microsoft Excel developed for the simultaneous quantification of ITC uptake and release experiments. Such analysis provides the avidity of membrane interactions, the membrane permeability, and additional electrostatic properties in terms of a partition coefficient, a lipid accessibility factor, and an effective charge number, respectively.

2 Materials

1. A complete set of uptake, release, and blank experiments requires approximately 40 mg phospholipid. Zwitterionic lipids such as 1-palmitoyl-2-oleoyl-*sn*-glycero-3-phosphocholine (POPC) and negatively charged lipids such as 1-palmitoyl-2-oleoyl-*sn*-glycero-3[phospho-*rac*(1-glycerol)] (POPG) or 1-palmitoyl-2-oleoyl-*sn*-glycero-3-phospho-L-serine (POPS) are commonly used to mimic uncharged mammalian cell and negatively charged bacterial membranes, respectively. High-purity lipids (>99 %) are commercially available (e.g., from Avanti Polar Lipids, Genzyme, or Lipoid).
2. A basic set of uptake and release experiments typically requires 0.2–0.3 μmol solute (e.g., peptide).
3. All reagents (buffer, salt, additives, etc.) should be of highest purity possible. Use water having a resistivity of at least 18 MΩ cm for buffer preparation, and use this buffer in all experiments.
4. The assay depends on the availability of a properly calibrated and well-maintained high-sensitivity isothermal titration microcalorimeter. All published uptake and release experiments were performed on MicroCal VP-ITC and iTC_{200} systems (GE Healthcare; www.microcal.com), but other instruments with comparable technical specifications are also available (see www.tainstruments.com).

5. Unilamellar lipid vesicles are produced by extrusion or sonication. To obtain large unilamellar vesicles (LUVs; diameter typically 100–500 nm), we use a LiposoFast extruder (Avestin), but other types such as a Lipex extruder (Northern Lipids) and mini-extruders (Avanti Polar Lipids) may also be used. Polycarbonate membranes for extrusion come in different pore sizes (50–400 nm). To obtain small unilamellar vesicles (SUVs; diameter ~30 nm), we employ indirect sonication, which uses no microtip probe but instead a water-filled vessel (e.g., Bandelin BR 30) into which a glass vial containing the sample is submerged during sonication.
6. The vesicle size distribution after extrusion or sonication can be determined by dynamic light scattering (DLS) using, for instance, a Zetasizer Nano S90 (Malvern).
7. For a quantitative analysis and interpretation of uptake and release data for ionic solutes, you need:
   (a) Microsoft Excel 3.0 or later with the Solver add-in. Solver performs nonlinear least-squares fitting of experimental data on the basis of a theoretical model using an iterative method [24]. Solver was developed by Frontline Systems and has been included in every distribution of Microsoft Excel since 1990. We recommend using Windows Excel 2010 (or 2003) since, in our experience, the 2007 version is rather slow. In the newest version of Excel for Mac (Excel 2011), Solver is not automatically installed but can be downloaded free of charge from http://www.solver.com/mac/dwnmac2011solver.htm.
   (b) NITPIC, an automated peak-shape analysis and baseline adjustment program [25] for ITC. NITPIC allows automated integration of all peaks in a consistent and efficient way without user interference and bias. The program can be downloaded from http://biophysics.swmed.edu/MBR/software.html.
   (c) Copy of the fitting program *ITC Uptake and Release Mole Fraction.xls* or *ITC Uptake and Release Mole Ratio.xls*, which can be obtained upon request from the authors.

3 Methods

3.1 Preparation of Solutions for a Set of Uptake, Release, and Blank Experiments

1. The following protocol is described for a membrane-interacting peptide but is equally applicable to other solutes such as proteins or small molecules.
2. Prepare 700 μL of a 300 μM peptide stock solution in buffer (10 mM Tris-HCl, 100 mM NaCl, pH 7.4). Prepare 1.8 mL of a 20 μM peptide solution by adding 120 μL of the 300 μM peptide stock solution to 1.68 mL buffer.

3. Dissolve 30 mg POPC in 1.5 mL chloroform and 10 mg POPG in 0.5 mL chloroform in separate glass vials to make 20 mg/mL stock solutions. Always use glass, stainless steel, or Teflon to prepare and transfer lipids in organic solvents.
4. Using a Hamilton syringe, prepare a lipid mixture of POPC and POPG at a molar ratio of 3:1 by adding 0.5 mL POPG stock solution to 1.5 mL POPC stock solution.
5. Take two 5-mL glass vials with lids attached, label the first with Vial 1 (for uptake and blank experiments) and the second with Vial 2 (for release experiment), and weigh them for their tare weights.
6. Transfer 1.5 and 0.4 mL of the lipid mixture to Vial 1 (30 mg) and Vial 2 (8 mg), respectively. Pass a gentle stream of nitrogen over the solutions under the chemical hood until the solvent evaporates. Place the vials in a vacuum desiccator (at 10^{-2}–10^{-3} mbar) and let dry for several hours or overnight.
7. Take the vials out of the desiccator and seal them immediately. Weigh the sealed vials and subtract the tare weights (**step 4**) to obtain the net weights of the lipid films.

3.2 Preparation of Lipid Vesicles for Uptake and Blank Experiments

1. Resuspend the dry lipid film in Vial 1 by adding 1.31 mL buffer (at room temperature) to yield a final lipid concentration of 40 mM and hydrate the mixture for 30 min. Vortex for at least 5 min (*see* **Note 1**).
2. To produce LUVs, perform five freeze–thaw cycles by immersing the lipid suspension alternately in liquid nitrogen for 10 s to freeze and in a 25–30 °C water bath for 2 min to thaw. This procedure is necessary in the preparation of vesicles for release experiments where the peptide needs to distribute uniformly over both inner and outer membrane leaflets. Assemble two stacked polycarbonate membranes with a pore size of 100 nm in the extruder. Attach a 1-mL gas-tight syringe to one end and tighten. Load the lipid dispersion into this syringe, push the plunger until a droplet appears at the other end, then attach a second 1-mL gas-tight syringe to the other end and tighten. Perform 35 extrusion steps (*see* **Note 2**).
3. Alternatively: To produce SUVs, place the sealed vial containing the multilamellar dispersion (obtained in **step 1**) in a water-cooled ultrasonic vessel. Sonicate for 50 min at maximum amplitude using a pulse duration of 10 s followed by a pause of 1 s. Make sure that the sample is kept cool and the vial is placed in the middle of the vessel at least 2 mm above the base. A clear solution should result (*see* **Note 3**).
4. Check the size distribution of the LUVs or SUVs by DLS. Pipette 1 μL lipid vesicles in 1 mL buffer, transfer to a cuvette, and measure. A mean diameter of about 120 or 30 nm is typically

obtained upon extrusion through 100-nm pore filters or sonication, respectively.

5. Add 250 μL of 40 mM lipid vesicles (**step 2** or **3**) to 250 μL buffer to make 500 μL of 20 mM lipid vesicles.

3.3 Uptake Experiment

1. In an uptake experiment, the peptide is taken up into the membrane upon titration with lipid vesicles.
2. For a MicroCal VP-ITC, fill the injection syringe (~300 μL) with 20 mM lipid vesicles and the sample cell (1.4 mL) with 20 μM peptide. In the case of MicroCal iTC_{200}, similar concentrations apply, but it requires only ~70 μL lipid and ~300 μL peptide to fill the injection syringe and the sample cell, respectively.
3. Set the total number of injections to 30, the injection volume to 10 μL, and the spacing between injections to 6 min. A standard ITC uptake titration consists of an initial 1-μL injection followed by a series of 10-μL injections. However, more sophisticated injection schedules with nonconstant injection volumes are possible (e.g., 1 × 1 μL, 3 × 3 μL, 4 × 5 μL, and 27 × 10 μL). This allows not only a good resolution of peaks at the start of the titration, particularly since the heats at this point are largest and most variable, but also a measurement of the small heats of dilution towards the end of the titration with a reasonably good signal-to-noise ratio. Set up the other parameters as recommended by the instrument's manufacturer. Allow sufficiently long spacing times between injections to accommodate complete return of the signal to the baseline. While spacings of 240–300 s are sufficient for many fast-equilibrating systems, slowly equilibrating processes may require up to 60 min spacing times (*see* **Note 4**).
4. At the start of the ITC run, monitor the height of the power peak so that it does not exceed the limits of the detection range (for VP-ITC, between 0 and 35 μcal/s). Adjust the injection volume if necessary.
5. After the titration, clean the sample cell thoroughly according to the manufacturer's instructions.

3.4 Blank Experiment

1. To ensure the absence of contaminants that can contribute additional reaction heats, a blank experiment is performed, where a pure lipid vesicle suspension is titrated into buffer.
2. Refill the syringe with the pure lipid vesicle dispersion and load the sample cell with the same buffer used to prepare the peptide solution.
3. Apply the same parameters used for the uptake experiment. Reduce the spacing times as appropriate. If peaks are small and

constant during the first few injections, you may stop the experiment.

4. Clean both sample cell and syringe thoroughly.

3.5 Preparation of Preloaded Vesicles for Release Experiments

1. To prepare unilamellar vesicles prebound with peptide, add 524 μL of the 300 μM peptide solution to the dry lipid film in Vial 2. Vortex vigorously for at least 5 min to yield a final lipid concentration of 20 mM lipid mixture.
2. Perform freeze–thaw cycles followed by either extrusion or sonication as described in Subheading 3.1.
3. Check the size distribution of the peptide-preloaded LUVs or SUVs by DLS.

3.6 Release Experiment

1. In a release experiment, vesicles are first preloaded with peptide on both leaflets and are then titrated into buffer. This results in the release of the peptide from the lipid membrane, in which the extent of desorption of the peptide crucially depends on its ability to translocate across the membrane.
2. Fill the injection syringe with the suspension containing the peptide-preloaded vesicles and load the sample cell with buffer.
3. Set up the run and the injection parameters similar to those given for the uptake experiment (**steps 2** and **3** of Subheading 3.3). For proper normalization of reaction heats during subsequent data analysis, enter the lipid concentration (i.e., 20 mM) as the concentration in the syringe rather than the peptide concentration.
4. Clean the sample cell and syringe thoroughly after titration.

3.7 Data Analysis

Prepare the .itc files generated from the uptake and release experiments (*see* **Note 5**). This section focuses on the simultaneous analysis of uptake and release data; additionally, evaluation of either uptake or release data alone is described in Subheading 4.

1. Load an .itc file containing uptake data into NITPIC by clicking *File: Read ITC Data: Execute.* To check the quality of peak integration, click any data point on the "Isotherm" panel. To export the integrated heats of reaction, click *Save DAT and XP* (*see* **Note 6**). A typical NITPIC window is shown in Fig. 1. Repeat this step for a .itc file containing release data.
2. Open the fitting program *ITC Uptake and Release Mole Fraction.xls* or *ITC Uptake and Release Mole Ratio.xls*, which refers to *u*ptake and *r*elease evaluated on the basis of a mole fraction (X) partition coefficient or a mole ratio (R) partition coefficient, respectively (*see* **Note 7**). For the present purpose, we take *ITC Uptake and Release Mole Fraction.xls.* A screenshot of the main worksheet ("Fit") of the program with uptake

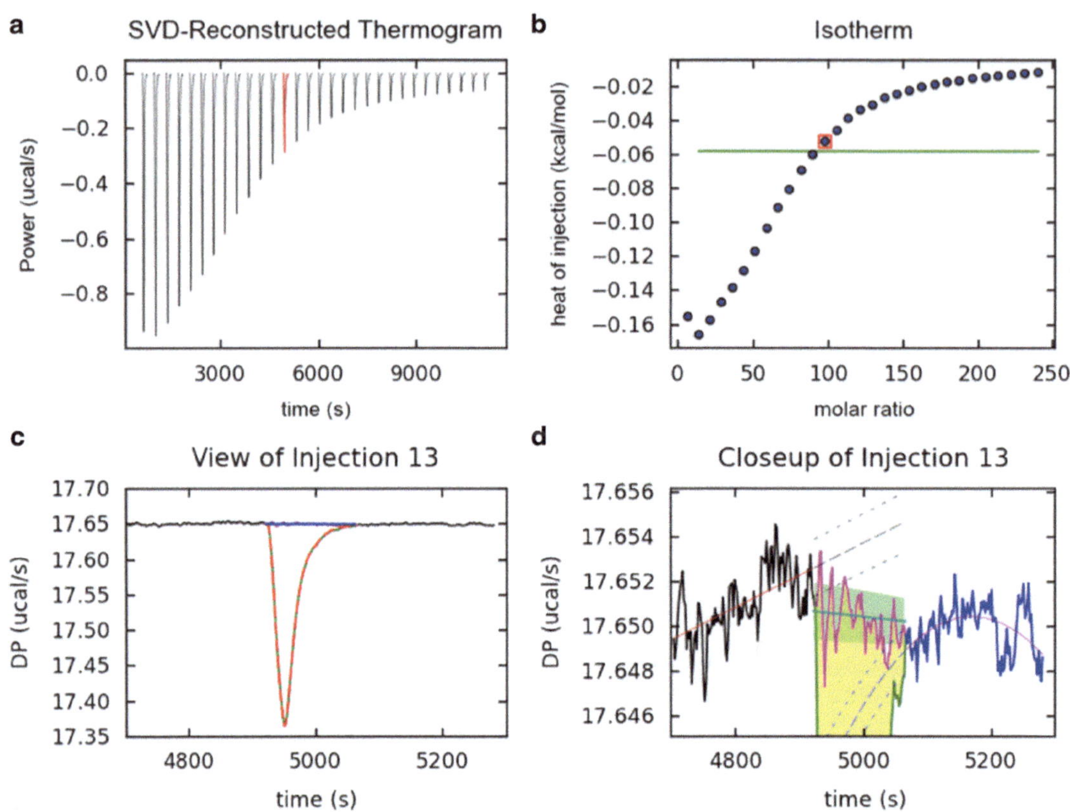

Fig. 1 NITPIC output window for ITC uptake data of the cell-penetrating peptide penetratin. (**a**) Thermogram reconstructed by singular value decomposition (SVD), a computational method for peak-shape analysis. (**b**) Integrated heats of reaction. (**c**) Example view of an injection peak (e.g., injection 13). (**d**) Zoomed-in view of an injection peak where the area highlighted in yellow corresponds to the reaction heat. An in-depth description of the NITPIC software is found elsewhere [25]

and release data of penetratin is shown in Fig. 2 (for a description of the other sheets, *see* **Note 8**). The main worksheet contains two panels. The upper panel displays the normalized heat of reaction as a function of the injection number. Shown are experimental ITC data (red) with error bars (green) and predicted values (blue), which depend on the fitting parameters and the implicitly given surface potential of the membrane calculated using Gouy–Chapman theory. The lower panel depicts the charge density at the membrane surface versus the injection number. It contains the values calculated from the membrane surface potential according to Gouy–Chapman theory (blue) and a second, independent expression for the charge density (red) derived from its definition. In the upper left corner of the worksheet, the fitting parameters are provided in column C and are flanked by their lower and upper bounds in columns B and D, respectively. The experimental settings are given below in column B.

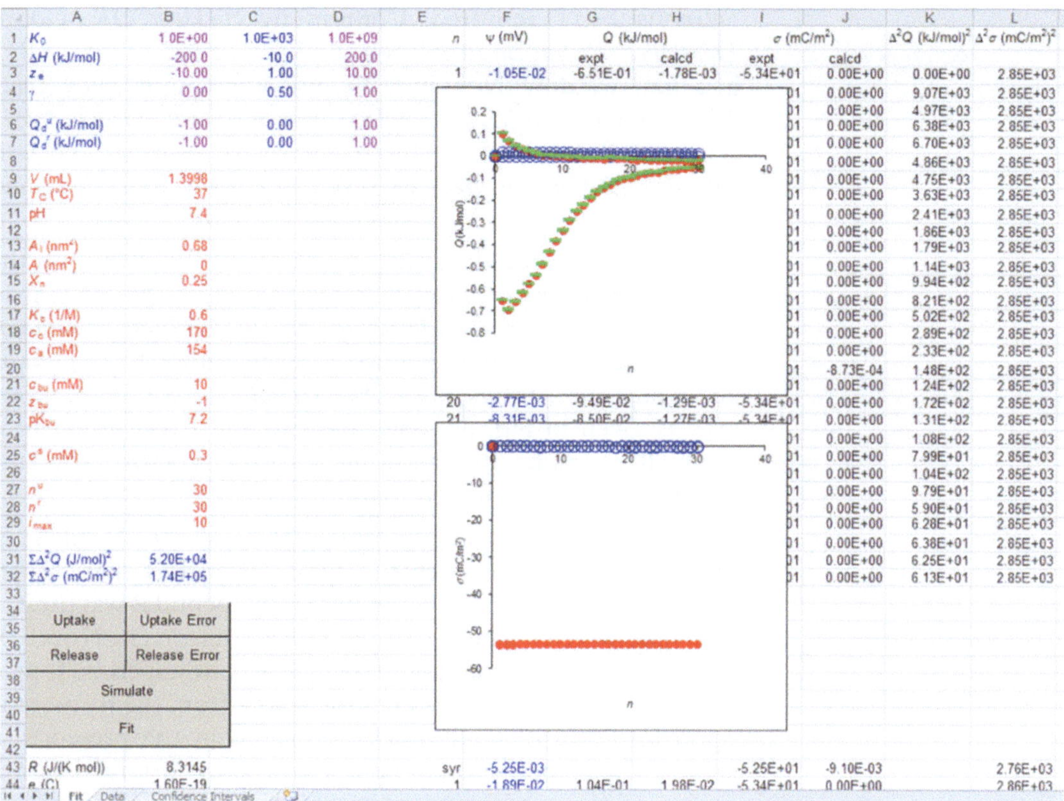

Fig. 2 Main worksheet of the spreadsheet program for the analysis of uptake and release experiments showing data of penetratin after data loading but before fitting. Cells B9–B29 contain the experimental settings, while cells B31 and B32 give the sums of squared residuals (SSRs) for the heat, Q, and the electrostatic surface density, σ, respectively. The values of the fitting parameters are provided in cells C1–C7 and are flanked by their lower and upper bounds in columns B and D, respectively. *Gray* buttons are used for loading data into the analysis program (*Uptake*, *Uptake Error*, *Release*, and *Release Error*), simulating data using the current set of parameter values (*Simulate*), and fitting data (*Fit*). The *upper panel* shows the normalized heat of reaction as a function of the injection number. This contains the experimental ITC data (*red*) with error bars (*green*) as well as the predicted values (*blue*). Note that uptake heats are exothermic (i.e., negative), whereas release heats are endothermic (i.e., positive). The *lower panel* depicts the charge density at the membrane surface vs. the injection number. It contains the values calculated from the membrane surface potential according to Gouy–Chapman theory (*blue*) as well as those computed from a second, independent expression for the charge density derived from its definition (*red*)

3. Upon opening the program, messages may appear asking if you want to *Enable content* and to *Activate macros* contained in the program. Confirm by clicking. If the latter message does not show up, the security level of your Excel installation is probably too high. You can fix this by opening any Excel file and clicking *Developer: Macro Security* and choosing *Disable all macros with notification* in the *Macro Settings* category (*see* **Note 9**).
4. This step applies only the first time you use the fitting program on a given computer. Click *Developer: Add-Ins* and make sure

Solver is NOT activated. Confirm by clicking OK. Go to *Developer*: *Visual Basic* to open the Visual Basic Editor (VBE) in a new window. Now in VBE, not Excel itself, click *Tools*: *References* and make sure *MISSING*: *SOLVER.XLA* is NOT activated. Return to Excel, go to *Developer*: *Add-Ins*, and activate *Solver*. Then go back to VBE, click *Tools*: *References*, and activate *Solver*. Close VBE and return to Excel.

5. Check the Solver settings [24]. Click *Data*: *Solver* to open the *Solver Parameters* window. Settings should be the same as in Fig. 3a except for *Set Objective* and *By Changing Variable Cells*, which may contain symbols one can ignore. Click *Options* and adjust the settings in the *All Methods* and *GRG Nonlinear* tabs as given in Fig. 3b. Save and exit the Excel fitting program. Reopen it, again allowing it to use macros. Now, the program is ready to fit your data.
6. Click the *Uptake* button (*see* **Note 10**). Load the .dat file generated by NITPIC (cf. **step 1**). Uptake refers to a classical partitioning experiment in which solute (e.g., peptide) was titrated with lipid vesicles. Then, click the *Release* button (*see* **Note 10**) and load the corresponding .dat file. A release experiment provides the heats of desorption of the solute from the membrane upon injection of lipid vesicles prepared in the presence of solute. The combination of both assays allows for the determination of the membrane permeability to the solute, represented by the lipid accessibility factor, γ (for assessment of either uptake or release data alone, *see* **Notes 11** and **12**).
7. Load standard error data from the peak integration process performed by NITPIC. To load the uptake error data, click *Uptake Error*. In the pop-up window, select *All Files* and open the .errordat file generated by NITPIC. Do the same with the release error data by clicking the *Release Error* button. In the upper panel of the "Fit" sheet, the error bars are indicated in green.
8. Specify lower and upper bounds (i.e., the minimum and maximum allowed values; *see* **Note 13**) for the fitting parameters. In the worksheet, columns A and B correspond to the lower and upper bounds, respectively. From top to bottom, these are:
   (a) Bounds on the intrinsic partition coefficient, K_0 (*see* **Note 14**)
   (b) Bounds on the molar transfer enthalpy, ΔH (*see* **Note 15**)
   (c) Bounds on the effective charge number, z_e (*see* **Note 16**)
   (d) Bounds on the lipid accessibility factor, γ (*see* **Notes 12** and **17**)
   (e) Bounds on the heat of dilution for the uptake experiment, Q_d^u (*see* **Note 18**)
   (f) Bounds on the heat of dilution for the release experiment, Q_d^r (*see* **Note 18**)

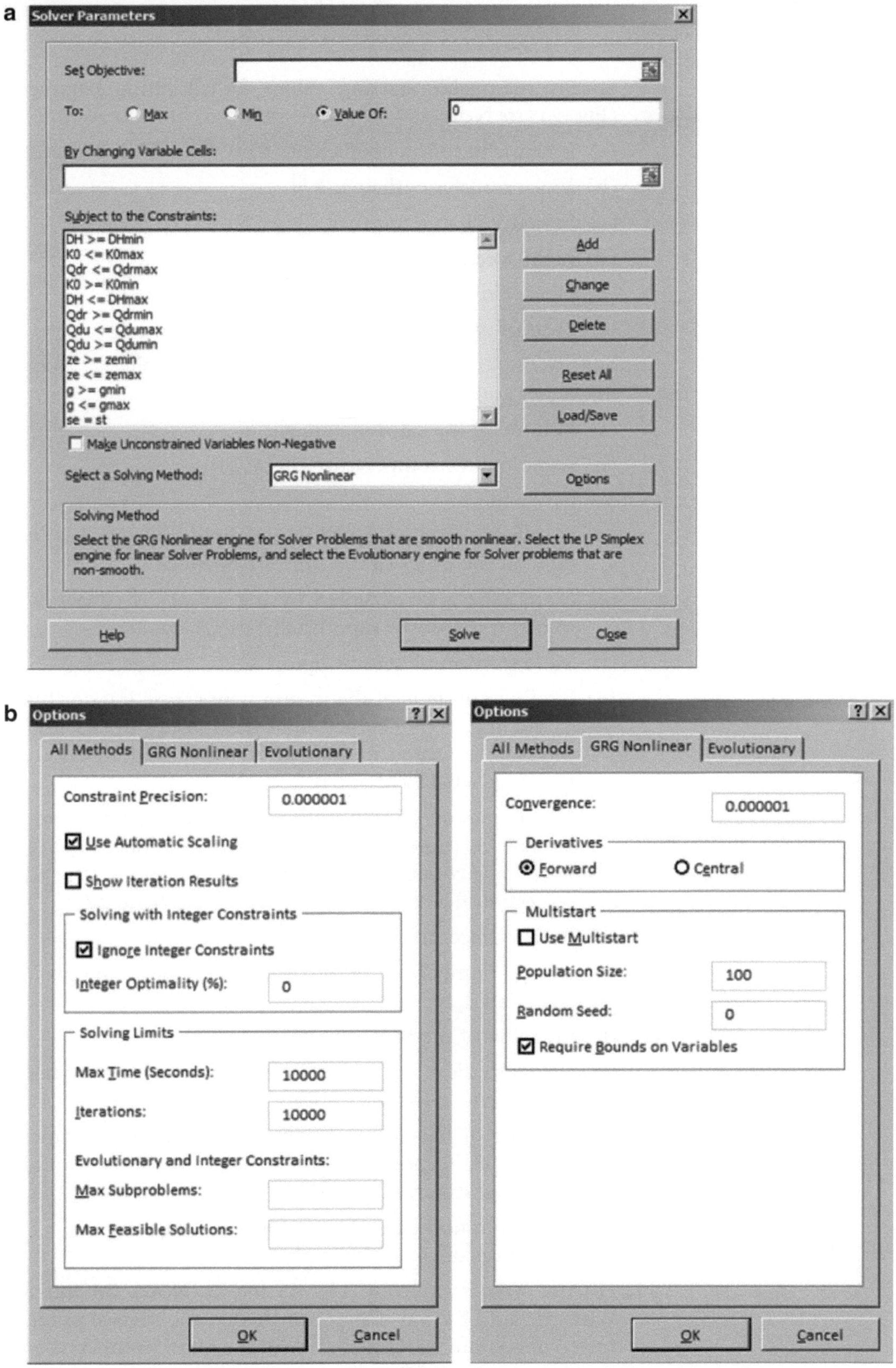

a

b

Fig. 3 Recommended Solver settings. (**a**) *Solver Parameters* window. (**b**) *All Methods* and *GRG Nonlinear* tabs in the *Options* window. For proper functioning of the program, make sure all settings are as given in these panels

9. Specify reasonable starting values for all fitting parameters (Fig. 2) (*see* **Note 19**), which from top to bottom are the:
   (a) Intrinsic partition coefficient, K_0
   (b) Molar transfer enthalpy, ΔH
   (c) Effective charge number, z_e (*see* **Note 20**)
   (d) Lipid accessibility factor, γ (*see* **Notes 12** and **17**)
   (e) Heat of dilution for the uptake experiment, Q_d^u
   (f) Heat of dilution for the release experiment, Q_d^r
10. Specify the known experimental parameters (Fig. 2), including the:
   (a) Volume of the calorimeter cell, V
   (b) Experimental temperature (in °C), T_C
   (c) pH value
   (d) Membrane area occupied by one lipid molecule, A_l (typically on the order of 0.7 nm^2; *see* **Note 21**)
   (e) Membrane area occupied by the solute, A (depends on solute; if it adsorbs superficially, use $A=0$)
   (f) Mole fraction of anionic lipid (X_n)
   (g) Binding constant of the monovalent cation in your buffer to the negatively charged lipid headgroup, K_c (for POPG and Na^+, this value amounts to $K_c=0.6/M$, for POPG and K^+ it is $K_c=0.15/M$ [26])
   (h) Concentrations of monovalent cations and anions, c_c and c_a, respectively (note that these two concentrations might differ since some counterions (either cations or anions, depending on your buffer) come with the buffer (*see* **Note 22**))
   (i) Buffer concentration, c_{bu}
   (j) Charge number of the buffer when the buffering group is protonated, z_{bu}
   (k) pK value of the buffering group, pK_{bu}
   (l) Concentration of solute in the syringe during release experiment, c^s (irrelevant for uptake experiment)
   (m) Number of injections of uptake and release experiments, n^u and n^r, respectively (the maximum number of data points per experiment is 40)
   (n) Maximum number of iterations per fitting step, i_{max} (10 is a good default)
11. Hit the *Simulate* button (Fig. 4a and *see* **Notes 23** and **24**). With the aid of Gouy–Chapman theory, the surface potential (given in column F) is calculated for the set of parameters specified in **steps 9** and **10** without, however, changing these values. In the lower panel, you now see if the requirements imposed

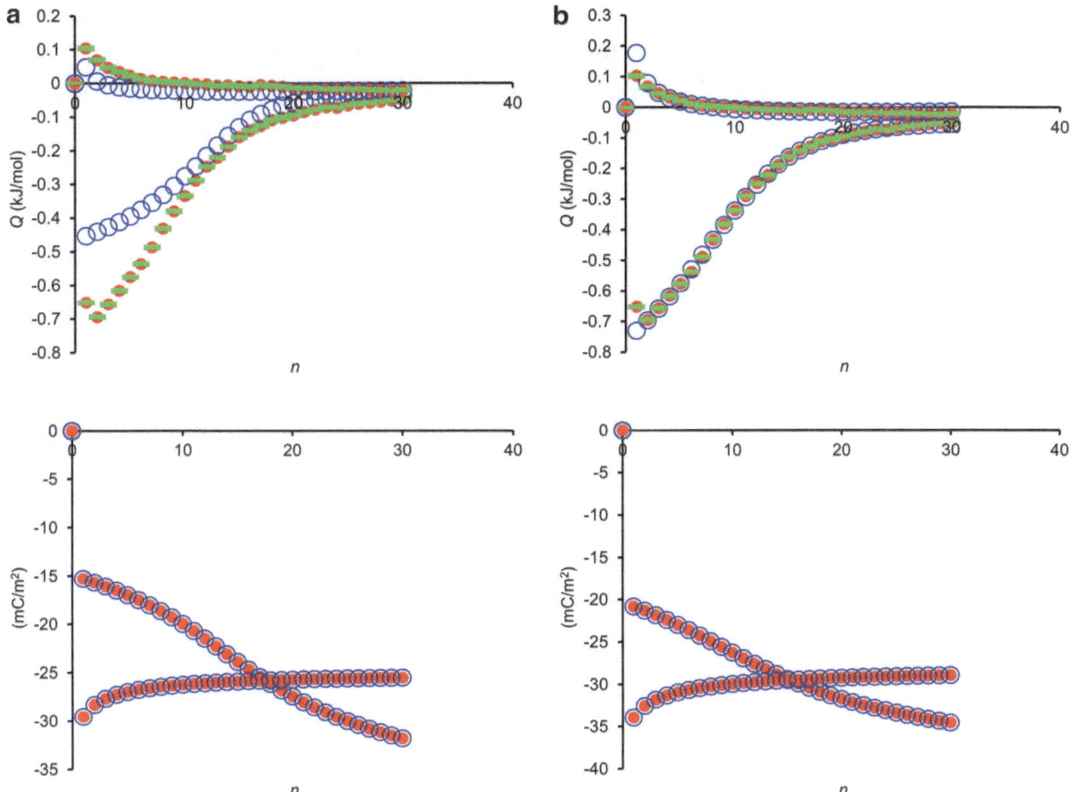

Fig. 4 Snapshots of uptake and release data analysis. *Upper panels* show the normalized heat of reaction as a function of the injection number, including the experimental ITC data (*red*) with error bars (*green*) as well as the predicted values (*blue*). Note that uptake heats are exothermic (i.e., negative), whereas release heats are endothermic (i.e., positive). *Lower panels* depict the charge density at the membrane surface versus the injection number. They contain the values calculated from the membrane surface potential according to Gouy–Chapman theory (*blue*) as well as those computed from a second, independent expression for the charge density derived from its definition (*red*). (**a**) Simulated data as obtained from **step 11** (Subheading 3.7). (**b**) Fitted data as obtained from **step 12** (Subheading 3.7)

by Gouy–Chapman theory are fulfilled, that is, if the blue and red symbols overlap. The upper panel draws the experimental ITC data (red) along with the simulation based on the parameter values given above (blue).

12. Click the *Fit* button (Fig. 4b and *see* **Note 24**) to initialize the fitting procedure. In contrast with the simulation carried out in **step 11**, fitting not only takes into account Gouy–Chapman theory but also adjusts the fitting parameters (whose starting values were defined in **step 9**) to find the best agreement between experimental and calculated data. This process may take a while but can be accelerated by choosing good starting values. Repeat this step by clicking *Fit* until the values of the fitting parameters and the sum of squared residuals (SSR; *see* **Notes 25** and **26**) no longer change. The two SSR values are

given in cells B31 and B32 (Fig. 2), where the former refers to the ITC data and the latter to Gouy–Chapman theory.

13. Repeat **steps 9**, **11**, and **12** using different starting values for the fitting parameters to check how reliable the results are. Ideally, the outcome of the fitting procedure should not depend on the starting values used, but in a few cases the fitting program may become trapped in local minima, particularly when poor starting values are chosen. The goodness of fit is quantified by the two SSR values (*see* **Note 27**).
14. Read the best-fit parameter values for the system under investigation. Most importantly, K_0 is the sought measure of membrane affinity, while γ reflects the ability of the solute to translocate across lipid bilayers under the conditions used in ITC uptake and release assays (*see* **Note 12**). For the penetratin–POPC/POPG system [14, 15], the values obtained for K_0, ΔH, z_e, and γ are ~1,300, ~−50 kJ/mol, −4.8, and 0.58, respectively. The latter value indicates that penetratin does not translocate across lipid bilayers (*see* **Note 12**).

4 Notes

1. Multilamellar vesicles can be stored under Ar or N_2 at −20 °C for a few months.
2. For such bidirectional extrusion, use an odd number of extrusion steps to ensure that the desired material does not end up in the initial syringe where unwanted particles that fail to cross the membranes might have been trapped.
3. LUVs are stable for a week under Ar or N_2 at 4 °C, whereas SUVs are best used immediately, although they may be kept under Ar or N_2 at room temperature for a few days. Freezing must be avoided for both LUVs and SUVs because it would rupture unilamellar vesicles.
4. Spacing times can be adjusted while the titration is in progress.
5. This protocol was designed and developed using datasets acquired on VP-ITC and iTC_{200} calorimeters from MicroCal/GE Healthcare. A global analysis of more than one set of uptake and release experiments is possible but requires modification of the Excel spreadsheet and macros as well as installation of the Premium Solver software, which can be purchased commercially.
6. NITPIC produces several files. The .dat file corresponds to the automated peak-analysis data, while the .error-dat file contains the standard error of the integrated peaks. Another file with extension .xp is specific for SEDPHAT, a powerful platform for global multi-method data analyses. However, this does not

contain the fitting model required for analyzing uptake and release data.

7. In a system obeying the laws of ideal mixing, solute partitioning between the membrane and the interfacial aqueous phase is described by the mole fraction partition coefficient. Use *ITC Uptake and Release Mole Fraction.xls* to analyze such data. By contrast, the mathematically simplest case of non-ideal mixing is based on the assumption that the mole ratio partition coefficient is constant, for which case you need to use *ITC Uptake and Release Mole Ratio.xls*. Both program versions take into account electrostatic effects to describe the partitioning of ionic solutes into initially uncharged or negatively charged lipid membranes.
8. The "Data" sheet contains the integrated heats generated by NITPIC with the corresponding concentrations of the components (i.e., lipid and peptide) for each injection as well as the standard error data. The "Confidence Intervals" sheet provides for the calculation of the confidence intervals of the fitting parameters (*see* **Note 27**).
9. The paths and items described here for setting up the macro security and activating the Solver add-in apply to Microsoft Excel 2010 for Windows. They may differ in other versions of Excel.
10. Always load data into a new (i.e., empty) copy of a fitting program.
11. Given the limitation to perform only one type of experiment, it should be considered that an uptake experiment generally produces larger heats, thus making data analysis more straightforward and reliable. By contrast, a release experiment may be helpful if the solute is not well soluble. In such a case, where only low solute concentrations are possible, one may prepare a mixture of lipid vesicles and solute and dilute this solution into buffer. To analyze either uptake or release data, use a fixed value for γ (**steps 8** and **9**; *see* **Notes 12** and **17**).
12. If translocation of the solute across the membrane takes place, it can freely distribute between both the outer and the inner leaflets of the vesicles on the one hand and both external and internal aqueous solutions (i.e., the vesicle lumen), on the other hand. The fraction of the total lipid accessible to the solute from the outside is quantified by the lipid accessibility factor, γ. If the solute can equilibrate across the membrane, $\gamma = 1.0$. If the membrane is impermeable to the solute, $\gamma = 0.5$ for LUVs ($d \geq 100$ nm) and $\gamma = 0.6$ for SUVs ($d \approx 30$ nm) since 60 % of the lipid molecules reside in the outer leaflet of highly curved SUVs. If $0.5 < \gamma < 1.0$ for LUVs or $0.6 < \gamma < 1.0$ for SUVs, only partial translocation occurred during the equilibration

time between injections. Such nonequilibrium situations should be interpreted with care [27].

13. Bounds on the fitting parameters prevent Solver from searching for a match between experimental data and calculated values based on parameter values that are physically meaningless or impossible. Recommendations for the choice of bounds on each variable are found in the corresponding notes. For most cases, the best-fit parameter values should be within the bounds that are given by default. If, however, after the fitting process, the value of any fitting parameter equals one of the respective bounds, this limiting bound should be extended within a physically meaningful range. Moreover, bounds can also be used to fix a fitting parameter at a desired value during the fitting procedure. This enables one to exclude one or more parameters from the fit (*see* **Note 17**).

14. The so-called intrinsic partition coefficient K_0 quantifies the partition equilibrium between the *interfacial* aqueous phase and the bilayer phase [9, 13]. Note that, being an equilibrium constant, K_0 is always >0.

15. The molar transfer enthalpy, ΔH, is the enthalpy change upon partitioning of 1 mol of solute from the aqueous phase into the bilayer phase. For endothermic or exothermic reactions, set the lower and upper bounds to zero, respectively.

16. The effective charge number, which may deviate from and often is smaller than the formal charge, is the solute valence that contributes to the surface potential [1]. Set the lower and upper bounds to zero for positively and negatively charged solutes, respectively.

17. Note that $\gamma > 0$, while $\gamma > 1$ indicates that the lipid concentration is higher than assumed. Since γ can be determined only from a combined analysis of uptake and release experiments together, you have to fix its value when only either uptake or release data are being analyzed (*see* **Notes 11** and **12**).

18. Q_d^u refers to the heat of dilution upon injecting lipid vesicles (uptake), while Q_d^r applies to the dilution of vesicles preloaded with solute (release). For endothermic and exothermic heats of dilution, set the lower and upper bounds to zero, respectively.

19. An easy way to check the suitability of the starting values is by visual comparison of the red and blue symbols corresponding to the experimental and calculated data, respectively, in the upper panel. Adjust the starting values for the fitting parameters in such a way that both datasets come close and conform to each other. However, this can be achieved only after the values of the membrane surface potential have been recalculated. To do this, click *Simulate* (**step 11**). Repeat these steps until a good approximation is achieved.

20. Use the formal charge as a starting value for z_e.
21. $A_l = 0.68$ nm^2 applies to POPC (1-palmitoyl-2-oleoyl-sn-glycero-3-phosphocholine) [28] but has also been previously used for 3:1 mixtures of POPC and POPG [14]. Other lipid area values are also available [29].
22. The values c_c and c_a account for additional salt in the solution as well as buffer counterions. Be aware that upon adjusting the pH of the buffer solution, the added acid or base contributes additional anions or cations, respectively. Moreover, more than one counterion might come with one buffer molecule (e.g., for phosphate buffer, the basic component Na_2HPO_4 contributes two Na^+ ions).
23. After simulation, the value displayed in cell B32 is usually smaller than 10^{-10}.
24. If i_{max} is reached before Solver finishes its calculations, it stops, and a pop-up window stating *Maximum number of iterations reached* appears. If this occurs in **step 11** or **12**, click *OK* and check the Solver settings again (**step 5**). For **step 12**, also check the suitability of the starting values of the fitting parameters again. If the pop-window still appears upon repeating **step 11** or **12**, set i_{max} (**step 10**) to a higher value.
25. A squared residual (SR) is the squared difference between a measured value and its corresponding calculated value. The SSR is the quantity Solver minimizes by adjusting the fitting parameters and thus serves as a measure of the goodness of fit.
26. If error data are loaded into the program, the squared residuals (SR) are weighted by dividing each SR by the corresponding squared standard error, which is the squared difference between the upper and lower error bounds. As a result, peaks suffering from large noise are weighted less in the fitting procedure.
27. In addition to determining the best-fit values of the adjustable parameters, it is advisable to estimate their confidence intervals. In nonlinear least-squares fitting, a robust approach for doing so consists in monitoring the changes in the SSR on perturbation of the corresponding parameter value, as described in detail elsewhere [24]. The "Confidence Intervals" sheet is described in a manual that can be obtained together with the fitting program from the authors.

Acknowledgments

We thank Sebastian Fiedler, Martin Textor, and Sebastian Unger (all University of Kaiserslautern) for helpful discussions and comments. This work was supported by the Phospholipid Research Centre, the Research Initiative Membrane Biology, and the Stiftung Rheinland-Pfalz für Innovation (grant 961-386261/969 to S.K.).

References

1. Seelig J (2004) Thermodynamics of lipid–peptide interactions. Biochim Biophys Acta 1666:40–50
2. Nakagaki M, Katoh I, Handa T (1981) Surface potential of lipid membrane estimated from the partitioning of methylene blue into liposomes. Biochemistry 20:2208–2212
3. Fernández-Vidal M, White SH, Ladokhin AS (2011) Membrane partitioning: "classical" and "nonclassical" hydrophobic effects. J Membr Biol 239:5–14
4. Silvius JR, Nabi IR (2006) Fluorescence-quenching and resonance energy transfer studies of lipid microdomains in model and biological membranes. Mol Membr Biol 23:5–16
5. Wang T, Cady S, Hong M (2012) NMR determination of protein partitioning into membrane domains with different curvatures and application to the influenza M2 peptide. Biophys J 102:787–794
6. Scheidt HA, Huster D (2008) The interaction of small molecules with phospholipid membranes studied by 1H NOESY NMR under magic-angle spinning. Acta Pharmacol Sin 29:35–49
7. Lee TZ, Mozsolits H, Aguilar MI (2001) Measurement of the affinity of melittin for zwitterionic and anionic membranes using immobilised lipid biosensors. J Pept Res 58:464–476
8. Heerklotz HH, Binder H, Epand RM (1999) A "release" protocol for isothermal titration calorimetry. Biophys J 76:2606–2613
9. Tsamaloukas A, Keller S, Heerklotz H (2007) Uptake and release protocol for assessing membrane binding and permeation by way of isothermal titration calorimetry. Nat Protoc 2:695–704
10. Aveyard R, Haydon DA (1973) An introduction to the principles of surface chemistry. Cambridge University Press, Cambridge
11. McLaughlin S (1977) Electrostatic potentials at membrane-solution interfaces. Curr Top Membr 9:71–144
12. McLaughlin S (1989) The electrostatic properties of membranes. Annu Rev Biophys Biophys Chem 18:113–136
13. Keller S, Heerklotz H, Blume A (2006) Monitoring lipid membrane translocation of sodium dodecyl sulfate by isothermal titration calorimetry. J Am Chem Soc 128: 1279–1286
14. Keller S, Böthe M, Bienert M et al (2007) A simple fluorescence-spectroscopic membrane translocation assay. Chembiochem 8:546–552
15. Bárány-Wallje E, Keller S, Serowy S et al (2005) A critical reassessment of penetratin translocation across lipid membranes. Biophys J 89: 2513–2521
16. Cambridge SB, Geissler D, Keller S et al (2006) A caged doxycycline analogue for photoactivated gene expression. Angew Chem Int Ed 45:2229–2231
17. Hagen V, Dekowski B, Nache V et al (2005) Coumarinylmethyl esters for ultrafast release of high concentrations of cyclic nucleotides upon one- and two-photon photolysis. Angew Chem Int Ed 44:7887–7891
18. Gilbert D, Funk K, Lechler R et al (2007) Caged capsaicins: new tools for the examination of TRPV1 channels in somatosensory neurons. Chembiochem 8:89–97
19. Geissler D, Antonenko YN, Schmidt R et al (2005) (Coumarin-4-yl) methyl esters as highly efficient, ultrafast phototriggers for protons and their application to acidifying membrane surfaces. Angew Chem Int Ed 44: 1195–1198
20. Keller S, Heerklotz H, Jahnke N et al (2006) Thermodynamics of lipid membrane solubilization by sodium dodecyl sulfate. Biophys J 90:4509–4521
21. Heerklotz H, Tsamaloukas AD, Keller S (2009) Monitoring detergent-mediated solubilization and reconstitution of lipid membranes by isothermal titration calorimetry. Nat Protoc 4:686–697
22. Krylova OO, Jahnke N, Keller S (2010) Membrane solubilisation and reconstitution by octylglucoside: comparison of synthetic lipid and natural lipid extract by isothermal titration calorimetry. Biophys Chem 150:105–111
23. Broecker J, Keller S (2010) Membrane translocation assayed by fluorescence spectroscopy. Methods Mol Biol 606:271–289
24. Kemmer G, Keller S (2010) Nonlinear least-squares data fitting in Excel spreadsheets. Nat Protoc 5:267–281
25. Keller S, Vargas C, Zhao H et al (2012) High-precision isothermal titration calorimetry with automated peak-shape analysis. Anal Chem 84:5066–5073
26. Tocanne J-F, Teissie J (1990) Ionization of phospholipids and phospholipid-supported interfacial lateral diffusion of protons in

membrane model systems. Biochim Biophys Acta 1031:111–142

27. Heerklotz H (2001) Membrane stress and permeabilization induced by asymmetric incorporation of compounds. Biophys J 81:184–195

28. Altenbach C, Seelig J (1984) Ca^{2+} Binding to phosphatidylcholine bilayers as studied by deuterium magnetic resonance. Evidence for the formation of a Ca^{2+} complex with two phospholipid molecules. Biochemistry 23: 3913–3920

29. Nagle JF, Tristram-Nagle S (2000) Structure of lipid bilayers. Biochim Biophys Acta 1469: 159–195

Part IV

Membrane Biogenesis and Dynamics

Chapter 17

Analyzing Membrane Dynamics with Live Cell Fluorescence Microscopy with a Focus on Yeast Mitochondria

Dirk Scholz, Johannes Förtsch, Stefan Böckler, Till Klecker, and Benedikt Westermann

Abstract

With the availability of increasing numbers of fluorescent protein variants and state-of-the-art imaging techniques, live cell microscopy has become a standard procedure in modern cell biology. Fluorescent markers are used to visualize the dynamic processes that take place in living cells, including the behavior of membrane-bound organelles. Here, we provide two examples of how we analyze the membrane dynamics of mitochondria in living yeast cells using wide field and confocal microscopy: (1) Long-term observation of mitochondrial shape changes using mitochondria-targeted fluorescent proteins and (2) monitoring the behavior of individual mitochondria using a mitochondria-targeted version of a photoconvertible fluorescent protein.

Key words Fluorescent proteins, Live cell imaging, Mitochondria, Organelle dynamics, *Saccharomyces cerevisiae*, Yeast

1 Introduction

Most membrane-bound organelles of eukaryotic cells are highly dynamic. Depending on intrinsic and extrinsic stimuli they fuse and divide, and they are moved in the cell by cytoskeleton-dependent transport mechanisms. Live cell fluorescence microscopy is an important method to study these processes [1]. It is being used to examine a great variety of different organisms and cell types, including unicellular microorganisms, fungi, mammalian and plant tissue culture cells, small translucent animals such as nematode worms or zebrafish larvae, and others. Live cell fluorescence microscopy can be used to capture fast events that occur within a second or less as well as cellular processes that take several hours or even days.

The choice of the imaging system depends on multiple parameters, including the size of the studied object (i.e., the dimensions

Doron Rapaport and Johannes M. Herrmann (eds.), *Membrane Biogenesis: Methods and Protocols*, Methods in Molecular Biology, vol. 1033, DOI 10.1007/978-1-62703-487-6_17,

of the labeled cellular structures and the thickness of the specimen), the speed of the process to be observed, the speed of data acquisition that can be achieved, the brightness of the signal, the sensitivity of detection, the available fluorescent markers, the sensitivity of the fluorescent dyes to photobleaching, and the viability of the specimen. When considering these parameters, it is obvious that each biological question and organism requires a special protocol that needs to be optimized.

In this chapter, we describe two procedures to illustrate the potential of live cell fluorescence microscopy for the study of membrane dynamics. While these examples are related to our own research, similar general principles and basic considerations also apply for the study of other organelles and cellular systems. We use the yeast *Saccharomyces cerevisiae* as a model organism to study the dynamic behavior of mitochondria [2, 3] and employ live cell fluorescence microscopy for high throughput screens as well as for mechanistic studies. The properties of yeast as a model organism for mitochondrial research [4] and straightforward procedures to analyze yeast mitochondrial morphology by staining with vital dyes, targeted fluorescent proteins, or immunostaining [5] have been described earlier in this series. Here, we describe two procedures for the analysis of mitochondrial dynamics: First the observation of mitochondrial shape changes over a long time period using mitochondria-targeted fluorescent proteins; and second the use of a mitochondria-targeted version of the photoconvertible Dendra2 protein to observe the behavior of individual mitochondria in yeast.

Numerous plasmids are available that express GFP or other fluorescent proteins fused to a mitochondrial targeting sequence in yeast (see for example [5–7]) and other organisms. Transformation of these plasmids into cells allows the staining of mitochondria independent of respiratory activity and the presence of a high membrane potential. Mitochondria-targeted fluorescent proteins produce virtually no background staining. They can be used with regular wide field fluorescence microscopes, spinning disc, or scanning confocal microscopes. We describe a procedure to image mitochondria in yeast cells that are trapped in a microfluidic chamber where they can be observed for 1 h or longer.

Classical in vivo mitochondrial fusion assays are based on labeling of mitochondria with differently colored fluorescent proteins or vital dyes in different cells. Upon fusion of these cells (induced by mating, viruses, polyethylene glycol treatment, etc.) intermixing of fluorescent labels indicates fusion. This strategy was successfully employed for yeast [8] and mammalian [9–12] and plant [13] tissue culture cells. However, this assay requires significant manipulations of cells before mitochondrial fusion can be analyzed. Furthermore, expression of marker proteins has to be shut-off after cell fusion in order to prevent import of newly synthesized fluorescent proteins that would yield false-positive results. More recently, a variety of switchable fluorescent proteins became available [14, 15] that can be used to observe the behavior of single organelles in living cells without the necessity of cell fusion or shut-off of protein expression. Mitochondria-targeted variants of switchable proteins are expressed in cells, fluorescence of individual mitochondria or mitochondrial subpopulations is converted by irradiation with a laser beam, and fusion can be observed by merging of converted and non-converted fluorescent mitochondria. Several examples illustrate that this strategy is useful to monitor mitochondrial fusion in living cells, including photoconversion of GFP at low oxygen concentrations in yeast [16], PAGFP in mammalian cells [17, 18], and Kaede in onion epidermal cells [19]. Here we describe the use of mitochondria-targeted Dendra2 to image mitochondrial fusion in yeast.

2 Materials

2.1 Long-Term Observation of Yeast Cells Expressing Mitochondria-Targeted Fluorescent Proteins

1. Yeast strain expressing a mitochondria-targeted fluorescent protein (*see* **Note 1**): We used the wild-type strain BY4742 (*MAT*α, *ura3*, *lys2*, *his3*, *leu2*) [20] transformed with plasmid pYX142-mtERFP. For construction of pYX142-mtERFP, the yeast-enhanced mRFP (yEmRFP) coding sequence [21] was amplified by PCR and cloned into the *Bgl*II and *Xho*I sites of pYX142-mtGFP thereby replacing the GFP coding sequence by yEmRFP [6]. The resulting plasmid allows expression of a mitochondrial matrix-targeted yEmRFP protein under control of the constitutive *TPI* promoter. The plasmid vector contains an *LEU2* marker and an *ARS/CEN* sequence for propagation in *Saccharomyces cerevisiae.*
2. Yeast growth medium (*see* **Note 2**). SD minimal medium (supplemented with Ura, His, Lys): 0.67 % Bacto-yeast nitrogen base without amino acids; 2 % glucose; 20 mg/L uracil; 20 mg/L histidine; 30 mg/L lysine. YPD medium: 1 % Bacto-yeast extract; 2 % Bacto-peptone; 2 % glucose.
3. Microfluidic chamber (*see* **Note 3**): CellASIC Onix™ Microfluidic Perfusion System (CellASIC Corp., Hayward, CA, USA), ONIX™ Microfluidic Plates (Y04C Yeast Perfusion Plate, 3.5–5 μm), and ONIX™ FG Software.
4. Live cell microscopy system (*see* **Note 4**). We used a Leica DMI 6000 wide field fluorescence microscope (Leica Microsystems, Wetzlar, Germany) equipped with an HCX PL APO 100×/1.40–0.70 oil objective, Leica DFC360FX camera (high speed kit), an incubator BL (PeCon GmbH, Erbach, Germany), and Leica LAS AF Software Version 2.1.0. Filter sets suitable for yEmRFP are N3 (excitation 546 ± 6 nm, dichroic mirror at 585 nm, emission 600 ± 20 nm) and for GFP L5 (excitation 480 ± 20 nm, dichroic mirror at 505 nm, emission 527 ± 15 nm).

2.2 Mitochondrial Dynamics in Yeast Cells Imaged with Photoconvertible Dendra2

1. Yeast strain expressing a mitochondria-targeted photoconvertible fluorescent protein (*see* **Note 5**): We used the wild-type strain YPH500 (*MAT*α, *ura3*, *lys2*, *ade2*, *trp1*, *his3*, *leu2*) [22] transformed with plasmid pYX232-mtDendra2. For construction of pYX232-mtDendra2, the Dendra2 coding sequence [23] was amplified by PCR and cloned into the *Bam*HI and *Hin*dIII sites of pYX232-mtGFP [6] thereby replacing the GFP coding sequence by Dendra2. The resulting plasmid allows expression of a mitochondrial matrix-targeted Dendra2 protein under control of the constitutive *TPI* promoter. The plasmid vector contains a *TRP1* marker and a *2 μ* origin for propagation in *Saccharomyces cerevisiae.*
2. Yeast growth medium (*see* **Note 2**). SD minimal medium (supplemented with Ura, His, Leu, Lys): 0.67 % Bacto-yeast nitrogen base without amino acids; 2 % glucose; 20 mg/mL adenine; 20 mg/L uracil; 20 mg/L histidine; 60 mg/L leucine; 30 mg/L lysine.
3. 1 % low melting point agarose in medium. Prepare freshly and keep at 45 °C until use.
4. Scanning confocal microscope (*see* **Note 6**). We used a Leica TCS SP5 system (Leica Microsystems, Wetzlar, Germany) in combination with an inverted Leica DMI 6000 CS Trino microscope equipped with an HCX PL APO CS 63.0×/1.40 oil UV objective and LAS AF SP5 MicroLab software (Leica). This system is equipped with a Diode UV laser (405 nm/50 mW), an Argon laser (458 nm/5 mW, 476 nm/5 mW, 488 nm/20 mW, 496 nm/5 mW, 514 nm/20 mW), a DPSS laser (561 nm/20 mW), and the respective acousto-optic tunable filters (AOTF).

3 Methods

3.1 Long-Term Observation of Yeast Cells Expressing Mitochondria-Targeted Fluorescent Proteins

A representative experiment showing mitochondrial movements and shape changes over a time period of 1 h in a wild-type yeast cell expressing mitochondria-targeted yEmRFP is shown in Fig. 1.

1. Grow a pre-culture of yeast cells in liquid SD medium overnight (*see* **Note 2**). Use these cells to inoculate a culture in rich medium and grow to the logarithmic growth phase.
2. Prepare the microfluidic chamber (*see* **Note 3**). When using CellASIC Y04C Yeast Perfusion Plates, fill channels 1–8 with 100 μL YPD medium, rinse channels 1–8 for 5 min at 8 psi pressure with YPD medium, and remove the medium. Fill channels 1–6 with 300 μL YPD medium; leave channel 7 for flow-through. Load channel 8 with ca. 10^7 cells (*see* **Note 7**) of the logarithmic culture. Load the microscopy chamber with cells

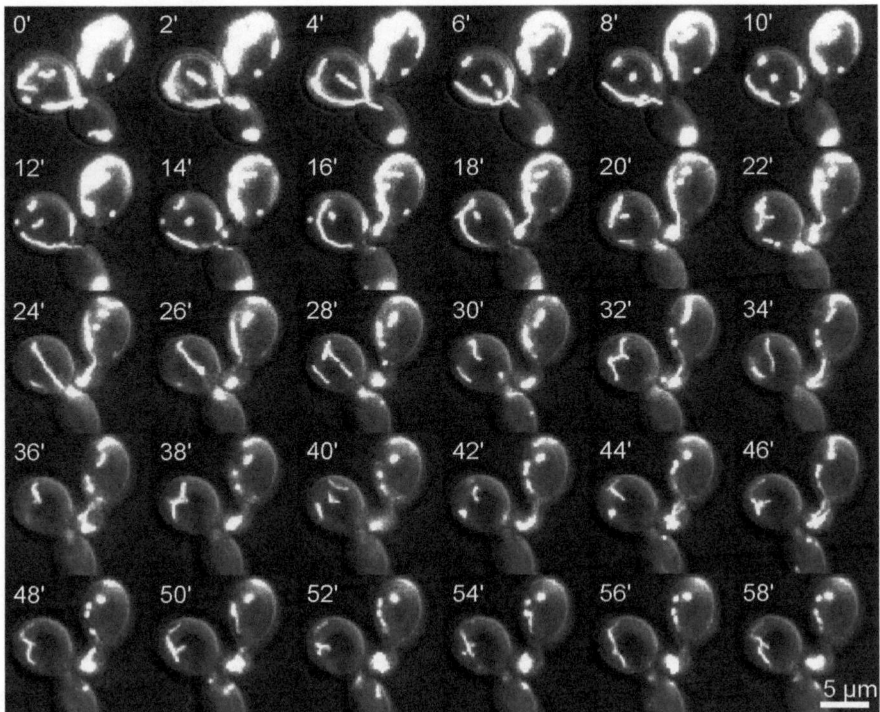

Fig. 1 Mitochondrial movements and shape changes in a wild-type yeast cell. Yeast cells expressing mitochondria-targeted yeast-enhanced mRFP were trapped in a microfluidic chamber, constantly supplied with medium, and observed by wide field fluorescence microscopy. Z-stacks consisting of ten *x*/*y* planes were taken every 2 min. Merged images of DIC and maximum intensity projections of red fluorescence are shown. Note the emergence and growth of a bud of the right cell and accumulation of mitochondria in the newly formed bud

from channel 8 until the desired cell density is reached (ca. 1–3 times for 20 s at 8 psi). To remove cells that have not been trapped rinse for 5 min at 8 psi with medium from either one of channels 1–6. Then supply fresh medium from either one of channels 1–6 at a constant flow rate at 3 psi until the end of the experiment.

3. View the cells in a wide field fluorescence microscope. Select a field of view with representative cells and take a z-stack consisting of differential interference contrast (DIC) and fluorescence images. Take z-stacks at regular intervals (*see* **Note 8**).

3.2 Mitochondrial Dynamics in Yeast Cells Imaged with Photoconvertible Dendra2

A representative experiment showing a mitochondrial fusion event in a wild-type yeast cell expressing mitochondria-targeted Dendra2 is shown in Fig. 2.

1. Grow yeast cells in liquid culture in SD medium to logarithmic growth phase (*see* **Note 2**).
2. If cell density is low, concentrate cells by a brief centrifugation in a microfuge.

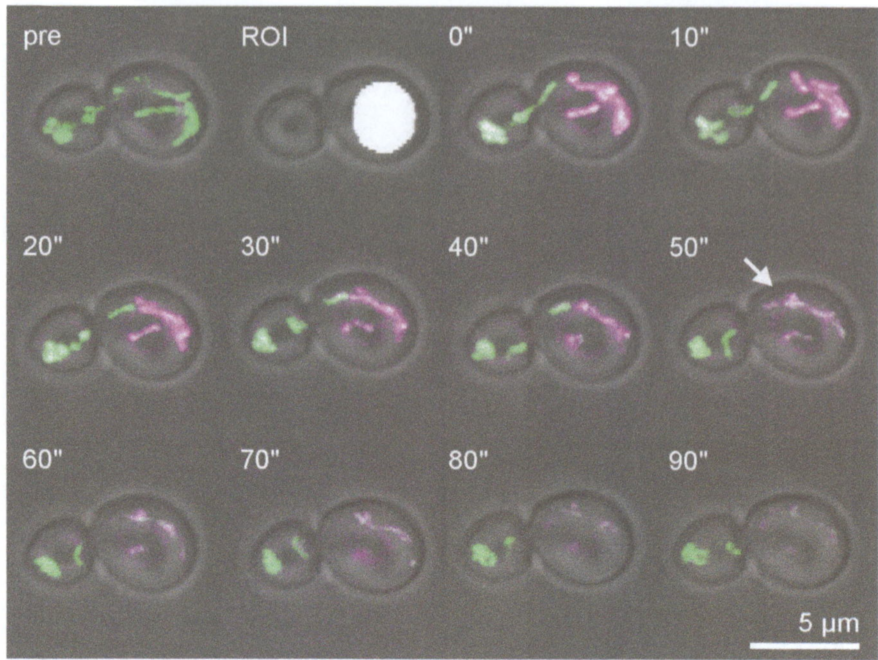

Fig. 2 Photoconversion of mitochondria-targeted Dendra2 in a wild-type yeast cell. Cells were grown to logarithmic growth phase in glucose-containing minimal medium and observed by DIC and confocal fluorescence microscopy. Photoconversion was performed by six z-stack scans of the region of interest (ROI). Mitochondrial behavior was observed every 10 s after photoconversion. Merged images of DIC and maximum intensity projections of green and red fluorescence are shown. Pre, fluorescence before photo-conversion; *ROI* region of interest (the UV-irradiated area is indicated in white, this image does not show fluorescence signals). The *arrow* points to a mitochondrial fusion event (merged *green* and *magenta* signals appear *white*)

3. Mix 10 µL yeast cell suspension with an equal volume of 1 % low melting point agarose, pipet 5 µL of this cell suspension onto a microscope slide, cover it with an 18 × 18 mm cover slip, and allow the agarose to solidify for ca. 5 min at room temperature.
4. View the cells in a scanning confocal microscope. First, focus on yeast cells using DIC and briefly check for the presence of green mitochondrial fluorescence and the absence of red fluorescence using wide field microscopy with low excitation intensity. Then, select a cell and take a z-stack consisting of DIC, green fluorescence (excitation 488 nm, emission 500–550 nm), and red fluorescence (excitation 561 nm, emission 575–700 nm). It is recommended to use bidirectional scanning and obtain DIC, green fluorescence, and red fluorescence images simultaneously. For the experiment shown in Fig. 2 the following settings were used: Diode UV laser, on; Argon laser (20 % power); DPSS laser, on; AOTF (405) 10 %; AOTF (488) 4 %; AOTF (561) 20 %. Data acquisition and photoconversion was performed in LiveDataMode.

5. Select a region of interest (ROI) (*see* **Note 9**) and start the photoconversion of Dendra2 with UV irradiation (405 nm) (*see* **Note 10**).
6. Take z-stacks as above (**step 4**) immediately after photoconversion and at several time points afterwards (*see* **Note 11**).

4 Notes

1. Here we used mitochondria-targeted yEmRFP in wild-type yeast. Similar protocols can be used for other fluorescent protein variants, organelles, strains, or microorganisms. We also made good experience with mitochondria-targeted GFP [6] and frequently analyze yEmRFP-labeled wild-type cells together with GFP-labeled mutant cells in the same sample (in this case, both strains must be of the same mating type to avoid mating and cell fusion). It is recommended to target fluorescent reporter proteins to the matrix, as foreign proteins targeted to other mitochondrial subcompartments may alter mitochondrial morphology when expressed at high levels. Although we have not analyzed this systematically, we observed that high-level expression of outer membrane targeted GFP sometimes induces fragmentation of mitochondria.
2. The use of minimal medium in the pre-culture allows efficient maintenance of plasmids in yeast cells and thus increases the proportion of cells expressing the fluorescent protein. The composition of minimal medium depends on the strain/plasmid combination that is used. When the cells do not contain additional plasmids, it is recommended to use rich medium during microscopy to keep the cells in good shape. Depending on the strain and experiment, other carbon sources can be used.
3. For short-term observations of yeast cells (i.e., a few minutes), it suffices to immobilize the cells in low melting point agarose in medium (*see* Subheading 2.2, **item 3** and Subheading 3.2, **step3**). For long-term observations, a variety of microfluidic chambers is available for different organisms and cell types.
4. Live cell fluorescence microscopy of yeast cells can be performed with any wide field, spinning disc, or scanning confocal microscope system. Wide field fluorescence microscopes are well suited for long-term observations, spinning disc confocal microscopy is the method of choice to capture fast events, and laser scanning confocal microscopes produce high resolution images. To obtain 3D information of whole cells, the microscope has to be motorized to move the stage along the *z*-axis. 3D image data can be reconstructed for viewing at any angle, or they can be displayed in 2D as a maximum intensity projection (*see* Figs. 1 and 2). Z-stacks taken with a wide field microscope can be further improved by deconvolution, a computational

technique that removes noise, scatter, glare, and blur [24]. An inverted microscope system is required when a microfluidic chamber is used.

5. We used mitochondria-targeted Dendra2 in wild-type yeast. Similar protocols can be used for other photoconvertible fluorescent proteins, organelles, strains, or organisms.
6. The experiment can be performed with any microscope system that allows photoconversion of an ROI in a confocal mode, and excitation and detection of the two fluorescent forms of Dendra2. Dendra2 can be activated by either UV, violet, or blue light. The excitation maximum of Dendra2 is 490 nm before activation and 553 nm after activation; the emission maximum is 507 nm before activation and 573 nm after activation.
7. The number of cells is not very critical; however, a too high cell density may cause problems when loading the cells. 10^7 cells approximately correspond to 0.3 mL of a logarithmically growing or 0.2 mL of a stationary culture with an OD_{600} of 1.0. This value might vary depending on strain and culture conditions. When cell density is low, cells can be concentrated by a brief centrifugation in a microfuge before loading of the microfluidic chamber.
8. The length and frequency of the time intervals depends on the speed of the process to be observed, the sensitivity of the reporter protein to photobleaching, and the number of x/y planes taken for each z-stack. When using deconvolution, the z-stacks should consist of as many x/y planes as possible to avoid artifacts.
9. Depending on the strain and experiment, it can be advantageous to perform photoconversion in a large area of the cell and follow the intermixing of the activated marker protein with the rest of the mitochondrial network, or alternatively to target single organelles to observe their fusion. Figure 2 shows an example of photoconversion of a large area within the mother cell and subsequent fusion with a single non-converted organelle.
10. If photoconversion involves a large volume of the cell, it should be performed in one or several serial scans of z-stacks.
11. Depending on the experiment, the time periods between the z-stacks can be varied from a few seconds up to several minutes.

Acknowledgment

This work was supported by the Deutsche Forschungsgemeinschaft through grants We 2174/4-2 and 5-1.

References

1. Stephens DJ, Allan VJ (2003) Light microscopy techniques for live cell imaging. Science 300:82–86
2. Merz S, Hammermeister M, Altmann K, Dürr M, Westermann B (2007) Molecular machinery of mitochondrial dynamics in yeast. Biol Chem 388:917–926
3. Westermann B (2010) Mitochondrial fusion and fission in cell life and death. Nat Rev Mol Cell Biol 11:872–884
4. Altmann K, Dürr M, Westermann B (2007) *Saccharomyces cerevisiae* as a model organism to study mitochondrial biology: general considerations and basic procedures. Methods Mol Biol 372:81–90
5. Swayne TC, Gay AC, Pon LA (2007) Fluorescence imaging of mitochondria in yeast. Methods Mol Biol 372:433–459
6. Westermann B, Neupert W (2000) Mitochondria-targeted green fluorescent proteins: convenient tools for the study of organelle biogenesis in *Saccharomyces cerevisiae*. Yeast 16:1421–1427
7. Swayne TC, Gay AC, Pon LA (2007) Visualization of mitochondria in budding yeast. Methods Cell Biol 80:591–626
8. Nunnari J, Marshall WF, Straight A, Murray A, Sedat JW, Walter P (1997) Mitochondrial transmission during mating in *Saccharomyces cerevisiae* is determined by mitochondrial fusion and fission and the intramitochondrial segregation of mitochondrial DNA. Mol Biol Cell 8:1233–1242
9. Chen H, Detmer SA, Ewald AJ, Griffin EE, Fraser SE, Chan DC (2003) Mitofusins Mfn1 and Mfn2 coordinately regulate mitochondrial fusion and are essential for embryonic development. J Cell Biol 160:189–200
10. Legros F, Lombes A, Frachon P, Rojo M (2002) Mitochondrial fusion in human cells is efficient, requires the inner membrane potential, and is mediated by mitofusins. Mol Biol Cell 13:4343–4354
11. Mattenberger Y, James DI, Martinou JC (2003) Fusion of mitochondria in mammalian cells is dependent on the mitochondrial inner membrane potential and independent of microtubules or actin. FEBS Lett 538:53–59
12. Ishihara N, Jofuku A, Eura Y, Mihara K (2003) Regulation of mitochondrial morphology by membrane potential, and DRP1-dependent division and FZO1-dependent fusion reaction in mammalian cells. Biochem Biophys Res Commun 301:891–898
13. Sheahan MB, McCurdy DW, Rose RJ (2005) Mitochondria as a connected population: ensuring continuity of the mitochondrial genome during plant cell dedifferentiation through massive mitochondrial fusion. Plant J 44:744–755
14. Chudakov DM, Matz MV, Lukyanov S, Lukyanov KA (2010) Fluorescent proteins and their applications in imaging living cells and tissues. Physiol Rev 90:1103–1163
15. Lukyanov KA, Chudakov DM, Lukyanov S, Verkhusha VV (2005) Innovation: photoactivatable fluorescent proteins. Nat Rev Mol Cell Biol 6:885–891
16. Jakobs S, Schauss AC, Hell SW (2003) Photoconversion of matrix targeted GFP enables analysis of continuity and intermixing of the mitochondrial lumen. FEBS Lett 554:194–200
17. Karbowski M, Arnoult D, Chen H, Chan DC, Smith CL, Youle RJ (2004) Quantitation of mitochondrial dynamics by photolabeling of individual organelles shows that mitochondrial fusion is blocked during the Bax activation phase of apoptosis. J Cell Biol 164:493–499
18. Molina AJA, Shirihai OS (2009) Monitoring mitochondrial dynamics with photoactivatable green fluorescent protein. Methods Enzymol 457:289–304
19. Arimura S, Yamamoto J, Aida GP, Nakazono M, Tsutsumi N (2004) Frequent fusion and fission of plant mitochondria with unequal nucleoid distribution. Proc Natl Acad Sci USA 101:7805–7808
20. Brachmann CB, Davies A, Cost GJ, Caputo E, Li J, Hieter P, Boeke JD (1998) Designer deletion strains derived from *Saccharomyces cerevisiae* S288C: a useful set of strains and plasmids for PCR-mediated gene disruption and other applications. Yeast 14:115–132
21. Keppler-Ross S, Noffz C, Dean N (2008) A new purple fluorescent color marker for genetic studies in *Saccharomyces cerevisiae* and *Candida albicans*. Genetics 179:705–710
22. Sikorski RS, Hieter P (1989) A system of shuttle vectors and host strains designed for efficient manipulation of DNA in *Saccharomyces cerevisiae*. Genetics 122:19–27
23. Chudakov DM, Lukyanov S, Lukyanov KA (2007) Tracking intracellular protein movements using photoswitchable fluorescent proteins PS-CFP2 and Dendra2. Nat Protoc 2:2024–2032
24. Wallace W, Schaefer LH, Swedlow JR (2001) A workingperson's guide to deconvolution in light microscopy. Biotechniques 31: 1076–1078

Chapter 18

Analysis of Protein Translocation into the Endoplasmic Reticulum of Human Cells

Johanna Dudek, Sven Lang, Stefan Schorr, Johannes Linxweiler, Markus Greiner, and Richard Zimmermann

Abstract

The development of small-interfering RNA (siRNA)-mediated gene-silencing strategies has made it possible to study the transport of precursors of soluble and membrane proteins into the endoplasmic reticulum (ER) of human cells. In these approaches, a certain target gene is silenced in the cell type of choice, followed by analysis of the effect of this silencing on the biogenesis of a single or set of precursor polypeptide(s) in cell culture or in cell-free assays involving semi-permeabilized cells and in vitro translations systems. These approaches allow for functional analysis of components of the ER-resident protein transport machinery as well as the elucidation of their potential cell-type variations and regulatory mechanisms. The gene-silencing and subsequent plasmid-based complementation carries the additional benefit of facilitating analysis of the consequences of disease-linked mutations in ER transport components.

Key words siRNA, Plasmid complementation, Transfection, Disease-linked mutations, Protein transport, ER-associated protein biogenesis, Endoplasmic reticulum, Human cells

1 Introduction

Traditionally, the early phase in the biogenesis of soluble and membrane proteins at the endoplasmic reticulum (ER) membrane of mammalian cells has been analyzed in cell-free assays comprising cytosolic extracts for in vitro synthesis of defined precursor polypeptides in combination with rough microsomes from canine pancreas or proteoliposomes with purified components [1, 2]. With the availability of small-interfering RNA (siRNA)-mediated gene-silencing strategies, it has become possible to study these fundamental cellular processes at the cellular level [3–5]. In these approaches, a certain target mRNA is degraded in the human cell type of choice (Fig. 1a), and the effect of this silencing on biogenesis of a defined set of precursor polypeptides is analyzed in cell culture or in cell-free assays that involve the traditional in vitro translation systems in combination with the so-called

Doron Rapaport and Johannes M. Herrmann (eds.), *Membrane Biogenesis: Methods and Protocols*, Methods in Molecular Biology, vol. 1033, DOI 10.1007/978-1-62703-487-6_18,

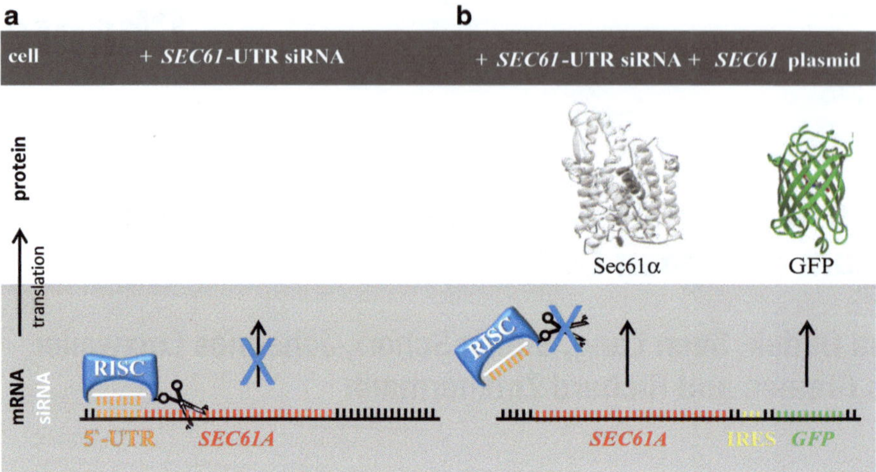

Fig. 1 Strategies for small-interfering RNA (siRNA)-mediated gene-silencing and plasmid complementation. (**a**) Effect of untranslated region (UTR)-targeted siRNA. (**b**) Effect of cDNA plasmid in the background of UTR-targeted siRNA. *IRES* internal ribosome entry site, *RISC* RNA-induced silencing complex

semi-permeabilized cells [6–13]. The latter approach also allows analysis of the stage of the translocation defect. In the case of signal peptide bearing precursor polypeptides, transport and membrane integration can be subdivided into three stages: targeting, initial membrane insertion (at the Sec61 complex), and completion of membrane integration or translocation. Initial insertion of precursor polypeptides into the Sec61 complex is assayed as modification by signal peptidase and/or oligosaccharyl transferase, completion of translocation by protease protection of the mature protein by the membrane (sequestration analysis). The gene-silencing and subsequent plasmid-based complementation experiments (Fig. 1b) have the advantage of facilitating analysis of the consequences of disease-linked mutations in transport components [13]. Compared to transport experiments with proteoliposomes, these strategies also carry the benefit that the role of single subunits of the transport machinery can be studied in a native context. Furthermore, these approaches allow for analysis of cell type-specific transport events and regulatory mechanisms.

First, a set of siRNAs directed against a certain target mRNA (including siRNAs directed against an untranslated region (UTR)) is evaluated in comparison to a negative control siRNA, such as a mismatch or scrambled sequence, with respect to silencing efficiencies and various biological readouts. The gene-silencing efficiencies are measured on the mRNA and protein level by quantitative methods, i.e., quantitative RT-PCR and quantitative western-blot analysis. To determine the suitable time point for protein transport experiments, however, it is crucial to determine the protein levels. Next, active siRNAs are titrated to determine the lowest possible levels.

For biological readouts, we routinely assess the effects of silencing of a specific gene on cell proliferation, cell viability, and cell and ER morphology in time-course experiments. These experiments identify a set of at least two active siRNAs (including a UTR-directed siRNA) and define the appropriate experimental time-window for the analysis of protein transport. The latter is particularly important in the case of essential transport components. Protein transport is analyzed in intact cells and in vitro, employing in vitro translation systems and semi-permeabilized cells prepared from the knock-down and control cells. As an ultimate control, a transport defect observed with at least one siRNA directed against the coding region of the target gene plus one UTR-targeting siRNA must be rescued by complementation of the UTR-siRNA-treated cells with a plasmid that allows expression of the respective cDNA (and is lacking the UTR) (Fig. 1). Ideally, the expression level of the cDNA of interest in the UTR-siRNA background should be comparable to the expression level of the target gene in completely untreated cells. Typically, we use internal ribosomal entry site (IRES)-green fluorescent protein (GFP) vectors that allow for quantitation of the plasmid transfection efficiency (Fig. 1b).

2 Materials

All plastic and glass wear has to be sterile and, therefore, is bought as such or is autoclaved or sterilized by the user in advance. Furthermore, gloves are worn throughout all experiments and cells are handled in laminar flow hoods. The handling can also be seen in a video [14].

2.1 Cells and Media

1. Routinely, we use HeLa cells (ATCC no. CCL-2) or PC3 cells (DSMZ no. ACC 465) (*see* **Note 1**). Alternatively, we have successfully employed murine NIH3T3 cells (ATCC no. CCL-92).
2. HeLa cells are cultivated in Dulbecco's Modified Eagle's Medium (DMEM) (Invitrogen Corporation, Carlsbad, CA, USA) containing 10 % fetal bovine serum (FBS) (Biochrom AG, Berlin, Germany) and 1 % penicillin/streptomycin (PAA Laboratories GmbH, Pasching, Austria) (*see* **Note 2**). Typically, FBS is pretreated for 30 min at 56 °C.

2.2 Transfections

1. When possible, we buy validated siRNAs from commercial sources (Invitrogen or Qiagen, Hilden, Germany) (*see* **Note 3**). As a negative control, we use the AllStars Negative Control siRNA from Qiagen. Dissolve the siRNAs (targeted siRNAs and control siRNA) in RNase-free deionized water to prepare a 20 μM stock solution. Mix using a vortex and observe solubilization visually. Store 50 μL aliquots at −20 °C.

2. Alternatively, plasmids that code for short-hairpin RNAs (shRNAs) can be used to create stable knock-down cell lines. However, this approach is not suitable for essential genes unless inducible promoters are used and we therefore do not discuss it here ([7, 8], also *see* **Note 4**). shRNA plasmids can be obtained from commercial sources (OriGene Technologies, Rockville, MD, USA).
3. To rescue the phenotype of *SEC61A1* silencing, the *SEC61A1* cDNA is inserted into the multiple cloning site (MCS) of a pCDNA3-IRES-GFP vector containing the cytomegalovirus (CMV) promoter, the MCS, the IRES, and the GFP coding sequence ([10–12], also *see* **Note 5**). The plasmids are purified using the Jet Star Plasmid Purification MIDI Kit according to the manufacturer's recommendations (Genomed GmbH, Löhne, Germany). Validated TrueORF cDNA plasmids are obtained from commercial sources, such as OriGene Technologies.
4. Many different transfection reagents are available from commercial sources. For HeLa cells, we routinely use HiPerFect from Qiagen for siRNAs and FuGENE HD from Promega Corporation (Madison, WI, USA) for plasmids (*see* **Note 6**).

2.3 Functional Assays

1. SPC buffer: 110 mM potassium acetate, 2 mM magnesium acetate, 20 mM HEPES/KOH, pH 7.2, adjusted at 2 °C.
2. Hepes buffer: 50 mM potassium acetate, 90 mM HEPES/KOH, pH 7.2, adjusted at 2 °C.
3. $CaCl_2$ solution: 200 mM $CaCl_2$ in deionized water.
4. Nuclease solution: 4,000 units of S7 nuclease (Roche Diagnostics GmbH, Mannheim, Germany)/mL of deionized water.
5. EGTA solution: 200 mM EGTA in deionized water.
6. TS-premix: 166 μg/mL BSA, 8.33 mM DTT, 0.83 mM ATP, 0.83 mM UTP, 0.83 mM CTP, 0.16 mM GTP, 10 mM magnesium acetate, 3.33 mM spermidine in 66.66 mM HEPES/KOH, pH 7.4.
7. Reagents for characterization of cells: BrdU Assay, WST Assay and LDH + assay (Roche Diagnostics GmbH) and Nuclear-ID™ Blue/Green cell viability reagent, Apoptosis/Necrosis Detection Kit (Enzo Life Sciences, Lausen, Switzerland).
8. PCR reagents: Superscript II, oligo-dT Primer (Invitrogen) and QIAamp RNA Blood Mini Kit, QuantiTect SYBR Green PCR Kit (Qiagen).
9. Western blotting reagents: Immobilon P (Millipore GmbH, Schwalbach, Germany), rabbit anti-phospho-eIF2α antibody (Cell Signaling Technology, Danvers, MA, USA), rabbit

anti-GAPDH antibody (Santa Cruz Biotechnology, Santa Cruz, CA, USA), mouse anti-β-actin antibody (Sigma Aldrich Chemie GmbH, Taufkirchen, Germany), and ECL™ Plex goat-anti-rabbit IgG-Cy5 conjugate, ECL™ Plex goat-anti-mouse IgG-Cy3 conjugate (GE Healthcare Life Sciences, Freiburg, Germany).

10. Reagents for protein transport: validated TrueORF cDNA plasmids coding for DDK-tagged model precursor proteins, murine anti-DDK antibody (OriGene Technologies), MG132, digitonin, phenylmethylsulfonyl fluoride (Merck Biosciences, Darmstadt, Germany), m^7GpppG, rRNasin Ribonuclease Inhibitor, SP6 RNA polymerase, T7 RNA polymerase, nuclease-treated rabbit reticulocyte lysate, TNT-coupled reticulocyte lysate (Promega Corporation), premium stabilized translation grade ^{35}S-methionine, EasyTag Express ^{35}S-cysteine/^{35}S-methionine Protein Labeling Mix (PerkinElmer LAS GmbH, Rodgau, Germany), and Proteinase K, RNase A, cycloheximide (Roche Diagnostics GmbH).

3 Methods

3.1 Cell Cultivation

Directly before transfection with siRNA, seed 5.2×10^5 HeLa cells in a 6-cm culture dish in DMEM containing 10 % FBS and 1 % penicillin/streptomycin and incubate at 37 °C in a humidified environment with 5 % CO_2 (final volume 3.9 mL).

3.2 Gene Silencing

Transfect the cells with targeted siRNAs, targeted UTR siRNA, or a negative control siRNA using HiPerFect Reagent according to the manufacturer's instructions (final concentration of siRNA: 20 nM).

1. Prepare siRNA-transfection mix freshly in a separate Eppendorf reaction cup prior to the actual transfection procedure. Add 20 μL HiPerFect transfection reagent to 4 μL of each siRNA (20 μM) that is dissolved in 80 μL of OptiMEM (Invitrogen Corporation). This mix is gently vortexed and incubated at room temperature for 10 min. Add the siRNA-transfection mix (0.104 mL) dropwise to the seeded 5.2×10^5 HeLa cells (3.9 mL).
2. After 24 h, change the medium (3.9 mL) and transfect the cells for a second time with fresh siRNA-transfection mix (0.104 mL).
3. At the time of cell harvest, cell number and viability are determined by employing the Countess® Automated Cell Counter (Invitrogen Corporation).
4. Silencing is evaluated by quantitative RT-PCR and western-blot analysis (*see* **Note 7**).

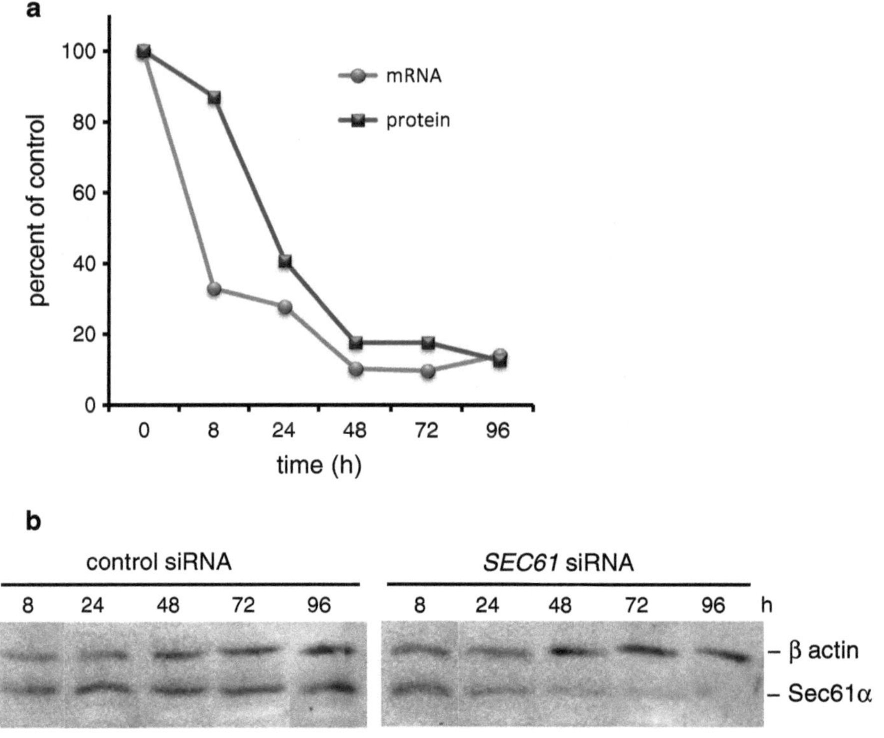

Fig. 2 Typical silencing results. HeLa cells were transfected with the indicated siRNAs. Silencing was evaluated by quantitative RT-PCR (**a**) and by quantitative western-blot analysis using anti-Sec61α antibody with anti-β-actin antibody as a control (**a**, **b**). The *SEC61A1* mRNA and Sec61α protein levels in *SEC61A1* siRNA-treated cells are given in % of control siRNA-treated cells

3.3 Evaluation of Gene Silencing by Quantitative RT-PCR

Cells are cultured as described above and harvested, and mRNA is isolated using the QIAamp RNA Blood Mini Kit. Reverse transcription of mRNA is performed with Superscript II RT and Oligo-dT_{12-18} primers; the resultant cDNA is purified using the PCR Purification Kit and adjusted to a final concentration of 50 ng/μL. Quantitative real-time PCR is performed with the QuantiTect SYBR Green PCR Kit and specific primers for *GAPDH* (fwd: 5′-AACGTGTCAGTGGTGGACCTG-3′; rev: 5′-AGTGG GTGTCGCTGTTGAAGT-3′) in a StepOne Plus 96-well system (Invitrogen Corporation). Each value is routinely determined in triplicate, a total of 50 ng cDNA per well is used in a 10 μL reaction volume and the primer concentration is 2 μM each. Relative gene expression in targeted siRNA-treated versus control siRNA-treated cells is calculated using the DDC_T-method [15] with *GAPDH* serving as internal control. Alternatively, *SOX* (Hs01053049_s1) and *BMI-1* (Hs00180411_m1) can be used as internal standards. A typical experiment is shown in Fig. 2a.

3.4 Evaluation of Gene Silencing by Quantitative Western Blotting

To quantify proteins of interest in lysate from cultured cells, 2×10^5 cells are used for western-blot analysis employing the respective primary antibodies from rabbit. The primary antibodies are visualized with an ECL™ Plex goat-anti-rabbit IgG-Cy5 conjugate and the Typhoon-Trio imaging system in combination with Image Quant TL software 7.0 (GE Healthcare). Routinely, we use antibodies that are directed against GAPDH or β-actin as controls for the fact that similar cell equivalents are compared (*see* **Notes 8** and **9**). A typical experiment is shown in Fig. 2a, b.

3.5 Complementation

To rescue the phenotype of *SEC61A1* silencing, the *SEC61A1* cDNA is inserted into the MCS of a pCDNA3-IRES-GFP vector containing the CMV promoter, the MCS, the IRES, and the GFP coding sequence (*see* **Note 10**). At 48 h after the first siRNA transfection, exchange the medium for a second time and transform the cells with either an empty vector or the *SEC61A1* expression plasmid using FuGENE HD according to the manufacturer's protocol (final ratio of vector to FuGENE HD is 4 μg vector to 16 μL FuGENE HD).

1. Prepare plasmids freshly in a separate Eppendorf reaction cup prior to the transfection procedure. Add 16 μL FuGENE HD transfection reagent to 4 μg of each plasmid that is dissolved in 80 μL of OptiMEM. This mix is gently vortexed and incubated at room temperature for 10–15 min. Add the plasmid mix (0.1 mL) dropwise to the HeLa cells (3.9 mL) (*see* **Note 6**).
2. Transfection efficiency can be visualized as GFP fluorescence and should be above 80 %.

3.6 Defining the Experimental Time-Window for Protein Transport Experiments

1. Cell proliferation, viability, and apoptosis/necrosis are analyzed in time-course experiments with commercially available assays following the manufacturer's instructions. We use the BrdU Assay in a 96-well format with 2.5×10^4 cells, the WST Assay in 12-well plates with 1×10^5 cells, and the LDH + assay according to the manufacturers' protocols. In addition, cells are microscopically evaluated using the Nuclear-ID™ Blue/Green cell viability reagent and the Apoptosis/Necrosis Detection Kit, respectively, according to the manufacturer's protocol.
2. Morphological integrity of cells and ER is analyzed by indirect immunofluorescence staining with an affinity-purified rabbit antipeptide antibody directed against the COOH terminal undecapeptide of the human Sec62 protein (plus an aminoterminal cysteine) and Alexa-Fluor-488- or Alexa-Fluor-594-coupled secondary antibody from goat (Invitrogen Corporation). We note that the anti-Sec62 antibody is specific for Sec62 under denaturing as well as native conditions (i.e., western blot and fluorescence microscopy signals are quenched after silencing of the *SEC62* gene). Cells are analyzed by

microscopy, such as on an Elyra SIM (Carl Zeiss- MicroImaging, Göttingen, Germany).

3. ER stress is analyzed by western blotting using antibodies directed against proteins overproduced after induction of the unfolded protein response. More immediate and sensitive assays are directed at the detection of the CHOP mRNA and the spliced form of the XBP1 mRNA, respectively, by quantitative RT-PCR and the detection of phosphorylated eIF2α by western-blot analysis [12].
4. On the basis of the results from these experiments, the silencing time for subsequent protein transport experiments is determined. In general, we aim for a maximum silencing efficiency at the protein level in combination with a minimum impact on cell growth, viability, morphology, and the unfolded protein response. The idea is to minimize secondary effects as far as possible.

3.7 Cellular Assays for Protein Translocation

ER protein transport is evaluated in silenced cells either under steady-state conditions, i.e., by cell fractionation into cellular supernatant, cytosol, and organelles/membranes plus subsequent SDS-PAGE and western blotting, or in pulse/chase experiments [7], i.e., short radioactive labeling of newly synthesized polypeptides and their chase to their final destination in ER, plasma membrane, or extracellular space (*see* **Note 11**). The steady-state experiments are analyzed as described for the evaluation of silencing efficiency, employing primary antibodies directed against substrate polypeptides of the ER-resident protein transport machinery. Alternatively, we transfect the cells with plasmids that code for certain model substrate precursor proteins and a tag for easy detection, such as the DDK-tag. Because precursor polypeptides in the cytosol are prone to degradation by the proteasome, the experiments with intact cells are also carried out after incubation of the cells with the proteasome inhibitor MG132 for the last 6–8 h. In our hands, HeLa cells tolerate a final concentration of 10 μM MG132 for this time period (*see* **Notes 12** and **13**). For optional cell transfection, *see* Subheading 3.5.

3.8 Preparation of Semi-permeabilized Cells for In Vitro Protein Translocation

For subsequent use in in vitro translocation reactions, semi-permeabilized cells are prepared from identical cell numbers according to a published procedure [6] (*see* **Note 11**). A typical experiment is shown in Fig. 3. After the actual translocation assay, their concentrations are confirmed by SDS-PAGE and protein staining with Coomassie Brilliant blue.

1. Wash the cells that are present in a 6 cm culture dish with 4 mL PBS and incubate with 300 μL of trypsin solution (0.05 % trypsin, 0.02 % EDTA in PBS) for 3 min at room temperature. Add 2 mL of soybean trypsin inhibitor solution (125 μg/mL

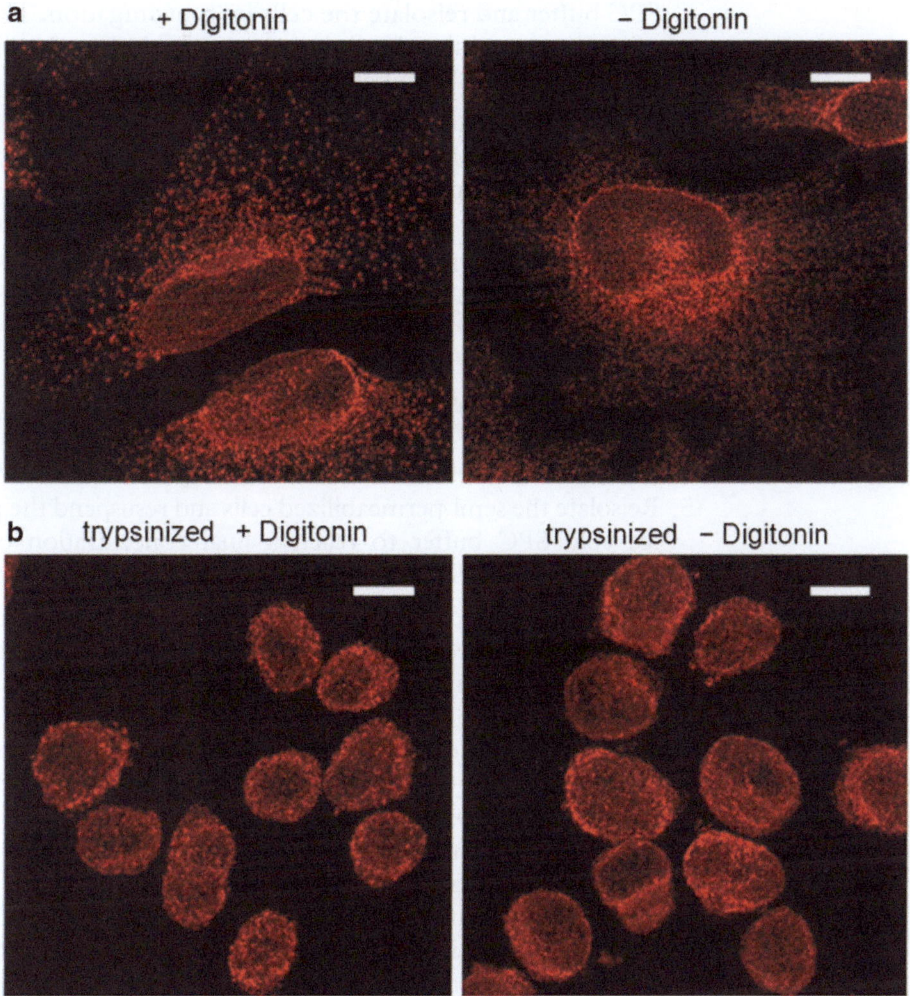

Fig. 3 Morphology of semi-pemeabilized cells. HeLa cells were either grown on 25 mm cover slips in 6 cm culture plates, fixed with paraformaldehyde and then semi-permeabilized with digitonin or not (**a**), or grown in 6 cm culture plates, trypsinized and semi-permeabilized with digitonin or not as described in methods and, subsequently fixed with paraformaldehyde (**b**). After treatment of all samples with saponin in PBS, indirect immunofluorescence staining was performed with an affinity-purified rabbit antipeptide antibody directed against the COOH-terminal undecapeptide of the human Sec62 protein and Alexa594-coupled secondary antibody. Cells were analyzed on an Elyra structured illumination microscope (SIM). Representative images are shown in maximum intensity projection. The *bar* corresponds to 10 μm. The images that are shown in (**b**) represent the semi-permeabilized cells as they are present in cell-free protein transport

in SPC buffer). Suspend the cells and transfer the cell suspension to a 15 mL Falcon tube.

2. Determine the cell number and pellet the cells by centrifugation (3 min at 1,000 × *g* and 4 °C). Resuspend the cellular pellet in 6 mL ice cold SPC buffer. Per one million cells 1 μL of digitonin solution (40 mg/mL of DMSO) is added and the mixture is incubated for 5 min on ice. Add 8 mL of ice cold

SPC buffer and reisolate the cells by centrifugation. The pellet of semi-permeabilized cells is resuspended in 14 mL of ice cold Hepes buffer and incubated on ice for 10 min.

3. After centrifugation, the semi-permeabilized cells are resuspended in 100 μL of SPC buffer, transferred to an Eppendorf reaction cup, supplemented with 0.5 μL $CaCl_2$ solution plus 1 μL of nuclease solution, and incubated for 12 min at room temperature. Stop the nucleic acid degradation by the addition of 2 μL of EGTA solution and reisolate the semi-permeabilized cells by centrifugation (3 min at 6,000 × *g* and 4 °C). Resuspend the pellet in 1.5 mL of ice cold SPC buffer.
4. Determine the concentration of the semi-permeabilized cells and the extent of permeabilization by staining with Trypan blue. The permeabilization efficiency should be close to 100 %.
5. Reisolate the semi-permeabilized cells and resuspend the pellet in ice cold SPC buffer to reach a final concentration of semi-permeabilized cells of 40,000/μL. Keep the suspension on ice.
6. Aliquots of the semi-permeabilized cells are subjected to SDS-PAGE and subsequent staining with Coomassie Brilliant blue and western-blot analysis, respectively.

3.9 In Vitro Protein Translocation Assays

Precursor polypeptides are synthesized in reticulocyte lysate (nuclease-treated rabbit reticulocyte lysate or TNT-coupled reticulocyte lysate) in the presence of [^{35}S]methionine plus buffer, canine pancreatic rough microsomes (positive control), or semi-permeabilized cells for 60 min at 30 °C (co-translational transport experiment). Alternatively, precursor polypeptides are synthesized in reticulocyte lysate in the presence of [^{35}S]methionine for 15 min at 30 °C. After 5 min of incubation with RNase A and cycloheximide at 30 °C, buffer, canine pancreatic rough microsomes or semi-permeabilized cells are added and the incubation is continued for 30 min (posttranslational transport experiment) (*see* **Note 14**).

For sequestration analysis, samples are divided into three aliquots and incubated in the absence or presence of proteinase K (final concentration: 170 μg/mL) for 60 min at 0 °C. In the third aliquot, Triton X-100 is present at a final concentration of 0.1 %. Protease treatment is terminated by the addition of phenylmethylsulfonyl fluoride (100 mM). All samples are analyzed by SDS-PAGE and phosphorimaging (Typhoon-Trio imaging system). Image Quant TL software 7.0 is used for quantifications. A typical experiment is shown in Fig. 4.

1. In order to prepare mRNA for programming protein synthesis in nuclease-treated rabbit reticulocyte lysate, supplement 30 μL of TS-premix in an Eppendorf reaction cup with 1 μg of plasmid DNA (*see* **Note 15**) and 2.5 μL of m^7GpppG (5 mM in HEPES/KOH, pH 7.4), 1.5 μL of recombinant RNasin,

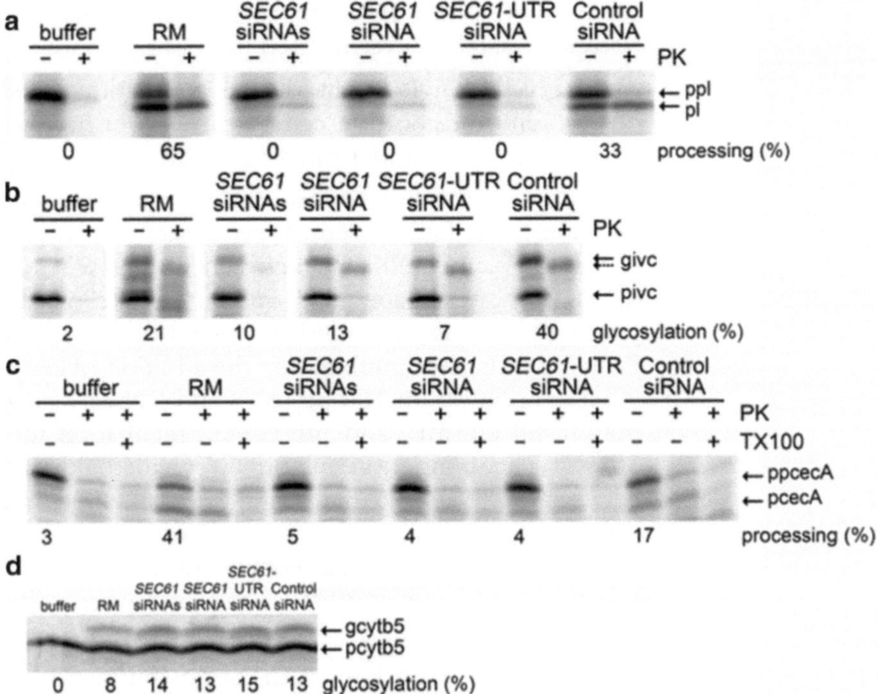

Fig. 4 Typical effect of silencing on ER protein translocation. HeLa cells were transfected with the indicated siRNAs. The indicated precursor polypeptides were imported into the indicated semi-permeabilized cells under co- (**a**, **b**) or posttranslational (**c**, **d**) transport conditions. Canine pancreatic microsomes (RM) served as positive control, SPC buffer as negative control. The membranes were subjected to sequestration analysis (**a**–**c**), followed by SDS-PAGE and phosphorimaging. (**a**) Preprolactin (ppl); (**b**) pre-MHC class II invariant chain (pivc), a membrane protein; (**c**) preprocecropin A (ppcecA); (**d**) hybrid cytochrome b5 (cytb5-ops28). The g indicates a glycosylated polypeptide. Only the areas of interest for single gels are shown

(40 U/μL), 1 μL of SP6 or T7 polymerase (20 U/μL), and enough sterile deionized water, to bring the final volume to 50 μL. Incubate for 3 h at 40 °C, then freeze in liquid nitrogen, and store at −80 °C.

2. For co-translational transport of precursor proteins in nuclease-treated rabbit reticulocyte lysate, prepare a mixture of 50 μL reticulocyte lysate, 2 μL TL-premix minus methionine (or minus cysteine) 2 μL rRNAsin, between 4 and 12 μL of transcription product, and 4 μL [^{35}S]-methionine (1,000 Ci/mmol) (or 4 μL EasyTag Express [^{35}S]-methionine/[^{35}S]-cysteine protein labeling mix) in an Eppendorf reaction cup. This mixture is divided into 9 aliquots of 15.5 or 17.5 μL (depending on the volume of transcription product that was used) and supplemented with buffer, rough microsomes plus buffer, or various semi-permeabilized cells plus buffer, to a final volume of 25 μL in an Eppendorf reaction cup. Routinely, we use between 2 and 8 μL of membrane suspension per 25 μL

reaction (*see* **Note 16**). Incubate for 60 min at 30 °C. The transport reaction is terminated by cooling on ice.

3. Alternatively, coupled transcription/translation and simultaneous protein translocation can be carried out in TNT-coupled reticulocyte lysate under similar conditions.
4. For posttranslational transport of precursor proteins in nuclease-treated rabbit reticulocyte lysate, the translation mixture is supplemented with between 75 and 95 μL of sterile deionized water (depending on the volume of transcription product that was used) and incubated for 15 min at 30 °C. Protein synthesis is inhibited by the addition of cycloheximide (final concentration: 100 μg/mL) and RNaseA (final concentration: 80 μg/mL) and subsequent incubation for 5 min at 30 °C. The translation mixture is divided into aliquots of between 15.5 and 17.5 μL and supplemented with buffer, rough microsomes plus buffer, or various semi-permeabilized cells plus buffer, to a final volume of 25 μL (*see* **Note 17**). Incubate for 30 min at 30 °C. The transport reaction is terminated by cooling on ice.
5. For sequestration analysis of co- or posttranslational transport reactions, prepare three solutions and chill on ice: 162.5 mM sucrose in deionized water, 340 μg/mL Proteinase K in sucrose solution, 0.2 % Triton X-100 in Proteinase K solution. For each transport reaction have ready on ice one Eppendorf reaction cup with 5 μL of sucrose solution, one cup with Proteinase K-solution, and one cup with Proteinase K/Triton X-100-solution. Then 5 μL of ice cold transport reaction are added to each cup and the cups are incubated for 60 min on ice (i.e., at 0 °C). The sequestration reactions are stopped by the addition of 1 μL of phenylmethylsulfonyl fluoride (100 mM in ethanol) and incubation for 5 min at 0 °C. After heating in SDS-PAGE sample buffer for 5 min at 95 °C, all samples are subjected to SDS-PAGE and phosphorimaging (Typhoon-Trio imaging system in combination with the Image Quant TL software 7.0) (*see* **Note 17**).

4 Notes

1. In principle, any mammalian cell may be used in this approach; however, in our hands adherent cells are better suited than non-adherent cells.
2. The serum quality is crucial for cell viability, particularly in combination with transfection reagents. Routinely, we test new batches of serum for their ability to support cell growth after transfection with control siRNA and control plasmid.

Furthermore, different cells may have different serum requirements. Therefore, prior to using a novel cell type, we test different sera in combination with siRNA or plasmid transfection.

3. Although siRNAs can be designed by available algorithms [3], we routinely use either validated siRNAs or predesigned siRNA. Both can be acquired from commercial sources. We note, however, that the validation is typically done by the companies at the mRNA level only.
4. Gene silencing by shRNA was successfully used for SRP14 in HeLa and HEK293T cells [7], demonstrating a role of the signal recognition particle (SRP) in ER protein transport at the cellular level. However, the same strategy failed in HeLa cells for SRP54 and SRP72 [8], i.e., in this case, no protein transport defect was observed. SRP is a cytosolic ribonucleoparticle involved in targeting of nascent precursor polypeptides and ribosomes to the ER [9].
5. In our hands, plasmids for transfection of HeLa cells are best purified from E. coli strain DH5α.
6. Transfection conditions must be optimized for each siRNA, each plasmid, and each cell line.
7. At the protein level, the maximum silencing effect is typically seen 72–96 h after the first transfection, while it usually takes less than 48 h for the mRNA to be efficiently degraded (Fig. 2). We note that in the case of protein complexes we have observed knock down of all subunits of the complex at the protein level in the absence of degradation of the non-silenced mRNAs [12]. Therefore, these effects can be assigned to protein degradation.
8. Quantitative western blotting requires that the whole assay is carried out in the linear range, i.e., it must be established for every single primary antibody/secondary antibody combination that quantitative differences are accurately monitored. This monitoring is best carried out by running serial dilutions of a cell suspension next to a protein mass standard for the protein of interest on SDS-PAGE. After transfer of the proteins to Immobilon P, the protein is visualized by fluorescence imaging. Typically, the fluorescence is proportional to the mass of the protein of interest only over a certain range of the serial dilution of the cell suspension. Ideally, an amount of cells corresponding to the middle of the linear range is used in routine assays. It is recommended that every once in a while, part of the serial dilution is run on the same gel as the samples from a gene-silencing or complementation experiment.
9. We note that employing primary antibodies in the western-blot evaluation of knock-down efficiencies allows for simultaneous evaluation of the antibodies. Subsequently, the antibodies are

also evaluated with targeted siRNA- vs. control siRNA-treated cells in immunofluorescence microscopy. As a rule, we use only those antibodies for fluorescence microscopy and immunocytochemistry that show a significant differential effect for targeted siRNA- vs. control siRNA-treated cells.

10. The gene-silencing and successful plasmid rescue experiments allow for the future analysis of rationally designed or disease-linked mutations in ER transport components [11, 13].
11. These transport assays can also be used for cells that were derived from knockout mice [12]. In this case, however, one has to take potential adaptive or compensatory mechanisms into consideration that may increase or decrease the effect of the knockout on ER protein translocation.
12. Be aware that MG132 induces the unfolded protein response. Therefore, the drug has to be used with caution.
13. Special consideration has to be given to reproducibility [16]. Therefore, experiments should routinely be carried out for different batches of cells.
14. Routinely, we study membrane integration of the tail-anchored model protein cytb5-ops28 which does not employ ER membrane proteins in its membrane integration [12] and therefore controls for overall effects on the integrity of the ER membrane, because its N-glycosylation involves oligosaccharyl transferase which depends on a lipid-linked oligosaccharide.
15. Typically, we use linearized plasmids for in vitro transcription for subsequent translation of the transcription products. This tends to increase the yield of mRNA. Alternatively, appropriate PCR products can be used as templates for in vitro transcription. Note that some coupled transcription/translation systems require circular plasmids.
16. The suitable membrane concentration has to be titrated for each transport substrate in order to be in the linear range of the assay.
17. It is mandatory to carry out the sequestration analysis at 0 °C and to stop the protease prior to the addition of SDS-PAGE sample buffer. Either omission can lead to complete protein degradation.

Acknowledgment

This work was supported by the Deutsche Forschungsgemeinschaft (DFG).

References

1. Blobel G, Dobberstein B (1975) Transfer of proteins across membranes: I. Presence of proteolytically processed and unprocessed nascent immunoglobulin light chains on membrane-bound ribosomes of murine myeloma. J Cell Biol 67:835–851
2. Görlich D, Rapoport TA (1993) Protein translocation into proteoliposomes reconstituted from purified components of the endoplasmatic reticulum membrane. Cell 75:615–630
3. Pei Y, Tuschl T (2006) On the art of identifying effective and specific siRNAs. Nat Methods 3:670–676
4. Niwa R, Slack FJ (2007) Ins and outs of RNA interference analysis. In: Zuk D (ed) Evaluating techniques in biochemical research. Cell Press, Cambridge, MA, pp 34–36
5. Editorial (2003) Whither RNAi? Nat Cell Biol 5:489–490
6. Wilson R, Allen AJ, Oliver J et al (1995) The translocation, folding, assembly, and redox-dependent degradation of secretory an membrane proteins in semi-permeabilized mammalian cells. Biochem J 387:679–687
7. Lakkaraju AK, Scherrer MC, Johnson AE et al (2008) SRP keeps polypeptides translocation competent by slowing translation to match limiting ER-targeting sites. Cell 133:440–451
8. Ren GY, Wagner KW, Knee DA et al (2004) Differential regulation of the TRAIL death receptors DR4 and DR5 by the signal recognition particle. Mol Biol Cell 15:5064–5074
9. Zimmermann R, Eyrisch S, Ahmad M et al (2011) Protein translocation across the ER membrane. Biochim Biophys Acta 1808:912–924
10. Lang S, Erdmann F, Jung M, Wagner R, Cavalié A, Zimmermann R (2011) Sec61 complexes form ubiquitous ER Ca^{2+} leak channels. Channels (Austin) 5:228–235
11. Erdmann F, Schäuble N, Lang S et al (2011) Interaction of calmodulin with Sec61α limits Ca^{2+} leakage from the endoplasmic reticulum. EMBO J 30:17–31
12. Lang S, Benedix J, Fedeles SV et al (2012) Different effects of Sec61α-, Sec62 and Sec63-depletion on transport of polypeptides into the endoplasmic reticulum of mammalian cells. J Cell Sci 125:1958–1969
13. Schäuble N, Lang S, Jung M et al (2012) BiP-mediated closing of the Sec61 channel limits Ca^{2+} leakage from the ER. EMBO J 31(15):3282–3296. doi:10.1038/emboj.2012.189
14. Lang S, Schäuble N, Cavalié A et al (2011) Live cell calcium imaging in combination with siRNA mediated gene silencing identifies Ca^{2+} leak channels in the ER membrane and their regulatory mechanisms. J Vis Exp (53), e2730. doi:10.3791/2730
15. Linxweiler M, Linxweiler J, Barth M et al (2012) Sec62 bridges the gap from 3q amplification to molecular cell biology in non-small cell lung cancer. Am J Pathol 180:473–483
16. Editorial (2012) Significant statistics. EMBO Rep 13:280

Chapter 19

An Assay to Monitor the Membrane Integration of Single-Span Proteins

Katrin Krumpe and Doron Rapaport

Abstract

In vitro import experiments with isolated organelles are a powerful tool for investigation of the biogenesis of proteins. A key issue in such experiments is an assay to distinguish between correctly and incorrectly imported proteins. Here we describe an assay to monitor in vitro the proper membrane integration of single-span proteins. In this assay non-imported proteins are distinguished from correctly imported protein species by labelling of unprotected cysteine residues and a resulting migration shift in SDS-PAGE.

Key words In vitro import, Tail-anchored proteins, Signal-anchored proteins, IASD-assay, Mitochondria, Endoplasmic reticulum

1 Introduction

The majority of mitochondrial and ER proteins are synthesized on cytosolic ribosomes and transported into their target organelles by distinct pathways. Import of proteins in vitro into isolated cell organelles is a well established method to investigate such pathways in detail. It helped to elucidate several import pathways into mitochondria and the endoplasmic reticulum. However, a central requirement for such an assay is the ability to differentiate between proteins that are indeed imported into the corresponding organelle and those who associate with the surface of the organelle without being integrated into its structures. Hence in vitro import experiments require an assay that monitors whether the investigated protein acquired its correct localization and topology. In this chapter we describe an assay for the analysis of the import of single-span proteins of the outer mitochondrial membrane and the ER membrane. A family of such single-span proteins are the tail-anchored proteins. These proteins contain a single transmembrane domain close to the C-terminus whereas the bulk of the protein is facing the cytosol. Tail-anchored proteins might contain a short soluble C-terminus not longer than 30 residues that faces the luminal side

Doron Rapaport and Johannes M. Herrmann (eds.), *Membrane Biogenesis: Methods and Protocols*,
Methods in Molecular Biology, vol. 1033, DOI 10.1007/978-1-62703-487-6_19, © Springer Science+Business Media, LLC 2013

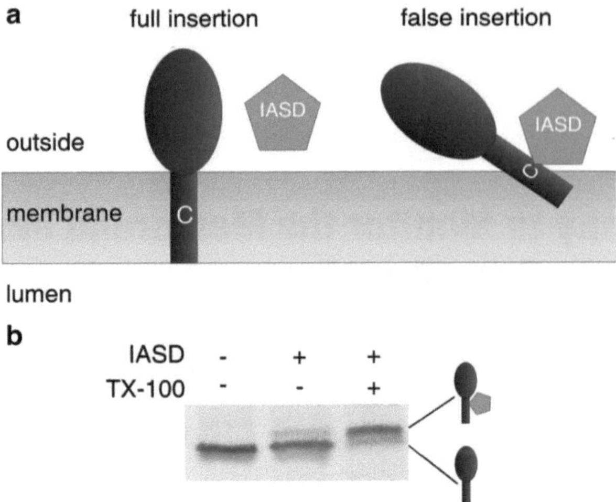

Fig. 1 The IASD-assay. (**a**) Schematic overview of the IASD-labelling of signal- or tail-anchored protein. Correctly membrane-inserted proteins are not accessible to IASD, while non-imported or only membrane-associated proteins can be labelled by IASD. (**b**) Autoradiography of Fis1 import into isolated mitochondria followed by an IASD-assay. The lower band (*see lanes 1* and *2*) corresponds to unlabelled imported protein, whereas the upper band (*see lanes 2* and *3*) corresponds to labelled non-imported protein

of the membrane [1, 2]. The majority of ER-resident TA-proteins are inserted into the mammalian ER membrane by the TRC40-containing pathway [3, 4] or into yeast ER by the equivalent GET machinery [5]. However, some ER TA-proteins, like mammalian cytochrome b5, were reported to insert into the ER membrane in an unassisted process [6]. A dedicated machinery for the insertion of mitochondrial TA-proteins was not identified so far. Signal-anchored proteins possess a transmembrane domain close to the N-terminus which is facing the luminal side of membrane. The larger C-terminal part of the protein forms a cytosolic domain [7, 8]. We and others have used an assay that is based on modification of a single cysteine residue as readout for the proper membrane integration of both signal- and tail-anchored proteins. The assay is based on the sulfhydryl-reactive reagent 4-acetamido-4′-((iodoacetyl)amino)stilbene-2,2′-disulfonic acid (IASD) [9]. A single cysteine reside should be introduced within the transmembrane region of an otherwise cysteine-less protein. IASD cannot modify this residue if the protein is properly imported, since in that case the membrane around the transmembrane region makes the cysteine residue inaccessible to the IASD. Only nonintegrated proteins can be labelled by the IASD reagent as the cysteine would be exposed to an aqueous environment (Fig. 1a). A modification by IASD would lead to a migration shift detectable in SDS-PAGE

(Fig. 1b). In this way correctly imported tail- or signal-anchored proteins can be reliably identified and distinguished from those molecules that are only membrane-associated. Two control reactions are performed to assure the dependency of the size shift on the exposure of the thiol groups to IASD. In the first IASD is omitted from the reaction mixture and in the second control reaction Triton X-100 is included to solubilize all membranes and thus to deliberately expose the thiol group. We present our results with the mitochondrial tail-anchored protein Fis1 as an example for usage of this assay (Fig. 1b).

2 Materials

2.1 In Vitro Import Reaction

1. Isolated and purified mitochondria or microsomes (*see* **Note 1**).
2. In vitro translated ^{35}S-labelled protein of interest containing only one cysteine in the transmembrane region (*see* **Note 2**).
3. Import buffer: 250 mM Sucrose, 10 mM MOPS, 80 mM KCl, 5 mM $MgCl_2$. Adjust the pH to 7.2 with KOH. Add 3 % BSA (fatty acid-free). The buffer can be stored at −20 °C.
4. 0.2 M ATP: ATP dissolved in H_2O, pH adjusted to 7.2 with KOH. Aliquots can be stored at −20 °C (*see* **Note 3**).
5. 0.2 M NADH: NADH dissolved in H_2O. Aliquots can be stored at −20 °C (*see* **Note 3**).
6. SEM-K80 buffer: 250 mM sucrose, 10 mM MOPS, 1 mM EDTA, 80 mM KCl. Adjust pH to 7.2 with KOH.

2.2 IASD-Assay

1. Labelling buffer: 0.6 M sorbitol, 20 mM HEPES. Adjust pH to 7.4 with KOH. Add 80 mM Tris–HCl pH 8.0, 4 M urea, 1 mM DTT. The buffer can be stored at −20 °C.
2. 5 mM IASD: Dissolve IASD in Labelling buffer. The solution can be stored at −20 °C (*see* **Note 4**).
3. 5 mM IASD + 1 % Triton X-100: Add the appropriate volume of a 20 % Triton X-100 stock solution to 5 mM IASD solution in order to obtain a final concentration of 1 % Triton X-100. This solution should be made fresh before use.

3 Methods

Before starting to work with radioactive material make yourself familiar with the safety procedures.

All procedures should be carried out on ice unless specified otherwise.

3.1 In Vitro Import of Radio-labelled Proteins

In the described method it is essential that the tail- or signal-anchored protein of interest contains only one cysteine, which needs to be located within the transmembrane domain. If the wild-type protein harbors additional cysteines in the cytosolic part, these residues should be mutated to other amino acids. The single cysteine can be introduced into the transmembrane region by site-directed mutagenesis, in the case that the transmembrane region does not contain any cysteines. Before starting the experiments one has to make sure that the modified protein is still functional and localized to its native destination.

1. Prepare mitochondria or microsomes in Import buffer: 30 μg mitochondria or microsomes in 100 μL Import buffer supplemented with 5 mM ATP and 2.5 mM NADH (*see* **Note 5**). Three samples are required for any specific import conditions.
2. Start import reaction by adding 6 μL of ^{35}S-labelled protein containing lysate (*see* **Note 6**). Mix well and shift the reactions to 25 °C (*see* **Note 7**).
3. Stop import reaction after the desired time by shifting samples back on ice and adding 400 μL SEM-K80 buffer (*see* **Note 8**).
4. Re-isolate mitochondria or microsomes by centrifugation: For mitochondria 10 min at 20,000 × *g*, 2 °C are sufficient; for microsomes 45 min at 100,000 × *g* at 2 °C are necessary. Carefully discard the supernatant.

3.2 IASD-Assay

1. The three pellets of the three samples of each import condition are resuspended in different solutions. One of the three reactions should be resuspended in 30 μL Labelling buffer, another one in 30 μL of 5 mM IASD, and the third in 30 μL of 5 mM IASD + 1 % Triton X-100.
2. Incubate all reactions 30 min on ice to allow sufficient time for solubilization of the membranes in those samples that contain Triton X-100 (*see* **Note 9**). Afterwards shift reactions to 25 °C for 20 min to label accessible cysteines.
3. Re-isolate mitochondria or microsomes from reactions without Triton X-100 by centrifugation as described in Subheading 3.1, **step 4**. Discard the supernatant carefully and resuspend pellets in Laemmli-buffer. Alternatively proteins can be re-isolated by a trichloroacetic acid (TCA)-precipitation.
4. To re-isolate proteins from Triton X-100 containing reactions perform a TCA-precipitation. Resuspend pellets in Laemmli-buffer.
5. Incubate all samples 10 s at 95 °C (*see* **Note 10**).
6. Load samples on an SDS-PAGE and either blot the gel onto nitrocellulose membrane or dry the gel on a paper (*see* **Note 11**).
7. Put an autoradiography film on top of the membrane/paper for a few days (*see* **Note 12**) and develop the film (*see* **Note 13**).

8. Interpretation of results (*see* also Fig. 1b): The control reaction without IASD determines the size of the unlabelled correctly imported protein, whereas the control reaction with IASD and Triton X-100 determines the size of the IASD-labelled species. The reaction with only IASD normally shows both bands. The lower corresponds to the imported protein (compare control reaction without IASD) and the upper corresponds to the non-imported protein (compare control reaction with IASD and Triton X-100). By comparing the ratio between these two bands from different reactions it is possible to analyze import efficiency of signal- and tail-anchored proteins under different conditions.

4 Notes

1. Isolated mitochondria or microsomes can be prepared in advance and stored in single use aliquots at −80 °C.
2. Commercially available systems like rabbit reticulocyte lysate can be used for synthesis of ^{35}S-labelled protein. Before using the protein in this assay it has to be engineered in a way that its only cysteine residue is located within the transmembrane section.
3. ATP and NADH should be frozen and thawed only once.
4. Avoid more than three freeze–thaw cycles by making aliquots.
5. Import buffer should be supplemented with ATP and NADH by adding the appropriate volume of the 0.2 M stock solutions just before use. Mitochondria or microsomes should be resuspended in the buffer very carefully by pipetting the liquid up and down in a cut tip. Do not vortex. The amount of organelles per reaction can be adjusted according to the experimental needs.
6. The required amount of ^{35}S-labelled protein containing reticulocyte lysate might vary. If the signal after autoradiography is too strong or too weak the volume of lysate can be adjusted accordingly. But the amount of the reticulocyte lysate should not be more than 10 % of the total. If a higher amount of lysate is needed the total volume of the reaction has to be scaled up.
7. The import temperature can be adjusted to the needs of the experiment. If import of the examined protein is too fast to see any kinetics at 25 °C, temperature can be reduced to 15 °C or the import can even be carried out on ice. Temperature can also be raised to more than 25 °C if required.
8. To analyze import kinetics the time points of import have to be adjusted to the characteristics of the imported protein. A good starting point is an import for 2, 5, and 10 min. If an import time of over 20 min is required, fresh ATP and NADH should

be added to the reaction mix every 20 min. Keep in mind that mitochondria or microsomes might get damaged after a longer incubation time at high temperatures.

9. Samples containing Triton X-100 can be mixed every 10 min to obtain a better solubilization of the membranes.
10. Proteins in Laemmli solution can be stored at −20 °C. But ^{35}S has a half-life of approximately 87 days, so the intensity of the radioactive signal will fade after a certain time. For that reason SDS-PAGE and autoradiography should be performed in the next couple of days.
11. When choosing the gel system keep in mind that the size shift achieved by labelling of one accessible cysteine is only 0.6 kDa. Thus, the separation capacity of the gel should be very high in the range of the labelled protein.
12. The membrane/paper has to be absolutely dry before putting on the film. Otherwise the membrane/paper will stick to the film and the whole experiment will be spoiled.
13. Exposure time depends on the signal intensity. In general exposure times are between 2 and 3 days for strong signals and 1–4 weeks for weak signals. If the first exposure does not result in an appropriate signal, adjust exposure time accordingly.

Acknowledgments

Our work is supported by the Deutsche Forschungsgemeinschaft (SFB766/TP B11 and RA 1048/4-1 to D.R.).

References

1. Borgese N, Brambillasca S, Colombo S (2007) How tails guide tail-anchored proteins to their destinations. Curr Opin Cell Biol 19:368–375
2. Wattenberg B, Lithgow T (2001) Targeting of C-terminal (tail)-anchored proteins: understanding how cytoplasmic activities are anchored to intracellular membranes. Traffic 2:66–71
3. Favaloro V, Spasic M, Schwappach B, Dobberstein B (2008) Distinct targeting pathways for the membrane insertion of tail-anchored (TA) proteins. J Cell Sci 121:1832–1840
4. Stefanovic S, Hedge RS (2007) Identification of a targeting factor for posttranslational membrane protein insertion into the ER. Cell 128: 1147–1159
5. Schuldiner M, Metz J, Schmid V, Denic W et al (2008) The GET complex mediates insertion of tail-anchored proteins into the ER membrane. Cell 134:634–645
6. Borgese N, Colombo S, Pedrazzini E (2003) The tale of tail-anchored proteins: coming from the cytosol and looking for a membrane. J Cell Biol 161:1013–1019
7. Shore GC, McBride HM, Millar DG et al (1995) Import and insertion of proteins into the mitochondrial outer membrane. Eur J Biochem 227:9–18
8. Waizenegger T, Stan T, Neupert W, Rapaport D (2003) Signal-anchor domains of proteins of the outer membrane of mitochondria: structural and functional characteristics. J Biol Chem 278: 42064–42071
9. Kemper C, Habib SJ, Engl G et al (2008) Integration of tail-anchored proteins into the mitochondrial outer membrane does not require any known import components. J Cell Sci 121: 1990–1998

Chapter 20

Methods to Study the Biogenesis of Membrane Proteins in Yeast Mitochondria

Daniel Weckbecker and Johannes M. Herrmann

Abstract

The biogenesis of mitochondrial membrane proteins is an intricate process that relies on the import and submitochondrial sorting of nuclear-encoded preproteins and on the synthesis of mitochondrial translation products in the matrix. Subsequently, these polypeptides need to be inserted into the outer and the inner membranes of the organelle where many of them assemble into multisubunit complexes. In this chapter we provide established protocols to study these different processes experimentally using mitochondria of budding yeast. In particular, methods are described in detail to purify mitochondria, to study mitochondrial protein synthesis, to follow the import of radiolabeled preproteins into isolated mitochondria, and to assess membrane association and the aggregation of mitochondrial proteins by fractionation. These protocols and a list of dos and don'ts shall enable beginners and experienced scientists to address the targeting and assembly of mitochondrial membrane proteins.

Key words Carbonate extraction, Isolation of mitochondria, Membrane insertion, Mitochondrial targeting, Protein import, Translation

1 Introduction

Mitochondria are "semiautonomous" organelles [1] that, unlike most other cellular compartments, cannot be formed de novo. They propagate by growth and division from preexisting mitochondria thereby gaining their constituents such as proteins, lipids, and metabolites both by the uptake from outside and by the synthetic production inside. Of about 700–1,000 different proteins that constitute mitochondria [2–4] only a small number (eight in budding yeast, 13 in human) are translated in the mitochondrial matrix. The vast majority of proteins are synthesized on ribosomes in the cytosol from where they are imported into mitochondria. A large fraction of mitochondrial proteins are membrane proteins. Whereas proteomics identified only about 25–50 different proteins in the outer membrane of fungal mitochondria [5, 6], the inner membrane of mitochondria has a very complex protein

Doron Rapaport and Johannes M. Herrmann (eds.), *Membrane Biogenesis: Methods and Protocols*, Methods in Molecular Biology, vol. 1033, DOI 10.1007/978-1-62703-487-6_20, © Springer Science+Business Media, LLC 2013

composition. For example, more than 100 membrane proteins are required to build and assemble the five complexes of the respiratory chain that are embedded into the inner membrane. The budding yeast *Saccharomyces cerevisiae* has proven to be a highly valuable model system to study the biogenesis of membrane proteins for several reasons such as: (1) budding yeast can grow by fermentation and tolerates respiration-incompetent mutants; (2) mitochondria can be easily purified from yeast and remain functional even after being frozen for years; (3) yeast mitochondria can be easily subfractionated by hypotonic swelling, protease protection, carbonate extraction, or by simple aggregation assays; (4) isolated mitochondria efficiently import in vitro preproteins from the surrounding buffer or synthesize mitochondrial proteins on their own ribosomes. This chapter provides detailed protocols for the analysis of the biogenesis of mitochondrial membrane proteins.

1.1 Protein Import into Mitochondria

The import of proteins into mitochondria is mediated by protein translocation complexes that are embedded into the outer and inner membranes of the organelle. Many excellent review articles describe this complex process in great detail [7–11]. With the exception of some tail-anchored outer membrane proteins, all mitochondrial precursor proteins reach the organelle via the TOM (translocase of the outer membrane of mitochondria) complex (*see* Fig. 1). The TOM complex consists of surface-exposed receptor subunits that recognize cytosolic precursor proteins and of a protein-conducting channel that transfers them further across the outer membrane. Outer membrane proteins can be directly integrated into the outer membrane either by the TOM complex or by assistance of the SAM/TOB complex that particularly serves the insertion of β-barrel proteins [12–14].

Proteins destined to the matrix of mitochondria contain amino-terminal targeting sequences, called presequences or mitochondrial targeting sequences (MTSs) that are threaded through the TOM and the TIM23 complexes in a process that requires a membrane potential across the inner membrane and the hydrolysis of ATP in the matrix. Following translocation into the matrix, presequences are mostly removed by processing peptidases [15–17]. Proteins of the inner membrane can either be arrested at the level of the TIM23 complex and released into the inner membrane [18], or integrate from the matrix side in an export-like process that can be catalyzed by the membrane insertase Oxa1 [19, 20] or the AAA protein Bcs1 [21]. Some polytopic inner membrane proteins such as carrier proteins employ a specific insertion complex, called the TIM22 translocase [11, 22, 23].

1.2 Membrane Insertion of Mitochondrial Translation Products

Seven of the eight translation products of yeast mitochondria are membrane proteins. These are the cytochrome *b* of the bc_1 complex, subunits 1–3 of cytochrome *c* oxidase (Cox1, Cox2, and Cox3), and subunits 6, 8, and 9 of the F_oF_1-ATPase (Atp6, Atp8, and Atp9).

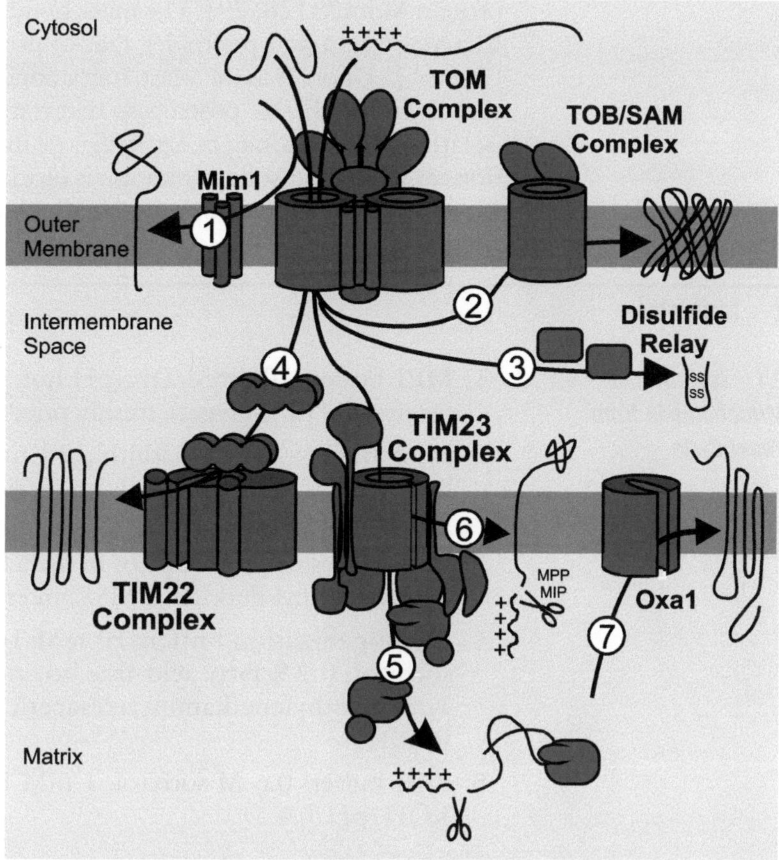

Fig. 1 Pathways of protein import into mitochondria. Some outer membrane proteins are integrated into the membrane from the cytosolic side (*route 1*) whereas others are first transported through the TOM complex before they insert into the outer membrane via the TOB/SAM complex (*route 2*). Some proteins of the intermembrane space (IMS) are imported in redox-dependent process by Mia40 and Erv1, also referred to as the disulfide relay system (*route 3*). Carrier proteins are inserted into the inner membrane by a specialized inner membrane translocase, the TIM22 complex (*route 4*). Preproteins with matrix targeting sequences are transported to the TIM23 complex and either imported into the matrix (*route 5*) or integrated into the inner membrane (stop-transfer pathway, *route 6*). Some inner membrane proteins of bacterial origin are inserted into the membrane in an export-like step from the matrix that can depend on the Oxa1 insertase (conservative sorting pathway, *route 7*). Oxa1 also catalyzes the membrane insertion of several mitochondrial translation products

In addition, the ribosomal protein Var1 is synthesized in mitochondria. Presumably as a consequence of their specialization on the production of membrane proteins, yeast mitochondrial ribosomes are physically associated with the inner membrane [24]. Membrane binding of the ribosomes is mediated by several components, for example, by the membrane insertase Oxa1 [25, 26], the ribosome receptor Mba1 [24, 27], and the inner membrane protein Mdm38 [28, 29]. The mitochondrial translation system is of bacterial origin and resembles that of prokaryotes in many aspects [30, 31]. Upon lysis of yeast mitochondria, ribosomes lose their capacity to synthesize proteins so that translation is studied either in isolated mitochondria (*in organello*) or in whole cells under conditions at which cytosolic translation is blocked (in vivo). Protocols for monitoring these processes are described in this chapter.

2 Materials

2.1 Isolation of Mitochondria from Yeast Cells

1. MP1 buffer: 100 mM Tris (pH not adjusted), 10 mM dithiothreitol (DTT). Prepare freshly prior to use.
2. MP2 buffer: 1.2 M sorbitol, 20 mM potassium phosphate buffer, pH 7.4, 3 mg zymolyase (Seikagaku Corporation) per g wet weight of cells.
3. PMSF: Freshly prepare prior to use a 0.2 M solution of phenylmethylsulfonyl fluoride (PMSF) in ethanol.
4. Homogenization buffer: 10 mM Tris–HCl, pH 7.4, 0.6 M sorbitol, 0.2 % fatty acid-free bovine serum albumin (BSA), 1 mM ethylenediaminetetraacetic acid (EDTA), 1 mM PMSF.
5. SEH buffer: 0.6 M sorbitol, 1 mM EDTA, 20 mM HEPES/KOH, pH 7.4.

2.2 In Vivo Labelling of Mitochondrial Translation Products

1. Yeast synthetic medium: 1.7 g/L yeast nitrogen base, 5 g/L ammonium sulfate, 20 mg/L adenine, 20 mg/L uracil, 20 mg/L tryptophan, 20 mg/L histidine, 100 mg/L leucine, 30 mg/L lysine. Supplement with the desired carbon source (glucose, galactose, glycerol, ethanol, or lactic acid) to a final concentration of 2 % (*see* **Note 1**). Adjust pH to 5.5 with KOH.
2. Cycloheximide: Prepare a cycloheximide stock solution of 7.5 mg/mL in water and store at −20 °C.
3. Nonradioactive ("cold") methionine: Prepare a methionine stock solution of 200 mM and store at −20 °C.
4. ^{35}S-methionine: 11 mCi/mL.
5. Puromycin: Prepare a puromycin stock solution of 1 mg/mL in water and store at −20 °C.
6. Lysis buffer: 1.8 M NaOH, 1 M β-mercaptoethanol, 1 mM PMSF.
7. Trichloroacetic acid (TCA): Adjust 72 g of TCA to 100 mL with water.
8. Cold acetone: Cool down acetone to −20 °C prior to the experiment.

9. Sample buffer: 2 % (w/v) sodium dodecylsulfate (SDS), 50 mM DTT, 10 % (v/v) glycerol, 0.02 % (w/v) bromophenol blue, 60 mM Tris–HCl, pH 6.8.

2.3 In Organello Labelling of Mitochondrial Translation Products

1. SH buffer: 0.6 M sorbitol, 20 mM HEPES/KOH, pH 7.4.
2. 1.5× *in organello* translation buffer: 375 μL of 2.4 M sorbitol, 225 μL of 1 M KCl, 22.5 μL of 1 M potassium phosphate buffer, pH 7.2, 19 μL of 1 M magnesium sulfate, 45 μL of 100 mg/mL fatty acid-free BSA (*see* **Note 2**), 30 μL of 200 mM ATP, pH 7.0, 15 μL of 50 mM GTP, 9.1 μL of amino acid stock solution (2 mg/mL of all proteinogenic amino acids except for tyrosine, cysteine, and methionine), 10 μL of 10 mM cysteine, 18.2 μL of 1 mg/mL tyrosine, 1.7 mg α-ketoglutarate, 3.5 mg phosphoenol pyruvate, and 0.5 mg/mL pyruvate kinase. Add distilled water to 1 mL. Prepare freshly each time.
3. Isolated yeast mitochondria at concentration of 10 mg/mL kept in single use aliquots at −80 °C (*see* **Note 3**).

2.4 Protein Import into Isolated Mitochondria

1. 2× import buffer: 6 % (w/v) fatty acid-free BSA, 1.2 M sorbitol, 160 mM KCl, 20 mM magnesium acetate, 4 mM potassium phosphate buffer, pH 7.2, 5 mM EDTA, 5 mM $MnCl_2$, 100 mM HEPES/KOH, pH 7.2 (*see* **Note 4**). Store in aliquots at −20 °C.
2. ATP: Prepare a 0.2 M solution in water and adjust the pH with KOH to 7.0. Make single use aliquots of 10 μL and store at −20 °C.
3. NADH: Prepare a 0.2 M solution in water. Make single use aliquots of 10 μL and store at −20 °C.
4. Creatine phosphate (CP): Prepare a 250 mM solution of CP in water. Store at −20 °C (*see* **Note 5**).
5. Creatine kinase (CK): Prepare a 20 mg/mL solution of CK in water. Make single use aliquots of 10 μL and store at −20 °C (*see* **Note 5**).
6. Succinate: Prepare a 200 mM solution and adjust the pH to 7.2 with KOH (*see* **Note 5**). Store at −20 °C.
7. Malate: Prepare a 200 mM solution and adjust the pH to 7.2 with KOH (*see* **Note 5**). Store at −20 °C.
8. Proteinase K (PK): Prepare a 10 mg/mL solution of PK in water and store in single use aliquots of 50 μL at −20 °C.
9. ^{35}S-labelled mitochondrial precursor protein prepared in an in vitro translation reaction based on rabbit reticulocyte lysate [33, 34] and kept as single use aliquots at −80 °C.
10. SH/KCl buffer: 0.6 M sorbitol, 20 mM HEPES/KOH, pH 7.2, 150 mM KCl. Store in 50 mL aliquots at −20 °C.

2.5 Carbonate Extraction

1. 100 mM Na_2CO_3, prepare freshly each time (*see* **Note 6**).

2.6 Aggregation Assay

1. Lysis buffer: 20 mM Tris–HCl, pH 7.4, 150 mM NaCl, 5 mM EDTA, 0.5 % Triton X-100.

3 Methods

3.1 Isolation of Mitochondria from Budding Yeast

S. cerevisiae is the standard source for mitochondria in many laboratories. The success of budding yeast is in part due to the fact that mitochondria can be prepared relatively easily and remain energized even upon long storage periods at −80 °C. In most protocols, enzymes like zymolyase are used to remove the cell wall. The resulting spheroplasts are then ruptured by douncing in glass homogenizers yielding in cellular lysate from which mitochondria can be purified by differential centrifugation. It is important to keep in mind that after douncing hydrolytic enzymes are released from the yeast vacuole so that all following steps should be carried out fast and on ice. This protocol was first described by the Schatz laboratory [35] and represents the standard procedure for the isolation of mitochondria in many laboratories. The resulting fraction contains biochemically highly active mitochondria but also other cellular membranes such as microsomes. Further purification steps will be required to get highly purified mitochondria [4, 36].

1. Grow yeast cells at 30 °C in a shaker for at least 36 h to an OD_{600} of 1.0–1.3 in the desired culture medium. During this time the cultures should never reach OD_{600} values larger than 1.5.
2. Harvest the cells by centrifugation for 5 min at 2,800 × *g* at room temperature.
3. Wash cells with H_2O and centrifuge again as in **step 2**.
4. Weigh the cells.
5. Resuspend the cells in 2 mL of MP1 buffer per g wet weight.
6. Incubate the cells for 10 min shaking at 30 °C.
7. Collect the cells by centrifugation for 5 min at 1,900 × *g* at room temperature.
8. Wash the cells in 100 mL of 1.2 M sorbitol and centrifuge as in **step 7**.
9. Resuspend the cells in 6.7 mL MP2 buffer per g wet weight including 3 mg zymolyase per g wet weight.
10. Incubate the cells while shaking at 30 °C for 30–60 min (*see* **Note 7**).
11. Perform the following steps at 4 °C.
12. Harvest the spheroplasts by centrifugation for 5 min at 1,100 × *g*.

13. Resuspend the spheroplasts in 13.4 mL of ice-cold homogenization buffer per g wet weight.
14. Rupture the cells with a cooled glass homogenizer with a glass pistil by douncing 10–12 times.
15. Remove cell debris, nuclei, and intact cells by centrifuging twice for 5 min at 1,900 × *g* (*see* **Note 8**).
16. Collect the mitochondria by centrifugation of the supernatant for 12 min at 17,700 × *g*.
17. Resuspend the pellet in 10 mL of SEH buffer (*see* **Note 9**).
18. Centrifuge the suspension for 5 min at 1,900 × *g*.
19. Take the supernatant and centrifuge it for 12 min at 17,700 × *g* to retrieve the mitochondria.
20. Resuspend the mitochondria in 0.3–1 mL of SEH buffer.
21. Determine the protein concentration via Bradford assay [37] and adjust it to 10 mg/mL with SEH buffer.
22. Make aliquots of 25–50 μL, freeze them in liquid nitrogen, and store them at −80 °C (*see* **Note 10**).

3.2 In Vivo Labelling of Mitochondrial Translation Products

Mitochondrial translation products can be easily labelled in cells by addition of radiolabelled amino acids such as ^{35}S-methionine. Cytosolic translation is blocked by cycloheximide prior to labelling.

1. Prepare an overnight culture of the desired yeast strains in synthetic media supplemented with the appropriate carbon source and auxotrophic amino acids/nucleotides.
2. Dilute the cells to an OD_{600} of 0.2 and let the cells grow to an OD_{600} of 0.5–1.0.
3. Take cells equivalent to 2 OD_{600} × mL.
4. Harvest the cells by centrifugation at 17,000 × *g* for 2 min at RT.
5. Wash the cells twice with 1 mL of synthetic medium without proteinogenic amino acids.
6. Centrifuge the cells again and resuspend them in 1 mL of synthetic medium without proteinogenic amino acids.
7. Add 6 μL of the amino acid mix, 12 μL of tyrosine, and 6.4 μL of cysteine.
8. Incubate the cells while shaking at 1,200 rpm for 5 min at 30 °C.
9. Add 20 μL of cycloheximide.
10. Incubate the cells while shaking at 1,200 rpm for 5 min at 30 °C.
11. Add 2 μL of ^{35}S-methionine.
12. Continue the incubation for 15 min.
13. Take 250 μL of the suspension and stop the labelling process by addition of 50 μL of lysis buffer and 10 μL of cold methionine.

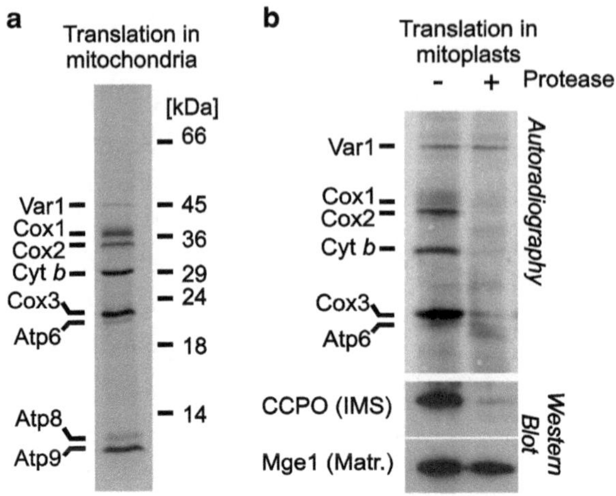

Fig. 2 Radioactive labelling of mitochondrial translation products. (**a**) Translation products were radiolabelled in isolated mitochondria ("*in organello* translation") as described in Subheading 3.3. The positions of the eight different translation products are shown. It should be noted that the migration of the translation products strongly depends on the acrylamide/bisacrylamide composition of the SDS gel. (**b**) Translation products were radiolabelled in mitoplasts, i.e., in mitochondria in which the outer membrane was ruptured by hypotonic swelling (*see* **Note 14**). Addition of protease after translation leads to a degradation of membrane proteins that expose domains to the IMS. In contrast, the ribosomal protein Var1 is inaccessible to protease and remains stable. The rupture of the mitochondria was controlled by Western blotting

14. Incubate samples for 10 min on ice.
15. Add 72 μL of TCA and vortex briefly.
16. Freeze the sample for at least 30 min at −20 °C.
17. Centrifuge at 16,000 × *g* for 15 min at 4 °C.
18. Discard the supernatant.
19. Wash the pellet: Add 1 mL of cold acetone (−20 °C) and invert the sample carefully.
20. Centrifuge at 16,000 × *g* for 15 min at 4 °C.
21. Discard the supernatant.
22. Resuspend the pellet in 30 μL of sample buffer.
23. Incubate the samples for 20 min while shaking (1,200 rpm) at 25 °C (*see* **Note 11**).
24. Analyze samples by SDS-PAGE and autoradiography.

3.3 *In Organello* Labelling of Mitochondrial Translation Products

Isolated mitochondria efficiently synthesize proteins that can be labelled by the addition of ^{35}S-methionine (Fig. 2). For translation and for the uptake of the amino acids from the buffer, highly energized mitochondria are a prerequisite. Therefore, the buffer contains, in addition to ATP and GTP, metabolic substrates and energy-regenerating systems. The importance of the different constituents of the buffer was tested in detail to optimize the labelling conditions [32].

1. Mix 20 μL of 1.5× translation buffer with 5–50 μg of mitochondria. Adjust volume to 29.5 μL.
2. Incubate the sample for 5 min at 25 °C.
3. Add 0.5 μL of ^{35}S-methionine to start the labelling process.
4. Incubate the sample for 20 min at 25 °C (*see* **Note 12**).
5. Stop the labelling process by addition of 5 μL cold methionine (*see* **Note 13**).
6. Incubate the sample for 3 min at 25 °C.
7. Centrifuge for 10 min at 25,000 × *g* (4 °C).
8. Discard the supernatant.
9. Add 1 mL of cold SH buffer to wash the mitochondria and centrifuge again for 10 min at 25,000 × *g* (4 °C) (*see* **Note 14**).
10. Resuspend the pellet in 30 μL of sample buffer.
11. Incubate the samples for 20 min while shaking (1,200 rpm) at 25 °C (*see* **Note 15**).
12. Analyze samples by SDS-PAGE and autoradiography.

3.4 Protein Import into Isolated Mitochondria

Many protocols were published for the import of proteins into isolated mitochondria (Fig. 3). Most of them use radiolabeled preproteins that were synthesized in reticulocyte lysate [33]. The chaperones of the lysate thereby keep preproteins import-competent. Alternatively, preproteins can be recombinantly expressed in bacteria and purified. When denatured in urea or guanidine hydrochloride these proteins can be imported; however, the efficiency of the import reaction is often low. Detailed protocols for the import of chemical amounts of proteins were published before [38].

1. Prepare one tube for each import reaction with 900 μL of SH buffer and keep it on ice.
2. Mix gently in the following order: 42 μL H_2O, 50 μL 2× import buffer, 1 μL ATP, 1 μL NADH, 50 μg (5 μL) mitochondria (*see* **Notes 4** and **15**).
3. Incubate the samples for 5 min at 25 °C.
4. Add 1 μL of radiolabelled precursor protein (*see* **Note 16**).
5. Incubate the sample for 20 min at 25 °C (*see* **Note 17**).
6. Add 900 μL of ice-cold SH buffer (*see* **Note 14**).
7. Add 10 μL of PK (*see* **Note 18**).
8. Incubate the sample for 30 min on ice.

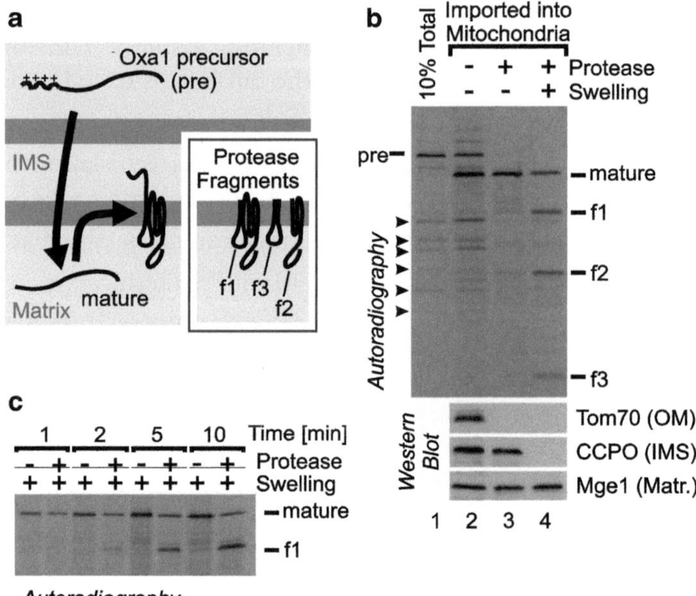

Fig. 3 Analysis of the topogenesis of a mitochondrial inner membrane protein. (**a**) Schematic representation of the Oxa1 import route. Oxa1 is a conservatively sorted inner membrane protein with five transmembrane segments. It is synthesized in the cytosol in a precursor form that is initially imported into the matrix from where it integrates into the inner membrane. Upon opening of the outer membrane and addition of PK (protease) characteristic fragments can be generated. (**b**) Import experiment with radiolabelled Oxa1. The precursor form (pre) of Oxa1 was synthesized in reticulocyte lysate (*lane 1*). Arrowheads depict N-terminally truncated versions of the proteins resulting from translational initiation reactions on internal AUG codons of the RNA. Upon incubation of the lysate with mitochondria, the preprotein is processed to its mature form (*lane 2*). When mitochondria are exposed to protease after the import reaction the precursor protein is removed but the imported mature form remains intact due to its inaccessibility to protease (*lane 3*). When the outer membrane is opened by hypotonic swelling of the mitochondria during the protease treatment, the imported protein is degraded and forms specific fragments that provide information on its topology in the inner membrane. Western blot signals for marker proteins are shown to control the mitochondrial fractionation: Tom70 is a protein of the outer membrane (OM), cytochrome *c* peroxidase (CCPO) is a protein of the IMS, and Mge1 is a protein of the matrix (Matr.). (**c**) Time course experiment with Oxa1. Oxa1 was imported into isolated mitochondria for the time periods indicated. Mitochondria were converted to mitoplasts by hypotonic swelling and protease was added as indicated. Note that at early time points Oxa1 is completely protected against protease and, hence, present in the matrix. Upon longer incubation periods, a fragment is formed by protease treatment indicating membrane integration of Oxa1 from the matrix

9. Add 10 μL of PMSF. Mix rapidly to avoid precipitation of PMSF.
10. Centrifuge for 10 min at 25,000 × *g* (4 °C).
11. Discard the supernatant carefully and add 500 μL of ice-cold SH/KCl containing 2 mM of PMSF (*see* **Note 19**).

12. Centrifuge for 10 min at 25,000 × *g* (4 °C).
13. Discard the supernatant carefully.
14. Resuspend the pellet in 30 μL of sample buffer and boil for 5 min at 96 °C (*see* **Note 20**).
15. Analyze samples by SDS-PAGE and autoradiography. As seen in the example shown in Fig. 3, imported matrix proteins that are proteolytically matured migrate more rapidly on an SDS-PAGE due to their smaller weight.

3.5 Carbonate Extraction

Alkaline extraction with carbonate solubilizes membrane-associated proteins but not integral membrane proteins. The stringency of the extraction thereby depends on the pH of the carbonate solution used. The higher the pH of the carbonate solution the more proteins are extracted from the membrane fraction. This method can be used to analyze proteins by Western blotting (steady state levels) or after an import experiment with radiolabeled precursor proteins (newly imported proteins or sorting intermediates).

1. Perform all steps at 4 °C.
2. Take 100 μg of isolated mitochondria and centrifuge them for 10 min at 25,000 × *g*.
3. Resuspend the pellet in 500 μL of Na_2CO_3.
4. Incubate the sample while shaking at 1,400 rpm for 30 min.
5. Separate the soluble and the membrane-bound proteins by centrifuging for 30 min at 186,000 × *g*.
6. Resuspend the pellet that contains integral membrane proteins in 30 μL of sample buffer and boil for 3 min at 96 °C.
7. Take the supernatant from **step 5** that contains soluble and peripheral membrane proteins and add 100 μL of 72 % TCA for protein precipitation.
8. Freeze the sample for at least 30 min at −20 °C.
9. Centrifuge at 25,000 × *g* for 15 min at 4 °C.
10. Discard the supernatant.
11. Wash the pellet: Add 1 mL of cold acetone (−20 °C) and invert the sample carefully.
12. Centrifuge at 25,000 × *g* for 15 min at 4 °C.
13. Discard the supernatant.
14. Resuspend the pellet in 30 μL of sample buffer.
15. Analyze both samples by SDS-PAGE and Western blotting.

3.6 Aggregation Assay

This simple fractionation assay separates aggregated proteins from soluble and membrane-embedded proteins. Due to their hydrophobicity, membrane proteins and their precursor forms tend to form aggregates, in particular if they are overexpressed.

1. Lyse 100 μg of mitochondria in 300 μL of lysis buffer by incubation on ice for 10 min. Vortex the sample briefly 3–4 times during the incubation time.
2. Perform a clarifying spin at 25,000 × *g* for 10 min (4 °C) (*see* **Note 21**).
3. Take 140 μL of the supernatant and precipitate proteins by addition of 500 μL of water and 140 μL of TCA (continue with **step 6**). This sample will be the total.
4. Centrifuge 140 μL of the remaining supernatant of **step 2** at 186,000 × *g* for 30 min (4 °C).
5. Separate pellet and supernatant. Resuspend the pellet in 30 μL of sample buffer and boil for 5 min at 96 °C; this sample contains aggregated proteins. Take the supernatant and precipitate proteins by addition of 500 μL of water and 140 μL of TCA. This sample contains soluble proteins.
6. Freeze the sample for at least 30 min at −20 °C.
7. Centrifuge at 25,000 × *g* for 15 min at 4 °C.
8. Discard the supernatant.
9. Wash the pellet: Add 1 mL of cold acetone (−20 °C) and invert the sample carefully.
10. Centrifuge at 25,000 × *g* for 15 min at 4 °C.
11. Discard the supernatant and resuspend the pellet in 30 μL of sample buffer. Boil for 5 min at 96 °C.
12. Analyze the three samples by SDS-PAGE and Western blotting or autoradiography.

4 Notes

1. The carbon source of the media has a strong influence on the amount of mitochondria in yeast cells. The largest yield of mitochondria will be reached with media which force cells to respire such as glycerol, ethanol, or lactate (or combinations thereof). Alternatively, galactose can be used for strains which are unable to respire or which grow poorly on non-fermentable carbon sources. It is not recommended to use glucose as carbon source since glucose represses the expression of many genes encoding mitochondrial proteins. For selection purposes marker amino acids or nucleotides can be omitted.
2. BSA is used to stabilize isolated mitochondria. It has a positive effect on the integrity of mitochondria, particularly during longer incubation periods of more than 30 min and at temperatures of 30 °C or more. It is important that only fatty acid-free BSA is used as BSA of lower quality can lead to the lysis of membranes.

3. Mitochondria should be thawed on ice and also be kept on ice until they are used. They should be used as soon as possible after thawing since a prolonged incubation even on ice strongly reduces their import and translation competence.
4. BSA is used to stabilize precursor proteins; however, not all precursors need BSA. Thus it can be omitted in many cases without any influence on the import efficiency. BSA should be omitted especially if the import reaction is followed by cross-linking or TCA precipitation.
5. The import of some mitochondrial proteins strongly depends on well-energized mitochondria and high levels of the mitochondrial membrane potential. To this end, CP (final concentration: 10 mM) and CK (final concentration: 100 μg/mL) can be added to the import reaction to improve the energy supply by regeneration of ATP. Likewise, malate and succinate can be added to the reaction (final concentration: 2 mM each) to further energize the mitochondria.
6. The pH of the carbonate solution is critical for the stringency of the extraction conditions. If the pH is not adjusted it is typically around 11.2–12.0. Under these conditions even some integral membrane proteins can be found in the supernatant, in particular tail-anchored proteins or integral membrane proteins with a small domain on one side of the membrane. To reduce the stringency, the carbonate solution can be adjusted to less alkaline pH values such as 9.5 or 10.5.
7. To test whether the cell wall was successfully digested, take twice 50 μL of the suspension and dilute one aliquot in 1.4 mL of H_2O and the other in 1.4 mL of 0.6 M sorbitol. The cells were efficiently converted to spheroplasts, if the OD_{600} of the H_2O sample is 10–20 % of that of the sorbitol sample.
8. It is important that no part of the pellet is collected together with the supernatant. If this cannot be avoided due to the soft consistency of the pellet, further centrifugation steps have to be performed.
9. Use a glass pipette to gently resuspend the mitochondria. Do not vortex the sample to avoid damage to the mitochondria.
10. Truncate the pipette tips with scissors before pipetting mitochondria to avoid rupturing of mitochondrial membranes.
11. Samples should not be boiled since the highly hydrophobic mitochondrial translation products aggregate at higher temperatures. These aggregates even remain unsoluble in SDS buffer. After boiling, especially the signals of Cox1, Cox3, and cytochrome *b* disappear on SDS gels. In order to prevent post-lysis degradation by PK, samples can be precipitated by TCA which inactivates the protease.

12. The mitochondrial translation works efficiently for up to 30–60 min. The labelling process can be performed at temperatures between 10 and 40 °C with an optimum from 25 to 35 °C.

13. The addition of an excess of nonradioactive methionine leads to a continued translation without labelling, thus, to a completion of the synthesis of the translation products and a decrease of the background signal.

14. Incubation of mitochondria under hypoosmotic buffer conditions will lead to a hypotonic swelling of the mitochondria which selectively opens the outer membrane. For swelling, the mitochondria-containing solution is adjusted to 60 mM sorbitol (final); after incubation for 20–30 min on ice, the mitochondria are converted to mitoplasts. Proteases such as PK may be added during this incubation period. Thereby the submitochondrial localization of proteins can be analyzed (Figs. 2 and 3). It is recommended that the opening of the outer membrane is controlled by Western blotting with antibodies against marker proteins.

15. Make sure that all components are properly mixed before you add the mitochondria. Do not vortex the sample after mitochondria have been added. Use truncated tips when pipetting mitochondria.

16. The amount of lysate that is used per import reaction strongly depends on its content of methionine residues, its quality, as well as the specific import efficiency of the precursor protein used. Usually, 1 % (v/v) of the lysate should be used per import reaction but up to 20 % can be used without major deleterious effects.

17. The length of the incubation time and the incubation temperature depend on the import substrate used. Some proteins are efficiently imported within 1–2 min; others need more than 30 min. To decrease the import rate, the temperature can be lowered to 16 °C. Temperatures up to 30 °C further increase the import rate.

18. The protease degrades all precursor proteins that are outside of the outer mitochondrial membrane. Thus, only those precursors are detected in the subsequent autoradiography which were successfully translocated across the outer membrane and hence are protected from protease treatment. Instead of PK, other proteases like trypsin can be used. However, higher concentrations of trypsin destabilize the outer membrane and can degrade mitochondrial proteins.

19. Add the PMSF to the SH/KCl buffer only prior to use since it is rapidly inactivated in aqueous solutions.

20. We observed that PK is sometimes not completely inactivated or removed by PMSF and SH/KCl treatment. Thus, it is strongly recommended to boil the samples immediately after addition of sample buffer. Otherwise, residual active PK will rapidly degrade the proteins of the sample during SDS-PAGE leading to a specific loss of proteins larger than 45–60 kDa.
21. During this clarifying spin incompletely dissolved mitochondrial debris is removed.

Acknowledgments

This work was supported by grants from the Deutsche Forschungsgemeinschaft and the Landesschwerpunkt für Membrantransport in Rheinland-Pfalz.

References

1. Federman M, Avers CJ (1967) Fine-structure analysis of intercellular and intracellular mitochondrial diversity in Saccharomyces cerevisiae. J Bacteriol 94:1236–1243
2. Mootha VK, Bunkenborg J, Olsen JV et al (2003) Integrated analysis of protein composition, tissue diversity, and gene regulation in mouse mitochondria. Cell 115:629–640
3. Elstner M, Andreoli C, Klopstock T et al (2009) The mitochondrial proteome database: MitoP2. Methods Enzymol 457:3–20
4. Sickmann A, Reinders J, Wagner Y et al (2003) The proteome of *Saccharomyces cerevisiae* mitochondria. Proc Natl Acad Sci USA 100: 13207–13212
5. Schmitt S, Prokisch H, Schlunck T et al (2006) Proteome analysis of mitochondrial outer membrane from *Neurospora crassa*. Proteomics 6:72–80
6. Zahedi RP, Sickmann A, Boehm AM et al (2006) Proteomic analysis of the yeast mitochondrial outer membrane reveals accumulation of a subclass of preproteins. Mol Biol Cell 17:1436–1450
7. Neupert W, Herrmann JM (2007) Translocation of proteins into mitochondria. Annu Rev Biochem 76:723–749
8. Chacinska A, Koehler CM, Milenkovic D et al (2009) Importing mitochondrial proteins: machineries and mechanisms. Cell 138: 628–644
9. Endo T, Yamano K (2009) Multiple pathways for mitochondrial protein traffic. Biol Chem 390:723–730
10. Koehler CM (2004) New developments in mitochondrial assembly. Annu Rev Cell Dev Biol 20:309–335
11. Rehling P, Brandner K, Pfanner N (2004) Mitochondrial import and the twin-pore translocase. Nat Rev Mol Cell Biol 5:519–530
12. Paschen SA, Waizenegger T, Stan T et al (2003) Evolutionary conservation of biogenesis of beta-barrel membrane proteins. Nature 426:862–866
13. Dukanovic J, Rapaport D (2011) Multiple pathways in the integration of proteins into the mitochondrial outer membrane. Biochim Biophys Acta 1808:971–980
14. Wiedemann N, Kozjak V, Chacinska A et al (2003) Machinery for protein sorting and assembly in the mitochondrial outer membrane. Nature 424:565–571
15. Hawlitschek G, Schneider H, Schmidt B et al (1988) Mitochondrial protein import: identification of processing peptidase and of PEP, a processing enhancing protein. Cell 53: 795–806
16. Witte C, Jensen RE, Yaffe MP, Schatz G (1988) MAS1, a gene essential for yeast mitochondrial assembly, encodes a subunit of the mitochondrial processing peptidase. EMBO J 7: 1439–1447
17. Vögtle FN, Wortelkamp S, Zahedi RP et al (2009) Global analysis of the mitochondrial N-proteome identifies a processing peptidase critical for protein stability. Cell 139:428–439
18. Van Loon APGM, Brändli AW, Schatz G (1986) The presequences of two imported

mitochondrial proteins contain information for intracellular and intra-mitochondrial sorting. Cell 44:97–100

19. Hell K, Neupert W, Stuart RA (2001) Oxa1p acts as a general membrane insertion machinery for proteins encoded by mitochondrial DNA. EMBO J 20:1281–1288
20. Herrmann JM, Neupert W, Stuart RA (1997) Insertion into the mitochondrial inner membrane of a polytopic protein, the nuclear encoded Oxa1p. EMBO J 16:2217–2226
21. Wagener N, Ackermann M, Funes S et al (2011) A pathway of protein translocation in mitochondria mediated by the AAA-ATPase Bcs1. Mol Cell 44:191–202
22. Sirrenberg C, Bauer MF, Guiard B et al (1996) Import of carrier proteins into the mitochondrial inner membrane mediated by Tim22. Nature 384:582–585
23. Koehler CM, Jarosch E, Tokatlidis K et al (1998) Import of mitochondrial carrier proteins mediated by essential proteins of the intermembrane space. Science 279:369–373
24. Ott M, Prestele M, Bauerschmitt H et al (2006) Mba1, a membrane-associated ribosome receptor in mitochondria. EMBO J 25: 1603–1610
25. Szyrach G, Ott M, Bonnefoy N et al (2003) Ribosome binding to the Oxa1 complex facilitates cotranslational protein insertion in mitochondria. EMBO J 22:6448–6457
26. Jia L, Dienhart M, Schramp M et al (2003) Yeast Oxa1 interacts with mitochondrial ribosomes: the importance of the C-terminal hydrophilic region of Oxa1. EMBO J 22:6438–6447
27. Preuss M, Leonhard K, Hell K et al (2001) Mba1, a novel component of the mitochondrial protein export machinery of the yeast Saccharomyces cerevisiae. J Cell Biol 153: 1085–1096
28. Frazier AE, Taylor RD, Mick DU et al (2006) Mdm38 interacts with ribosomes and is a component of the mitochondrial protein export machinery. J Cell Biol 172:553–564
29. Bauerschmitt H, Mick DU, Deckers M et al (2010) Ribosome-binding proteins Mdm38 and Mba1 display overlapping functions for regulation of mitochondrial translation. Mol Biol Cell 21:1937–1944
30. Christian BE, Spremulli LL (2011) Mechanism of protein biosynthesis in mammalian mitochondria. Biochim Biophys Acta 1819: 1035–1054
31. Rorbach J, Soleimanpour-Lichaei R, Lightowlers RN et al (2007) How do mammalian mitochondria synthesize proteins? Biochem Soc Trans 35:1290–1291
32. McKee EE, Poyton RO (1984) Mitochondrial gene expression in saccharomyces cerevisiae. I. Optimal conditions for protein synthesis in isolated mitochondria. J Biol Chem 259: 9320–9331
33. Pelham HRB, Jackson RJ (1976) An efficient mRNA-dependent translation system from reticulocyte lysates. Eur J Biochem 67: 247–256
34. Mokranjac D, Neupert W (2007) Protein import into isolated mitochondria. In: Leister D, Herrmann JM (eds) Mitochondria: practical protocols, vol 372. Humana Press, Totowa, NJ, pp 277–286
35. Daum G, Böhni PC, Schatz G (1982) Import of proteins into mitochondria: cytochrome b_2 and cytochrome c peroxidase are located in the intermembrane space of yeast mitochondria. J Biol Chem 257:13028–13033
36. Meisinger C, Sommer T, Pfanner N (2000) Purification of *Saccharomyces cerevisiae* mitochondria devoid of microsomal and cytosolic contaminations. Anal Biochem 287:339–342
37. Bradford MM (1976) A rapid and sensitive method for quantitation of microgram quantities of protein utilising the principle of protein-dye binding. Anal Biochem 72:248–254
38. Becker D, Krayl M, Voos W (2008) In vitro analysis of the mitochondrial preprotein import machinery using recombinant precursor polypeptides. Methods Mol Biol 457:59–83

Part V

Functional Analysis of Membrane Proteins

Chapter 21

Reconstitution of Mitochondrial Presequence Translocase into Proteoliposomes

Martin van der Laan, Ralf M. Zerbes, and Chris van der Does

Abstract

The isolation and functional reconstitution of large membrane protein complexes is an important step towards the biochemical characterization of such sophisticated molecular machines. Reconstitution is a multistep process that requires the mild solubilization of membrane protein complexes from native membrane preparations, the purification of the complexes from protein–detergent solutions, and their incorporation into artificial phospholipid vesicles through controlled detergent removal. The major challenge is to preserve the integrity and catalytic activity of the often fragile membrane protein assemblies during the entire procedure. Here we describe the protocols for a particularly intricate example, the functional reconstitution of the mitochondrial presequence translocase (TIM23 complex). This highly versatile and dynamic protein complex is the main protein translocation machinery of the inner mitochondrial membrane and mediates the import of precursor proteins with N-terminal presequences from the cytosol.

Key words Digitonin, Membrane protein complex, Mitochondria, Phospholipids, Protein import, Proteoliposomes, Reconstitution, TIM23 complex, Translocation

1 Introduction

Bioinformatic analysis of genome sequencing data predicts that ~30 % of all genes encode integral membrane proteins [1]. Moreover, ~50 % of all known drug targets represent membrane proteins, of which the largest group are the G-protein-coupled receptors [2]. Understanding the structure and function of membrane proteins is therefore a major focus of biomedical research. Due to their hydrophobic natural environment membrane proteins are barely water-soluble. As a consequence, their biochemical analysis has been and still is technically challenging. In order to purify a membrane protein it has to be extracted from its resident lipid bilayer with an organic solvent or (less harsh) with a suitable detergent. Whereas some types of experiments, like binding assays or structural analysis, can be performed with detergent-solubilized, purified membrane proteins, most functional studies rely on the

Doron Rapaport and Johannes M. Herrmann (eds.), *Membrane Biogenesis: Methods and Protocols*, Methods in Molecular Biology, vol. 1033, DOI 10.1007/978-1-62703-487-6_21,

reconstitution into an artificial membrane. Only such reconstituted systems are able to recapitulate the vectorial aspects of membrane protein activities. The importance of this feature becomes most obvious when looking at a channel protein or carrier that transports ions or organic compounds from one side of a membrane to the other. Two considerably different experimental setups have been used to deal with this problem: planar lipid bilayers [3] and artificial lipid vesicles, termed liposomes [4]. A major breakthrough in the analysis of protein-mediated membrane transport was the reconstitution of several transport activities from membrane extracts into proteoliposomes by Montal and coworkers in 1980 [5].

Nowadays, proteoliposome preparations are widely used to study membrane protein functions, but still there is no easy standard protocol for the functional reconstitution of a given membrane protein. For basically every class of membrane proteins an individual experimental procedure has to be established. A critical step of every reconstitution protocol is the solubilization of the membrane protein in an active state. For this purpose, a plethora of different detergents are commercially available that can be classified as either nonionic, anionic, zwitterionic, or bile acid salts. A relatively novel and particularly interesting group of detergents are the amphipols, synthetic polymeric surfactants that appear to have several advantages compared to "classical" detergents [6, 7]. Solubilization efficiency and stability of membrane proteins moreover depend on several physicochemical parameters, like the ionic strength and the pH of the buffer, as well as the incubation times and temperatures. An elegant and precise method to follow the solubilization of membrane proteins by isothermal titration calorimetry has recently been published [8]. However, in most cases technically less demanding procedures can be employed (see below). After solubilization membrane proteins must be reconstituted back into a lipid bilayer. In general, this is achieved by mixing of detergent-solubilized proteins and phospholipids and subsequent reformation of stable, protein-containing bilayers through the removal of the detergent(s) from this mixture. The concentration of the detergents must be reduced below the so-called critical micellar concentration (cmc), which is an intrinsic property of each individual detergent [9]. For detergents with a high cmc, this can be achieved either by rapid dilution of the samples into a detergent-free buffer or by dialysis. However, the most commonly used method is the adsorption of detergent molecules to hydrophobic surfaces, like polystyrene beads (see below). Finally, proteoliposomes have to be purified, the transmembrane topology of the reconstituted membrane proteins should be determined, and an activity assay has to be designed.

The reconstitution of a given membrane-bound process into proteoliposomes from purified components is the only way to directly determine the minimal requirements for the specific activity under investigation [10]. Such procedures may be relatively straight forward, when this activity is provided by a single polypeptide that can be recombinantly expressed and easily purified in large amounts. An additional level of complexity is added, when a membrane protein complex composed of multiple subunits has to be reconstituted. In principle such complexes may be rebuilt in vitro from individual membrane proteins that were expressed and purified separately. However, in most cases this turned out to be very laborious and inefficient, if not impossible. Therefore, it is necessary to purify the intact, functional protein complex directly from a native membrane preparation. As many membrane protein complexes are very labile upon solubilization, particularly mild detergents, like nonionic digitonin, are required. In the following we will describe the reconstitution of the membrane protein insertion activity of the presequence translocase of the inner mitochondrial membrane (TIM23 complex) from the baker's yeast, *Saccharomyces cerevisiae* [10–12]. Most mitochondrial proteins are nuclear-encoded and produced in the cytosol in the form of precursor proteins with specific mitochondrial targeting signals. The TIM23 complex imports precursor proteins with N-terminal, cleavable presequences into the inner mitochondrial membrane or into the matrix compartment, depending on the presence or absence of hydrophobic transmembrane segments in the polypeptide. These two distinct activities are carried out by different forms of the TIM23 complex. For the insertase activity the essential conserved core of the machinery associates with an additional inner mitochondrial membrane protein, termed Tim21. In contrast, matrix import depends on the cooperation of the TIM23 complex with the presequence translocase-associated import motor (PAM) that is located on the matrix side of the inner membrane. The insertase activity of TIM23 was reconstituted into proteoliposomes after purification of the complex via an affinity tag on the Tim21 protein from isolated mitochondria of an engineered yeast strain [10].

2 Materials

2.1 Detergent-Mediated Solubilization

1. Digitonin.
2. Ethanol (p.a. grade).
3. Glass centrifugation tubes (e.g., Corex).
4. Solubilization buffer without detergent: 20 mM Tris–HCl pH 7.4, 60 mM NaCl, 0.1 mM EDTA pH 8.0, 10 % (v/v) glycerol, 2 mM PMSF.
5. Solubilization buffers for digitonin titration: Same as in **item 4**; supplemented with 0.4–2.0 % (w/v) digitonin.

2.2 TIM23 Complex Purification from Isolated Mitochondria

1. Solubilization buffer: 20 mM Tris–HCl pH 7.4, 60 mM NaCl, 0.1 mM EDTA pH 8.0, 10 % (v/v) glycerol, 2 mM PMSF, 0.7 % (w/v) digitonin.
2. Mobicol columns with 35 μm filters (MoBiTec, Göttingen, Germany).
3. Immunoglobulin G (IgG) Sepharose beads.
4. Acetate buffer (prepare a 0.5 M acetic acid solution and adjust the pH to 3.4 with 0.5 M ammonium acetate).
5. Wash buffer: 20 mM Tris–HCl pH 7.4, 60 mM NaCl, 0.1 mM EDTA, 10 % (v/v) glycerol, 2 mM PMSF, 0.3 % (w/v) digitonin.
6. AcTEV protease (Life Technologies, Carlsbad, CA).
7. Ni^{2+}-NTA agarose beads.

2.3 Purification of Phospholipids

1. *Escherichia coli* polar lipid extract (Avanti Polar Lipids, Alabaster, AL); 25 mg/mL stock solution in chloroform.
2. Rotational evaporator.
3. Acetone (p.a. grade) supplemented with 2 mM DTT.
4. Glass centrifugation tubes (e.g., Corex).
5. Diethyl ether (p.a. grade) supplemented with 2 mM DTT.
6. Chloroform (p.a. grade).

2.4 Liposome Preparation

1. Purified *E. coli* polar lipid extract (prepared according to Subheading 3.3).
2. 1,1′,2,2′-Tetraoleoyl-cardiolipin (Avanti Polar Lipids, Alabaster, AL); 25 mg/mL stock solution in chloroform.
3. Chloroform (p.a. grade).
4. Rotational evaporator.
5. Liposome storage buffer: 20 mM Tris–HCl pH 7.2, 50 mM KCl, 2 mM DTT.
6. Tip sonicator.
7. Liquid nitrogen.
8. Mini-extruder equipped with a polycarbonate membrane filter of 400 nm pore size (Avanti Polar Lipids, Alabaster, AL).

2.5 Detergent Titration of Liposomes

1. Liposomes with a diameter of 400 nm (prepared according to Subheading 3.4).
2. Liposome buffer: 20 mM Tris–HCl pH 7.2, 50 mM KCl.
3. Detergent stock solutions: e.g., 5 % (w/v) digitonin, 10 % (w/v) dodecyl maltoside, or 10 % (v/v) Triton X-100.

2.6 Reconstitution of Purified TIM23 Complexes into Proteoliposomes

1. Elution fraction from a TIM23 complex purification as described in Subheading 3.2.
2. Liposomes in Liposome storage buffer prepared according to Subheading 3.4.
3. Buffer P: 20 mM Tris–HCl pH 7.2, 100 mM KCl, 1 mM DTT, 10 % (v/v) glycerol, 0.35 % (w/v) dodecyl maltoside, 0.15 % (w/v) digitonin.
4. Biobeads SM-2 (Bio-Rad, Hercules, CA).
5. Methanol (p.a. grade).
6. Ethanol (p.a. grade).
7. Buffer R: 20 mM Tris–HCl pH 7.2, 100 mM KCl, 5 mM $MgCl_2$.
8. Buffer R supplemented with 1.4 or 1 M sucrose, respectively.
9. 1 M $MgCl_2$ stock solution.

3 Methods

3.1 Detergent-Mediated Solubilization

The starting material for any solubilization trials should be a thoroughly resuspended membrane preparation. This can either be a crude membrane extract from cells obtained by ultracentrifugation or be purified cellular organelles, like in the case of the TIM23 complex, which is extracted from isolated mitochondria [13]. For crude membrane preparations obtained by ultracentrifugation, a good resuspension can be obtained using a Potter-Elvehjem pestle (Sigma Aldrich, St. Louis, MO). For every membrane protein complex the optimal detergent needs to be determined experimentally. Most important criteria are the solubilization efficiency, the stability of the protein complex of interest, compatibility with downstream applications (e.g., interference with enzymatic activities), and absorption of UV light (e.g., for protein detection during chromatography and in enzymatic assays). Typical detergent concentrations for initial solubilization trials range from 0.5 to 2 % with protein concentrations between 1 and 5 mg/mL. The detergent of choice for the solubilization of the TIM23 complex is the particularly mild nonionic digitonin, as none of the many other detergents tested in our laboratory leaves the TIM23 complex intact. Digitonin solubilization works best for membranes that have a relatively high sterol or cardiolipin content. As digitonin is rather difficult to remove from protein–lipid–detergent mixtures (see below), it is important to determine the minimal amount required for solubilization. Commercially available digitonin is a natural product that contains relatively large amounts of contaminants. Thus, the minimal digitonin concentration needed for solubilization may considerably differ from batch to batch and should ideally be determined again for each new batch.

3.1.1 Preparation of Digitonin Stock Solution

1. Suspend digitonin in ethanol (1 g digitonin per 20 mL ethanol) and boil on a heating plate while slowly stirring (magnetic stirrer) until the powder is dissolved.
2. Transfer the solution carefully to glass centrifugation tubes (e.g., Corex) and incubate at 4 °C over night. Digitonin will precipitate from the solution (*see* **Note 1**).
3. Centrifuge the samples at 6,000 × *g* and 4 °C for 15 min.
4. Dry the digitonin pellets at room temperature and dissolve in distilled water at a concentration of 5 % (w/v). To dissolve the digitonin boil the solution for ~10 min slowly stirring.
5. Incubate the solution at 4 °C over night and centrifuge at 6,000 × *g* and 4 °C for 15 min. Use the digitonin-containing supernatant as ~5 % detergent stock solution for further experiments (*see* **Note 2**). Store the stock solution at 4 °C.

3.1.2 Determination of Solubilization Efficiency

1. Dilute a suspension of isolated mitochondria to 2 mg/mL (total protein concentration) in Solubilization buffer without detergent.
2. The range of final digitonin concentrations to be tested is between 0.2 and 1.0 % (ideally in 0.1 %-steps). Prepare 2× digitonin solutions (0.4–2.0 %) in Solubilization buffer using the 5 % stock solution described in Subheading 3.1.1 (*see* **Note 3**).
3. Mix 500 μL of diluted mitochondria for every tested digitonin concentration with an equal volume of the corresponding 2× digitonin solution in Solubilization buffer in 1.5 mL reaction tubes (*see* **Note 4**).
4. Incubate samples for 30 min, slowly shaking head-over-head at 4 °C (*see* **Note 5**).
5. Centrifuge samples for 30 min at 16,000 × *g*. Solubilized membrane protein complexes will stay in the supernatant, while non-solubilized mitochondria and aggregated material will be pelleted (*see* **Note 6**).
6. Separate pellets and supernatants and analyze samples by SDS-PAGE and Western blotting according to standard protocols with antibodies against the proteins of interest (*see* **Note 7**).

3.2 Purification of TIM23 Complexes from Isolated Mitochondria

The starting material for the isolation of TIM23 complexes in the import motor-free form [10, 14] are isolated mitochondria from a yeast strain expressing a Tim21-Protein A fusion protein from the chromosomal *TIM21* locus under control of the native promoter. This strain was constructed by transformation of wild-type (YPH499) cells with a DNA fragment encoding the affinity tag together with a selectable marker amplified from plasmid pYM10 using the S2 and S3 primers described [15] with the respective overhangs for site-specific homologous recombination.

The C-terminal attachment of Protein A to Tim21 allows for the isolation of TIM23 complexes with IgG Sepharose beads. The fusion protein contains a short linker amino acid sequence (–RTLQVDGS–) followed by a Tobacco Etch Virus (TEV) protease recognition site (–ENLYFQG–) between the Tim21 sequence and the Protein A moiety, which allows for the elution of TIM23 complexes under particularly mild conditions through TEV protease cleavage. TIM23 complexes eluted in this way exhibit a very high purity [10]. Additional purification steps are not required (*see* **Note 8**). To obtain sufficient amounts of purified TIM23 complexes for reconstitution and subsequent activity assays, at least 20–25 mg (total protein amount) of isolated Tim21-Protein A mitochondria should be used. It is of crucial importance that during the entire purification procedure samples are kept cold. All centrifugation steps are carried out 4 °C; all incubations are done at 4 °C or on ice.

1. Pellet 20 mg isolated mitochondria (total protein amount) by centrifugation at 16,000 × *g* for 10 min.
2. Carefully resuspend pellets in 1 mL Solubilization buffer; add another 19 mL of Solubilization buffer to reach a final protein concentration of 1 mg/mL.
3. Shake samples head-over-head for 30 min.
4. Centrifuge at 16,000 × *g* for 15 min.
5. Transfer supernatant into a fresh tube. Take off 200 μL for subsequent analysis by SDS-PAGE and Western blotting ("Total" sample) (*see* **Note 9**).
6. Prepare two Mobicol columns: Use the plastic tool delivered with the columns to place 35 μm filters into the columns and put the columns into 2 mL collection tubes. Add 200 μL of a 50 % IgG Sepharose slurry in 20 % ethanol to each column, centrifuge 1 min at 100 × *g*, and discard flow-through. (The bed volume is 2× 100 μL.)
7. Add 0.5 mL Acetate buffer to each column, centrifuge 1 min at 100 × *g*, and discard flow-through (*see* **Note 10**) (perform this step two times).
8. Add 0.5 mL Solubilization buffer without detergent to each column, centrifuge 1 min at 100 × *g*, and discard flow-through (perform this step two times).
9. Seal the Mobicol columns with the provided bottom plugs. Close with a screw cap until use.
10. Take 0.5 mL of the mitochondrial solubilizate from **step 5** to resuspend the equilibrated beads in the two columns and add this suspension to the rest of the solubilizate. Mix well and distribute the mixture into 2 mL tubes (*see* **Note 11**).

11. Incubate the samples for at least 90 min at 4 °C, shaking head-over-head.
12. Spin the 2 mL tubes for 1 min at 100×*g* to pellet the beads, remove all supernatants, but leave ~0.4 mL on top of each pellet. Take off 200 μL from the supernatant for subsequent analysis by SDS-PAGE and Western blotting ("Unbound" sample).
13. Use the remaining ~0.4 mL of supernatant to resuspend the pellets and transfer these mixtures from the ten 2 mL cups to the two Mobicol columns one by one. (Remove the bottom plug from the columns before!) After every step centrifuge 1 min at 100×*g* and discard flow-through.
14. Wash each column ten times with 0.5 mL Wash buffer, after every step centrifuge 1 min at 100×*g*, and discard flow-through.
15. After the last washing step, spin 1 min at 200×*g* to remove residual buffer from the column. Close the columns again with the bottom plug.
16. Add 250 μL Wash buffer and 8 μL TEV protease (10 Units/μL) to each column, close with screw cap, and incubate over night, shaking at 800 rpm in a thermomixer.
17. Next morning, add 10 μL of Ni^{2+}-NTA beads equilibrated 2× with H_2O and 2× with Wash buffer to each column (*see* **Note 12**).
18. Incubate for 30 min, shaking at 800 rpm in a thermomixer.
19. Remove the bottom plugs and transfer the columns into 1.5 mL reaction tubes. Collect the elution fractions by centrifugation for 1 min at 500×*g*. Take off 20 μL from the supernatant for subsequent analysis by SDS-PAGE and Western blotting ("Elution" sample) (*see* **Note 13**).
20. Pool the two elution fractions (total volume ~0.5 mL) and keep the purified TIM23 complexes on ice until used for reconstitution.
21. Mix "Total", "Unbound", and "Elution" samples 1:1 with 2× SDS sample buffer. Generally, 30 μL per gel lane of this mixture is subjected to SDS-PAGE and Western blotting with antibodies against the proteins of interest.

3.3 Purification of Phospholipids

To obtain functional reconstitution of a membrane protein complex, the phospholipid composition in which the complex is reconstituted plays a crucial role. Either different total phospholipid extracts from natural sources (self-prepared or purchased) or completely custom-made mixtures of synthetic phospholipids may be used. The latter approach is certainly more expensive and in most cases not really necessary. The main advantage of using synthetic phospholipid mixtures is that they are chemically defined and do not exhibit natural variations in terms of, e.g., fatty acid chain lengths or the degree of saturation. Independent of the source of lipids, it is generally recommended to reconstitute a given membrane protein complex into a phospholipid mixture that closely resembles that of its native membrane environment. The phospholipid composition of the inner mitochondrial membrane, which is the resident membrane of the TIM23 complex in vivo, is similar to that of the plasma membrane of Gram-negative bacteria. The main difference is that the inner mitochondrial membrane contains higher amounts of cardiolipin [16, 17]. Thus, for the reconstitution of the TIM23 complex, an *Escherichia coli* polar lipid extract supplemented with 10 % (mol/mol) synthetic 1,1′,2,2′-tetraoleoyl-cardiolipin was used [10]. Commercially available natural lipid extracts often contain residual amounts of membrane proteins, particularly such that are very small and hydrophobic. Such contaminations can be a serious problem, especially as some of these residual proteins may render the membrane of the formed liposomes permeable to ions or other small molecules. In the case of the TIM23 complex a membrane potential ($\Delta\psi$) across the proteoliposome membrane is a prerequisite for its activity. Therefore, it was very important to obtain a highly pure phospholipid mixture for liposome formation. Thus, the purchased *E. coli* polar lipid extract was subjected to an additional purification procedure (*see* **Note 14**).

1. *E. coli* polar lipid extract (1 g) in chloroform is concentrated in a rotational evaporator to a maximal volume of 5 mL (*see* **Note 15**).
2. Supplement 150 mL of ice-cold acetone (from the −20 °C freezer) with 2 mM DTT in a glass flask with a glass stopper (*see* **Note 16**).
3. Slowly add the 5 mL phospholipid solution in single drops to the cold acetone/DTT, while stirring on ice (*see* **Note 17**). A white precipitate will appear. Stir the mixture at least 4 h (preferentially over night) in the dark under a nitrogen atmosphere. (Blow nitrogen gas over the mixture before closing the flask with the stopper.)
4. Transfer the mixture into glass centrifugation tubes (e.g., Corex) (*see* **Note 1**) and centrifuge at 2,500 × *g* and 4 °C for 10 min.
5. Remove the supernatant and dry the pellet under a stream of nitrogen. Don't forget to dry the walls of the tubes as well!
6. Dissolve the pellet in 200 mL diethyl ether containing 2 mM DTT. Transfer the mixture again into a glass flask with a stopper and stir for 20 min under a nitrogen atmosphere at room temperature.
7. Transfer the mixture back into glass centrifugation tubes and centrifuge at 2,500 × *g* and 4 °C for 10 min. Attention: The phospholipids are now present in the supernatant!

8. Take off the supernatant and evaporate the solvent in a rotational evaporator at room temperature. A phospholipid film will form in the flask.
9. Dissolve the phospholipid film in 10 mL chloroform and again evaporate close to dryness (*see* **Note 18**).
10. Wash the film with 1 mL ethanol and evaporate the solvent in a rotational evaporator at room temperature. Again evaporate only close to dryness and remove the remaining ethanol with a stream of nitrogen.
11. Dissolve the purified phospholipid mixture in chloroform to a final concentration of 25 mg/mL and store in a glass tube under a nitrogen atmosphere at –20 °C. Close the tube with a glass stopper and wrap parafilm around the stopper.

3.4 Preparation of Liposomes

1. Mix purified and synthetic phospholipids in chloroform as required in a glass tube.
2. Slowly dry the lipid mixture in a rotational evaporator. Residual amounts of solvent may be removed with a nitrogen stream (*see* **Note 19**).
3. Add Liposome storage buffer to the lipid film. The final lipid concentration in the Liposome storage buffer should be ~20 mg/mL. Incubate at room temperature for 30 min; occasionally shake the flask or pipet carefully up and down to get the phospholipids completely suspended (*see* **Note 20**).
4. Transfer the phospholipid suspension to a plastic tube and homogenize with a Branson tip sonicator (5 min, duty cycle 50 %, output control stage 5) under a stream of nitrogen (*see* **Note 21**).
5. Freeze the phospholipid suspension in liquid nitrogen and then leave it at room temperature to thaw slowly. Repeat this at least four times (*see* **Note 22**).
6. Pass the suspension of now mainly unilaminar and relatively large liposomes 11 times through a mini-extruder equipped with a polycarbonate membrane filter of 400 nm pore size (*see* **Note 23**).
7. Store liposomes on ice until used for reconstitution. Single use aliquots may be frozen in liquid nitrogen and stored at –80 °C, but keep in mind that freezing and thawing change the average diameter of the liposomes.

3.5 Titration of Liposomes with Detergent

When increasing amounts of a classical detergent, like Triton X-100 or dodecyl maltoside, are added to a liposome suspension, the detergent is soaked up into the liposome membrane and the liposomes slightly swell. This swelling can be detected as an increase of the optical density of the suspension measured, e.g., at a wavelength of 540 nm (OD_{540}) (Fig. 1). At a certain detergent

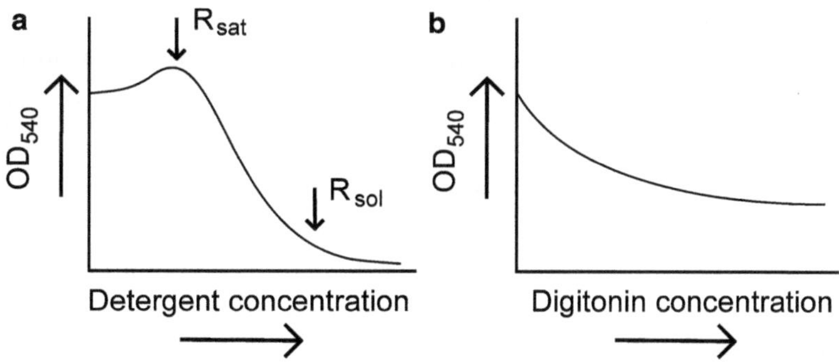

Fig. 1 Schematic representation of detergent titrations on liposomes, either with a classical detergent, like dodecyl maltoside or Triton X-100 (**a**), or with digitonin (**b**). R_{sat} saturation constant, R_{sol} solubilization constant, OD_{540} optical density at 540 nm

concentration the bilayer is saturated with detergent and the optical density reaches a maximum. The concentration required to reach this point is termed the saturation constant (R_{sat}) [18] (*see* **Note 24**). When more detergent is added, liposomes start to disintegrate and an equilibrium between detergent-saturated lipid vesicles and mixed detergent/phospholipid micelles is obtained. The equilibrium constant depends on the amount of detergent; further solubilization of the liposomes is measured as a rapid decrease of OD_{540}. At a certain detergent concentration, virtually all liposomes are solubilized and the OD_{540} of the suspension approaches the value of the fully micellar solution. This detergent concentration is termed the solubilization constant (R_{sol}) (*see* **Note 24**). To optimize the reconstitution efficiency and the direction of reconstitution, detergent–liposome mixtures from different parts of the solubilization curve (Fig. 1a) should be tested. Addition of the purified membrane protein (complex) in detergent solution either to liposomes that have been pretreated with a detergent concentration resembling R_{sat} or to almost completely solubilized liposomes is done most commonly. In the latter case, the membrane protein (complex) is generally reconstituted in equal amounts in both possible orientations, meaning that half of the proteins will acquire a "physiological" orientation, in which they can bind a ligand or substrate added from the outside of the proteoliposomes. If detergent-saturated liposomes are used, the preferred topology of the reconstituted proteins will be such that the largest hydrophilic domains remain on the outside of the proteoliposomes (*see* **Note 25**). Some membrane protein (complexes) may however deviate from these rules, and the reconstitution efficiency and orientation should always be determined experimentally.

1. Prepare liposomes with a diameter of 400 nm and with the desired phospholipid composition according to Subheading 3.4.
2. Dilute the liposome suspension to 8 mg/mL with Liposome buffer.

3. Prepare twofold concentrated detergent solutions in Liposome buffer, one for each concentration to be tested (*see* **Note 26**).
4. Add 1 mL of each 2× detergent solution to 1 mL liposome suspension and mix carefully by vortexing.
5. Incubate samples for 20 min on ice, occasionally vortex, or invert the tubes (*see* **Note 27**).
6. Measure OD_{540} of all samples and calculate from these data R_{sat} and R_{sol}.

3.6 Reconstitution of Purified TIM23 Complexes into Proteoliposomes

After all the preparations described in Subheadings 3.1–3.5, the stage is now set for the actual reconstitution of the purified membrane protein complex into the liposomes. The critical point is to remove the detergents from the reaction mixture in a controlled manner. In this way, detergent-solubilized membrane protein complexes will become incorporated with their hydrophobic domains into the slowly reforming and/or stabilizing phospholipid bilayers. The bilayer replaces the detergent that has previously kept the membrane protein complexes in solution and prevents their aggregation. The removal of the detergents must be as efficient as possible, because residual detergent molecules within the phospholipid bilayer will increase the permeability of the liposome membrane for small molecules, which becomes a major problem, if, e.g., ion gradients across the proteoliposome membrane have to be established and maintained during the course of activity assays [10]. As mentioned in the introduction, there are several methods to lower the detergent concentration in a reaction mixture below the cmc. The probably mildest and most often used method is the adsorption of detergents to polystyrene beads with a high surface area (e.g., Biobeads SM-2, Bio-Rad, Hercules, CA). This method was also successfully employed for the reconstitution of TIM23 complexes, but with slight modifications of the standard protocols (*see* **Note 28**). It is crucial that all steps of the procedure are carried out on ice or at 4 °C.

1. Mix 0.5 mL of the elution fraction from a TIM23 complex purification (as described in Subheading 3.2) with 0.5 mL of 4 mg/mL liposomes preincubated for 20 min in Buffer P (as described in Subheading 3.5) in a 2 mL reaction tube. Incubate this mixture for 30 min, shaking slowly head-over-head (*see* **Note 29**).
2. Pretreat the Biobeads according to the instructions of the supplier: Wash two times each with methanol, ethanol, distilled water, and Buffer R (*see* **Note 30**).
3. Add 0.5 mL Buffer R and 50 mg of Biobeads equilibrated with Buffer R (*see* **Note 31**) and incubate the mixture for 2 h, shaking slowly head-over-head or on a rolling bench (*see* **Note 32**).
4. Let Biobeads sediment and transfer the supernatant to a fresh tube with another 50 mg of equilibrated Biobeads.

Wash the Biobeads from the first incubation step with 100 μL Buffer R and add the supernatant to the sample (*see* **Note 33**). Incubate this mixture for 2 h, shaking slowly head-over-head or on a rolling bench.

5. Let Biobeads sediment and transfer the supernatant to a fresh tube with another 100 mg of equilibrated Biobeads. Wash the Biobeads from the previous incubation step with 100 μL Buffer R and add the supernatant to the sample. Important: At this stage, adjust the $MgCl_2$ concentration of the sample to ~10 mM. Incubate the mixture over night, shaking slowly head-over-head or on a rolling bench (*see* **Note 34**).
6. The next morning let the Biobeads sediment again and carefully collect the supernatant into an ultracentrifugation tube. Wash the Biobeads from the final incubation step with 100 μL Buffer R and add this wash fraction to the sample in the tube (*see* **Note 35**).
7. Collect the proteoliposomes from the sample by ultracentrifugation at 200,000 × *g* for 30 min.
8. Discard the supernatant and resuspend the proteoliposome pellet carefully and thoroughly in 1 mL Buffer R using a syringe.
9. Repeat **steps** 7 and **8**.
10. Completely resuspend the proteoliposome pellet in 200 μL 1.4 M sucrose in Buffer R (*see* **Note 36**).
11. Overlay with 500 μL 1 M sucrose in Buffer R and subsequently 200 μL Buffer R (without sucrose) to form step gradients (*see* **Note 37**).
12. Centrifuge the step gradients for 45 min at 225,000 × *g*.
13. Take out the milky proteoliposome fractions carefully and completely from the interface between the 1 M sucrose phase and top buffer phase.
14. Dilute the proteoliposome fraction at least five times with buffer R (*see* **Note 38**).
15. Repeat **steps** 7 and **8** two times.
16. Resuspend your gradient-purified proteoliposomes in 250 μL buffer R and keep on ice until used for activity assays or other downstream applications (*see* **Note 39**).

4 Notes

1. When handling solutions of phospholipids and/or detergents in organic solvents avoid using plastic tubes, as the hydrophobic molecules will adsorb to their surfaces. Always use glass tubes.
2. The precipitate contains various contaminants still present in the purchased powder, but also small amounts of digitonin.

Therefore, the true concentration of the stock solution will likely be lower than 5 % and the actual working concentrations have to be determined for each new digitonin solution. During storage at 4 °C the digitonin stock solution will become cloudy after a few weeks due to further precipitation. It is advisable to remove these precipitates by centrifugation and keep in mind that the active concentration will drop slowly over time. We therefore recommend to prepare fresh digitonin stocks at least every 2–3 months.

3. As the important parameter for solubilization efficiency is the protein:detergent and lipid:detergent ratio rather than the absolute detergent concentration, always the same amount of total mitochondrial proteins must be used for the solubilization step. We obtain best results with a total protein concentration of 1 mg/mL.
4. In order to assure optimal mixing of the sample during the solubilization step, the reaction tube used should be filled approximately two third, e.g., 1 mL solubilization mixture in a 1.5 mL tube. Use of smaller volumes is not recommended.
5. During solubilization avoid excessive foam formation, e.g., do not shake too fast. Moreover different detergent/phospholipid combinations require different time scales for solubilization; this has to be optimized.
6. These centrifugation conditions should be selected for digitonin solubilization of isolated mitochondria. With other membrane preparations (e.g., *E. coli* membranes) and/or different detergents, ultracentrifugation for 30–45 min at 100,000 × *g* will be necessary.
7. The amount of material to be loaded on SDS gels depends on the quality of the antibodies used for immunodecoration. For the TIM23 complex, the minimal amount of digitonin determined in this way to be required for efficient solubilization was 0.7 % (w/v).
8. Other affinity tags may be used as well. When tagging membrane proteins for purification, it should however be considered that small tags may be buried within detergent micelles during solubilization. This problem can be overcome by the introduction of flexible amino acid linkers between the C-terminus of the protein of interest and the N-terminus of the affinity tag.
9. The samples that are taken off during the purification for analysis by SDS-PAGE and Western blotting are kept on ice during the entire procedure.
10. At every step IgG Sepharose beads should be resuspended carefully, but pipetting up and down is not recommended, as the beads will stick to conventional pipette tips. Mixing can be achieved by either adding solutions with extra speed or

resuspending with a micro pipette tip slipped on a conventional one. Sepharose beads are too big to enter the micro tip.

11. The binding step may in principle be performed in one large tube, but we have obtained better results, when we distributed the 20 mL solubilizate into 10× 2 mL tubes for incubation. To minimize the loss of Sepharose beads at these steps, we use two times 0.5 mL of the solubilizate to rinse the two columns after the transfer of the beads and add this back to the rest.
12. The AcTEV protease from Life Technologies contains a histidine tag that allows its immobilization on Ni^{2+}-NTA beads prior to elution of the target protein complex.
13. Another elution step of the columns with 0.4 mL of 1× SDS sample buffer (without any β-mercaptoethanol or other reductants as they will dissociate the IgG light chains) may be performed. This denaturing elution step will elute also the uncleaved Tim21-Protein A fusion proteins from the column and allow to estimate the efficiency of the TEV protease elution.
14. All steps should be carried out in a fume hood as the organic solvents are toxic. Make sure that you dispose your waste properly!
15. Total *E. coli* phospholipid extracts are usually purchased as 25 mg/mL stock solutions. However, in the subsequent step the acetone/chloroform ratio is critical (should be ~30:1); therefore, the concentration step is required.
16. The DTT is added to prevent oxidative damage, mainly of the unsaturated fatty acid side chains in the phospholipid mixture.
17. Put a stirring bar into the glass flask containing the acetone/DTT, place the flask into a box filled with ice, and put the box onto a magnetic stirrer in the cold room.
18. The phospholipid solution in chloroform should now be clear and transparent and may be light brownish.
19. A common mistake is to "overdry" the lipids at this step. The lipid mixture should be dried very carefully, so that the resulting lipid film appears as regular concentric layers that look like waves of sand in the desert. When the lipids are dried too harsh, they become clumpy and the surface of the lipid film appears rather rough and irregular, looking like images of the moon surface.
20. The phospholipid suspension will look milky and light brownish, but should not have any clumps. Remaining clumps can be removed using a Potter-Elvehjem pestle. The suspension contains already phospholipid bilayer structures, but they are stacked together with irregular aggregates.
21. Always keep the phospholipid suspension on ice during sonication, as otherwise the sample will heat up and phospholipids may

precipitate. It is important that the tip does not touch the tube wall. Instead of a tip sonicator, a water bath sonicator can be used as well (settings depend on the bath sonicator used, but in general 5–10 min at full strength is enough). After sonication, the suspension is still milky, but clearly more transparent, as the lipid aggregates have been dissolved into multilaminar phospholipid vesicles of different sizes.

22. During freezing and thawing liposomes are ruptured and either reseal or fuse with each other. In this way mainly unilaminar vesicles are generated from multilaminar structures, while the average diameter of the vesicles becomes larger.

23. The number of passages through the extruder must not necessarily be 11, but should in any case be uneven. With an even number of passages, all the contaminations and aggregates that are held back by the membrane filter will be pushed back into the suspension.

24. The R_{sat} and R_{sol} values depend on the type of detergent, the composition of the liposome membrane, and on the concentration of liposomes used. For comparability reasons, detergent titrations should always be carried out with the same concentration of liposomes, e.g., 4 mg phospholipids/mL. This concentration must then be maintained for the actual reconstitution experiments.

25. In the case of the import motor-free TIM23 complex, the largest hydrophilic portion is the preprotein receptor domain on the intermembrane space side of the inner membrane [11, 12]. This domain should be on the outside of TIM23 proteoliposomes to allow for the initiation of preprotein insertion. Therefore, the strategy of choice was to pretreat liposomes with a detergent concentration close to R_{sat} before addition of the purified protein complex. The detergent used for the solubilization and purification of the membrane protein complex is usually also applied for the pretreatment of liposomes. However, in the case of digitonin, the only detergent that allows for the solubilization and purification of intact TIM23 complexes, it was not possible to obtain R_{sat} and R_{sol} values, as the titration curve strongly differed from that of virtually all other detergents (Fig. 1b). Indeed liposomes pretreated with digitonin at any concentration tested did not incorporate added TIM23 complexes. Therefore, we had to determine R_{sat} values for multiple detergents and to test, if the measured detergent concentrations leave intact the solubilized, purified TIM23 complexes in buffer containing 0.3 % of digitonin (Wash/elution buffer of the purification), when both solutions were mixed in a 1:1 ratio. As a readout for complex integrity we used blue native PAGE [19], and immunoblotting with antibodies against TIM23 complex subunits [10]. We found

that 0.35 % dodecyl maltoside, a concentration slightly higher than the measured R_{sat} value (0.3 %), fulfilled this requirement. Finally, in addition to 0.35 % dodecyl maltoside, a small amount of digitonin (0.15 %) was added to the buffer for pretreatment of the liposomes (Buffer P) to avoid a too strong dilution of the digitonin upon 1:1 mixing of the purified TIM23 complex with liposomes in Buffer P [10].

26. Detergent concentrations to be tested are usually between 0.1 and 2 %; 0.1 %-steps should be tested initially to determine the relevant range for a specific detergent. Subsequently, 0.05 %-steps may be tested within this range to obtain a more precise curve.
27. Depending on the detergent, it takes some time to reach equilibrium (detergent-saturated liposomes vs. mixed micelles), therefore at least 20 min incubation time is recommended.
28. In the case of digitonin, adsorption to Biobeads SM-2 is a very slow and inefficient process in standard buffers, but can be greatly enhanced by the addition of 10 mM $MgCl_2$ [10]. We have described an easy and broadly applicable test for the efficiency of detergent removal elsewhere, which is based on the determination of the residual solubilization capacity of a detergent solution towards a reference liposome suspension after incubation with Biobeads SM-2 [10].
29. In parallel, we always perform a mock reconstitution with the Wash buffer of the TIM23 complex isolation only. This is the buffer also used for the TEV elution step of the purification (*see* Subheading 3.2).
30. Biobeads that do not sediment to the bottom of the tube at one of these steps should be discarded. If many beads stay afloat gently tap with your finger against the tube to enhance sedimentation.
31. The addition of half a volume of Buffer R (without any detergents or glycerol) reduces the viscosity of the sample and greatly facilitates the sedimentation of Biobeads. The moderate dilution of the detergents further supports the reconstitution process. Weighing Biobeads on a balance is not trivial: access of fluid should be removed, e.g., with a tissue, but at the same time Biobeads should not fall completely dry. You simply have to practice this! During the course of a single reconstitution experiment the Biobeads can stay in the described Buffer R or similar standard buffers. Again Biobeads that fail to sediment should not be used anymore. Storage of equilibrated Biobeads for more than a week is recommended only in 20 % ethanol.
32. For longer incubations in the head-over-head shaker or on the rolling bench (e.g., with the small tubes put into a 50 mL Falcon tube) it is strongly recommended to wrap the reaction tubes with parafilm.

33. The transfer of supernatants is best carried out using a Pasteur pipet. Avoid to transfer Biobeads from one tube to the next. The additional 100 μL washing step minimizes the loss of material due to unspecific sticking to the beads or the tube wall. However, some binding of lipids and hydrophobic proteins to the polystyrene surface of the beads cannot be overcome. 20–30 % loss of material is to be expected.
34. At this stage, the $MgCl_2$ concentration in the sample is ~2 mM due do the stepwise addition of Buffer R that contains 5 mM $MgCl_2$. The total volume of the sample is ~1.7 mL. Thus, you have to add ~14 μL of a 1 M $MgCl_2$ stock solution to get a final concentration of ~10 mM. During the first two incubation steps most of the dodecyl maltoside has already been removed and proteoliposome formation is more or less completed. However, there is still a lot of digitonin in the samples that needs to be removed as well. The third incubation step with an increased amount of Biobeads and a high $MgCl_2$ concentration is therefore required to obtain sealed, largely digitonin-free proteoliposomes for downstream applications.
35. It is very important to avoid the transfer of Biobeads to the ultracentrifugation tubes as they may damage the tube during the run. Tubes should be carefully balanced with Buffer R.
36. Avoid the generation of air bubbles during resuspension. It is very important that the proteoliposome pellet is completely resuspended. Clumps will not float in the step gradient and material will be lost. Take your time with this step!
37. Pipet the layers very carefully and avoid mixing of the solutions. The sharper the separation of the layers, the more focused the proteoliposome fraction will be in the gradient. During the centrifugation step the proteoliposomes will migrate upwards in the step gradient according to their buoyant density and will accumulate at the very top of the 1 M sucrose phase. Aggregated and clumped material will be in the pellet after ultracentrifugation.
38. The proteoliposome fraction taken from the gradient contains a fairly high concentration of sucrose. If the samples are not sufficiently diluted, the proteoliposomes will not be efficiently pelleted in the following step.
39. The topology of the reconstituted membrane protein (complexes) should now be determined. This can, e.g., be done by protease accessibility assays or by differential labeling of specific amino acid residues with membrane-permeable and membrane-impermeable probes, respectively. Finally, an activity assay for the reconstituted proteins needs to be designed. This can be relatively straight forward, like in the case of carriers for small organic compounds that can be purchased in a

radiolabeled form, or rather complicated, like in the case of the insertase activity of the reconstituted import motor-free TIM23 complex. A very detailed description of this particular activity assay has been published earlier [10]. In brief, TIM23-containing proteoliposomes prepared in a potassium-based buffer system are diluted into a reaction buffer containing equimolar amounts of sodium ions. Addition of valinomycin renders the proteoliposome membrane selectively permeable for potassium ions, which will diffuse out of the proteoliposomes following their concentration gradient. As the sodium ions cannot pass the membrane under these conditions, a membrane potential ($\Delta\psi$) of physiological polarity is generated. Subsequently in vitro-synthesized, radiolabeled precursors of mitochondrial inner membrane-anchored proteins (substrates of the motor-free TIM23 complex) are added to the reaction mixture. If the reconstituted TIM23 complex is functional, these proteins will be inserted into the membrane, which can be assessed by carbonate extraction and by the protease resistance of the membrane-inserted domains of the precursors.

Acknowledgments

This work was supported by the Deutsche Forschungsgemeinschaft, Sonderforschungsbereich 746, and the Excellence Initiative of the German Federal and State Governments (EXC 294).

References

1. Krogh A, Larsson B, von Heijne G et al (2001) Predicting transmembrane protein topology with a hidden Markov model: application to complete genomes. J Mol Biol 305:567–580
2. Drews J (2000) Drug discovery: a historical perspective. Science 287:1960–1964
3. Andreoli TE (1974) Planar lipid bilayer membranes. Methods Enzymol 32:513–539
4. Reeves JP, Dowben RM (1969) Formation and properties of thin-walled phospholipid vesicles. J Cell Physiol 73:49–60
5. Darszon A, Vandenberg CA, Schonfeld M et al (1980) Reassembly of protein-lipid complexes into large bilayer vesicles: perspectives for membrane reconstitution. Proc Natl Acad Sci USA 77:239–243
6. Tribet C, Audebert R, Popot JL (1996) Amphipols: polymers that keep membrane proteins soluble in aqueous solutions. Proc Natl Acad Sci USA 93:15047–15050
7. Baneres JL, Popot JL, Mouillac B (2011) New advances in production and functional folding of G-protein-coupled receptors. Trends Biotechnol 29:314–322
8. Heerklotz H, Tsamaloukas AD, Keller S (2009) Monitoring detergent-mediated solubilization and reconstitution of lipid membranes by isothermal titration calorimetry. Nat Protoc 4:686–697
9. Le Maire M, Champeil P, Moller JV (2000) Interaction of membrane proteins and lipids with solubilizing detergents. Biochim Biophys Acta 1508:86–111
10. van der Laan M, Meinecke M, Dudek J et al (2007) Motor-free mitochondrial presequence translocase drives membrane integration of preproteins. Nat Cell Biol 9:1152–1159
11. Neupert W, Herrmann JM (2007) Translocation of proteins into mitochondria. Annu Rev Biochem 76:723–749

12. Chacinska A, Koehler CM, Milenkovic D et al (2009) Importing mitochondrial proteins: machineries and mechanisms. Cell 138: 628–644
13. Meisinger C, Pfanner N, Truscott KN (2006) Isolation of yeast mitochondria. Methods Mol Biol 313:33–39
14. Chacinska A, Lind M, Frazier AE et al (2005) Mitochondrial presequence translocase: switching between TOM tethering and motor recruitment involves Tim21 and Tim17. Cell 120:817–829
15. Knop M, Siegers K, Pereira G et al (1999) Epitope tagging of yeast genes using a PCR-based strategy: more tags and improved practical routines. Yeast 15:963–972
16. Kusters R, Dowhan W, de Kruijff B (1991) Negatively charged phospholipids restore pre-PhoE translocation across phosphatidylglycerol-depleted *Escherichia coli* inner membranes. J Biol Chem 266:8659–8662
17. Ridder AN, Kuhn A, Killian JA et al (2001) Anionic lipids stimulate Sec-independent insertion of a membrane protein lacking charged amino acid side chains. EMBO Rep 2:403–408
18. Paternostre MT, Roux M, Rigaud JL (1988) Mechanisms of membrane protein insertion into liposomes during reconstitution procedures involving the use of detergents. 1. Solubilization of large unilamellar liposomes (prepared by reverse-phase evaporation) by Triton X-100, octyl glucoside, and sodium cholate. Biochemistry 27:2668–2677
19. Wittig I, Schägger H (2008) Features and applications of blue-native and clear-native electrophoresis. Proteomics 8:3974–3990

Chapter 22

Single Channel Analysis of Membrane Proteins in Artificial Bilayer Membranes

Philipp Bartsch, Anke Harsman, and Richard Wagner

Abstract

The planar lipid bilayer technique is a powerful experimental approach for electrical single channel recordings of pore-forming membrane proteins in a chemically well-defined and easily modifiable environment. Here we provide a general survey of the basic materials and procedures required to set up a robust bilayer system and perform electrophysiological single channel recordings of reconstituted proteins suitable for the in-depth characterization of their functional properties.

Key words Bilayer, Membrane protein, Single channel analysis, Ion channel, Electrophysiology

1 Introduction

Channel forming membrane proteins are essential in electrical signaling and transmembrane transport of metabolic solutes as well as large molecules like proteins. Accordingly, the involved proteins range from classical ion channels with subnanometer pore sizes (e.g., KcsA ≈ 0.5 nm) to large, nanometer-sized pores (e.g., Tom40 ≈ 2 nm) [1, 2]. A common theme of these pore-forming membrane proteins is the fine-tuned regulation of their in vivo activity by a large range of effectors like membrane potential, pH, substrates, or regulatory subunits. For decades, cellular membranes have been studied by electrophysiological techniques such as the patch-clamp technique introduced by Neher and Sakmann [3] that allow ensemble or single channel measurements of channels in their natural membrane context including macromolecules that may influence channel functions. On the other hand, model membrane systems have proven to be useful tools for probing the molecular properties of lipid bilayers and membrane proteins in a chemically well-defined environment. Many of these model

[1] Philipp Bartsch and Anke Harsman have contributed equally to this work.

Doron Rapaport and Johannes M. Herrmann (eds.), *Membrane Biogenesis: Methods and Protocols*,
Methods in Molecular Biology, vol. 1033, DOI 10.1007/978-1-62703-487-6_22, © Springer Science+Business Media, LLC 2013

systems are based on freestanding artificial bilayers also referred to as black lipid membranes (BLM) that provide a dielectric isolation of two aqueous reservoirs which can be easily manipulated [4–9]. This especially benefits the study of membrane proteins from intracellular membranes as they are almost impossible to access with the patch-clamp technique.

Stable planar lipid bilayers can be established by the *painting technique* developed by Mueller et al. [4], in which a solution of lipid in hydrocarbon solvent is painted over a small aperture within a polytetrafluorethylene (PTFE) septum (⌀ ≈ 100 μm) or by lipid monolayer folding [10, 11]. Once a robust bilayer has been fabricated, purified membrane proteins may be incorporated by the use of several techniques [12, 13]. A method that can be applied for the study of most ion-permeable membrane proteins is based on the reconstitution of proteins into proteoliposomes which are then fused with the planar bilayer [14–17]. Finally, the permeation and gating properties as well as the regulation of integrated single transmembrane proteins, which may be ion channels, metabolite pores, or protein translocases, are studied at a single molecule level [18–20].

This chapter provides a practical guideline for the lipid bilayer technique. However, as we cannot cover all aspects in detail we recommend the following publications for further reading on technical background [21–24], definitions and explanations of single channel properties, and the theory underlying their analysis as well as more sophisticated single channel analysis techniques [25–29].

2 Materials

2.1 Bilayer Cuvette Assembly

1. Bilayer cuvette consisting of two half-chambers. For details see examples in Figs. 1 and 2 as well as **Note 1**. The half-chamber which in the fully assembled setup is situated closer to the operator is hereinafter denoted as *cis* chamber, while the other one is referred to as *trans*.
2. Casted PTFE film with a thickness of 25 μm and high dielectric strength (Chemfilm® DF1000, Saint-Gobain Performance Plastics Corporation, Hoosick Falls, NY, USA).
3. Expandable plastic film (e.g., Parafilm®).
4. Two customized glass windows, made of normal glass with a thickness of ≈1 mm.
5. Two sealing rings (made from the same material as the half-chambers).
6. High quality laboratory grease (glisseal HV, Borer Chemie AG, Zuchwil, Switzerland).

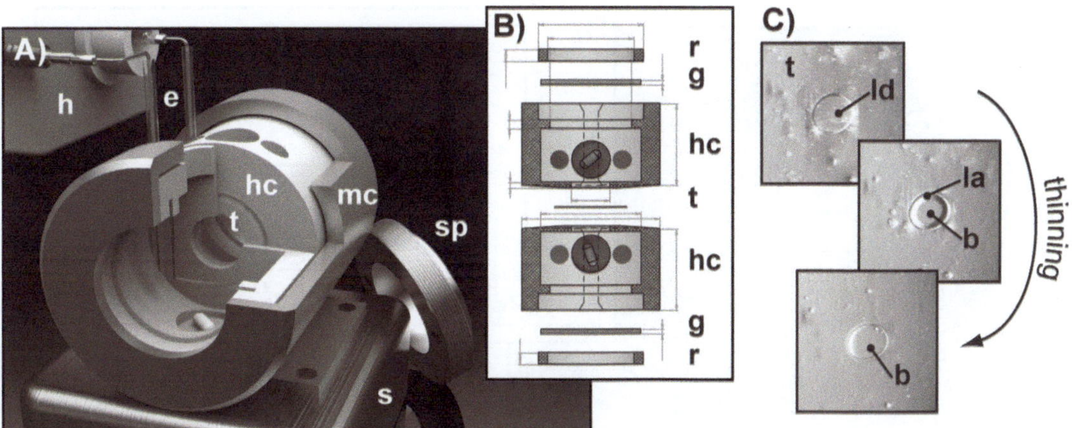

Fig. 1 Vertical bilayer cuvette. (**a**) Assembled bilayer cuvette with detached screw plug and cutaway view of the *cis* half-chamber. (**b**) Exploded view drawing without the metal cage. (**c**) Generation of a bilayer by lifting and lowering the buffer level thereby gradually thinning out the applied lipid. *b* bilayer, *h* headstage, *hc* half-chamber, *e* electrodes, *g* glass disc, *la* lipid annulus, *ld* lipid droplet, *mc* metal cage, *r* ring, *s* magnetic stirring plate, *sp* screw plug, *t* PTFE film

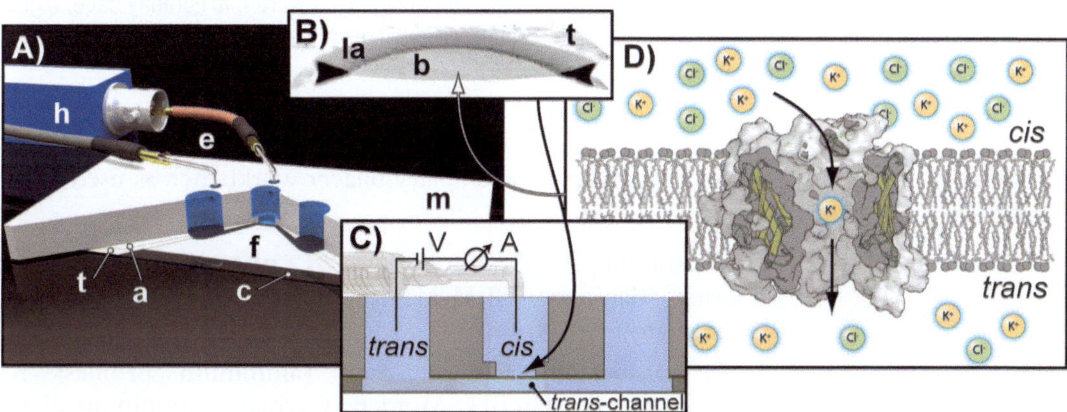

Fig. 2 Horizontal bilayer cuvette. (**a**) Cutaway view of an assembled bilayer cuvette. (**b**) Side-cut of a bilayer 3D reconstruction (LSM *YZ*-stack, thanks to Dr. Alf Honigmann). (**c**) Schematic setup of the bilayer chamber with indicated voltage source (*V*) and amperemeter (*A*). *Trans*-channel is not to scale. (**d**) Incorporated channel protein under asymmetric buffer conditions. *a* double sided adhesive foil, *b* bilayer, *c* cover glass, *e* electrodes, *f* double sided adhesive foil containing the cut-out representing the *trans*-channel, *h* headstage, *la* lipid annulus, *m* measurement chamber, *t* PTFE film

7. Screw-topped metal cage made by turning from solid aluminum to fit the half-chambers. A guidance fits notches of the half-chambers (marked by vertical dashed lines in Fig. 1b) thereby preventing involuntary rotation.
8. Tungsten carbide needles, tip diameter 1 μm (Fine Science Tools GmbH, Heidelberg, Germany).
9. Spark gap (self-built).

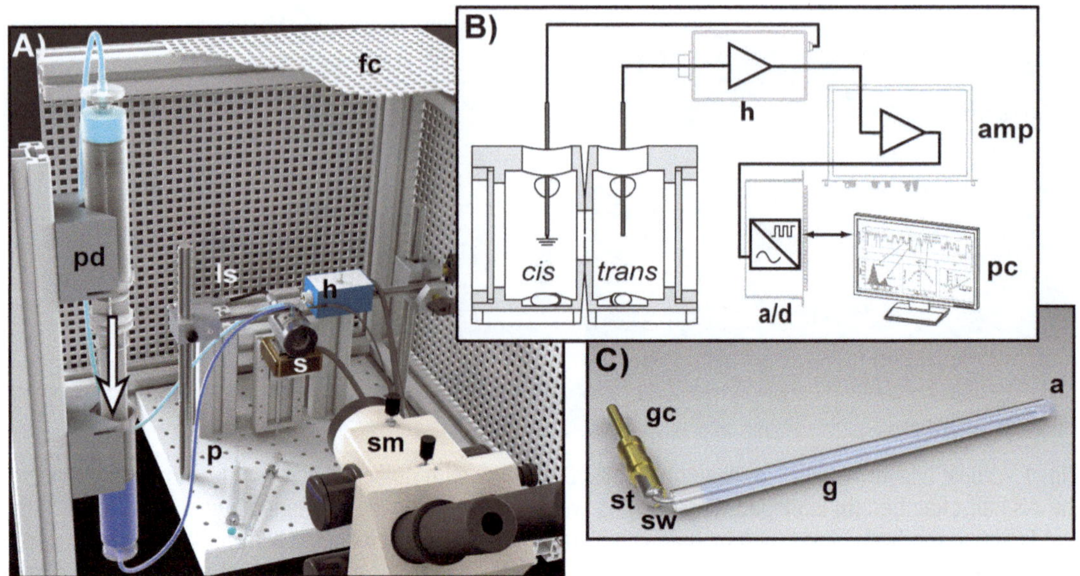

Fig. 3 (**a**) Setup of a bilayer workbench. (**b**) Equivalent circuit diagram of the data acquisition setup. (**c**) Close-up of an assembled electrode. *a* agarose, *amp* amplifier, *a/d* analog to digital converter, *fc* Faraday cage, *g* glass tube, *gc* gold connector, *h* headstage, *ls* light source, *p* vibration-cushioned plate, *pd* perfusion device, *s* stirring plate, *sm* stereo microscope, *st* solder tin, *sw* silver wire

2.2 Bilayer Workbench

Figure 3 illustrates an exemplary bilayer workbench as used in our laboratory.

1. Freestanding, vibration-cushioned plate (self-constructed using dampers: SLM-M1A, cplusw GmbH, Hamburg, Germany) (*see* **Note 2**).
2. Self-constructed Faraday cage (aluminum profiles: RK Rose + Krieger GmbH, Minden, Germany; aluminum plates: self-built).
3. Stereo microscope for optical control of the bilayer (*here*: Lomo MBC-10).
4. Two micro stirring bars.
5. Magnetic micro stirring plate (*see* **Note 3**).
6. Perfusion device for two syringes of 60 mL (CODAN pvb Medical GmbH, Lensahn, Germany, syringe mount: self-built). Each syringe is connected via plastic tubes to glass capillaries with an outer diameter of 1.55 mm and an inner diameter of 1.15 mm (Brand GmbH + Co. KG, Wertheim, Germany) (*see* **Note 4**).
7. Two micro syringes (10–50 μL) with blunt-ended and bended needles for applying lipid and sample, respectively.
8. Light source, e.g., small LED torch light.

2.3 Electrode Materials

1. Silver wire with a diameter of 0.5 mm (purity grade ≥99.9 %) (Carl Roth GmbH + Co. KG, Karlsruhe, Germany) which is soldered to gold connectors (and if necessary extension wires), appropriate for connection with the headstage (front and ground connectors).
2. Glass capillaries like those used for the perfusion device, cut to suitable length of about 1–2 cm.
3. 1.5 % agarose, 2 M KCl, 10 mM MOPS/Tris, pH 7.0; heat thoroughly until it is molten.
4. 3 M KCl solution.

2.4 Equipment for Data Acquisition and Analysis

1. Headstage (*here*: CV-5-1G, Molecular Devices, Sunnyvale, CA, USA) (*see* **Note 5**).
2. Patch-clamp amplifier (GeneClamp 500b, Molecular Devices, Sunnyvale, CA, USA) (*see* **Note 6**).
3. AD/DA converter (Digidata 1200, Molecular Devices, Sunnyvale, CA, USA) connected to a personal computer.
4. Recording software: Axon pClamp or similar software. Settings: Continuous recording mode, variable holding potential, low-pass Bessel filter frequency usually 1–5 kHz, and sampling frequency 10–50 kHz (*see* **Note 7**).
5. Analysis software: Single channel analysis software such as Clampfit (Molecular Devices, Sunnyvale, CA, USA) or others. We mostly use a self-written MATLAB-based program relying mainly on the algorithms published by Patlak [26].

2.5 Chemicals for Bilayer Fabrication and Measurements

1. 100 mg/mL L-α-phosphatidylcholine type IV-S (Sigma-Aldrich, St. Louis, MO, USA) in methanol/chloroform (1:1) stored at −20 °C (for long-term storage under argon) (*see* **Note 8**).
2. *n*-Decane (ReagentPlus, ≥99 %) (Sigma-Aldrich, St. Louis, MO, USA).
3. Buffer 1: 250 mM, 10 mM MOPS/Tris, pH 7.0 (*see* **Note 9**).
4. Buffer 2: 20 mM KCl, 10 mM MOPS/Tris, pH 7.0.
5. $CaCl_2$ solution: 2 M $CaCl_2$, 10 mM MOPS/Tris, pH 7.0.
6. Protein sample: purified target protein (*see* **Note 10**) reconstituted into proteoliposomes (*see* **Note 11**).

3 Methods

All solutions were prepared using ultrapure deionized water (18 MΩ at 25 °C).

3.1 Preparation of the Vertical Bilayer Chamber

1. Clean all chamber parts with acetone in an ultrasonic bath.
2. Insert glass windows into the half-chambers of the bilayer cuvette and loosely place Parafilm® onto them. Press rings in the half-chambers thereby fixing the glass discs and sealing the chambers with the Parafilm®. Crop excess Parafilm® that protrudes from the sealing rings and covers the glass discs.
3. Insert the first half-chamber into the metal cage with the glass window facing outward.
4. Thinly grease the PTFE adjacent side of both half-chambers.
5. Trim PTFE film to approximately fit the half-chamber, perforate it minimally by the tungsten carbide needle, and widen and smoothen the perforation by using the spark gap.
6. Place the prepared PTFE film onto the first half-chamber (avoid wrinkling of the film, as this might cause stray currents) and mount the second half-chamber.
7. Compress the two half-chambers by closing the metal cage until they are tightly sealed.

3.2 Preparation of Electrodes

1. The Ag wire of the electrodes is coated with AgCl by electrolysis. To this end, clean the silver wire (fine grade sandpaper or steel wool and ethanol) and submerge it, except for about 5 mm from the soldering joint, in 3 M KCl solution. Connect the electrode to the anode of an adjustable power supply unit and use another silver wire inserted into the same solution as cathode. Pass 5–10 mA/cm^2 for about 5–10 min until a matt grey layer is deposited on the wire. During this reaction gas bubbles (H_2) will rise from the cathode wire (*see* **Note 12**).
2. In order to minimize junction potentials between the electrode and the solution, Ag/AgCl electrodes are connected to the measurement solution via salt bridges. For this, insert the Ag/AgCl electrode into a precut glass tube (leaving a few millimeters space before the end of the glass tube). Dip its bottom end into the molten agarose solution and allow the fluid to fill the tube by capillary forces (*see* **Note 13**).
3. Connect the electrodes to the headstage (*see* **Note 14**) and insert their bottom ends into the buffer-filled bilayer chamber.

3.3 Fabrication of a Bilayer

1. The organic lipid solution is prepared freshly every day. For this purpose, transfer an aliquot of the lipid solution (e.g., 50 μL) to a glass test tube (*see* **Note 15**) and evaporate the solvent, for example, by using a diaphragm vacuum pump for 30 min. Redissolve the dried lipid film in *n*-decane to a final ratio of 50:100. Store the lipid solution for the day at RT.
2. In preparation for the formation of a vertical bilayer, mount the prepared bilayer chamber into the measurement setup and fill both half-chambers with buffer 1.

3. By use of a blunt-ended syringe, deposit a small volume of the lipid solution (approximately 1 μL) directly onto the aperture in the PTFE film from the *cis* as well as from the *trans* side of the bilayer (it should form a flat lipid droplet on the PTFE film around the aperture, Fig. 1c). Leave this lipid droplet on the PTFE film for at least 20 min to incubate. In the meantime, prepare the electrodes (*see* Subheading 3.1) and insert them into the bilayer cuvette.
4. After the incubation period, break the lipid droplet such that the two half-chambers are electrically connected and compensate for electrode offsets. Then, apply the smallest possible amount of fresh lipid onto the aperture and start thinning it out by repetitive lowering and raising of the buffer solution in the measurement chambers until a bilayer is formed like it is shown in Fig. 1c (a clean Pasteur pipette can be used for this step). Hereafter, adjust any possible output offset by applying a neutralizing DC voltage (*see* **Notes 16** and **17**).

3.4 Incorporation of Proteins into the Bilayer

1. Incorporation of membrane proteins into the lipid bilayer is achieved by osmotically induced swelling of proteoliposomes in immediate proximity of the lipid bilayer. This leads to a fusion of the vesicle and bilayer membranes (*see* **Note 18**). To this end, fill the *cis* chamber with buffer 1 (high salt) and the *trans* chamber with buffer 2 (low salt) and prepare a stable bilayer. This configuration is hereinafter referred to as "asymmetric conditions". Then use the second syringe to add the proteoliposomes (few microliters, depending on reconstitution success and concentration) as close to the bilayer as possible (*see* **Note 19**).
2. Wait for the integration of a pore-forming protein into the bilayer, while incubating with or without stirring for about 30 min with periodic controls of bilayer stability and quality in intervals of roughly 5 min. To improve the adhesion of vesicles to the bilayer you may add $CaCl_2$ to a final concentration of 10–20 mM (*see* **Note 20**). The integration of an ion-selective channel will result in a sudden shift of the measured current and potentially the appearance of current fluctuations representing single channel gating (*see* **Note 21**).
3. When you observe a current shift, try to gauge whether the activity inserted into the bilayer represents a single channel, as unbiased analysis requires individual channel proteins (*see* **Note 22**). If obviously multichannel incorporation occurred, form a new bilayer or restart with the fusion procedure from **step 1**.

3.5 Perfusion of the Bilayer Half-Chambers

1. In preparation for the following measurement, the buffer solution in the bilayer half-chambers is exchanged by perfusion to ensure precise ion concentrations. For this, switch off

the stirrer, insert the perfusion nozzle into the *trans* chamber, and perfuse the chamber solution with 20-chamber volumes of buffer 1 at a flow rate of approximately ≤10 mL/min (*see* **Note 23**).

2. Reload the perfusion syringes and repeat **step 1** with the other (*cis*) chamber. The resulting configuration with the same buffer in *cis* and *trans* is hereinafter referred to as "symmetric conditions".
3. Following sufficient perfusion to symmetric electrolyte conditions, the current in the absence of externally applied holding potentials should be zero per definition. If not, adjust the electrode offset by changing the zero current potential.

3.6 Recording of Single Channel Data: A Paradigm Case

1. First, current traces under symmetric buffer conditions are recorded at various holding potentials, which will provide the data for the analysis of, e.g., gating, conductance distribution, and open probability (*see* **Note 24**). To this end, start recording and apply a DC voltage gate-pulse of 60 s duration and variable amplitude with an idle period of 60 s (*see* **Note 25**).
2. In order to analyze the rectification of the channel or to display the voltage-dependent quality of an effector, voltage ramps are recorded under symmetric conditions. This means the holding potential is changed continuously at a constant increment of, e.g., 15 mV/s. Typically, voltage ranges of −100 to +100 mV are covered, but these parameters can be adjusted according to need.
3. For determination of the selectivity of a channel, the buffer solutions need to be changed to asymmetric conditions by perfusion (*see* Subheading 3.5) of one of the chambers to buffer 2. Then voltage ramps are recorded, as described in the previous step (**Note 26**).

3.7 Fundamental Single Channel Analysis (See Note 27)

1. In order to evaluate the *conductance* of the channel as well as the existence of subconductance states, gating transition amplitudes are analyzed. To this end, the difference in current passing the channel before and after a gating transition is calculated and divided by the applied holding potential to give the conductance of this gating event [27]. The resulting values may then be summarized in the form of conductance histograms, which depict the frequency distribution of gating transition amplitudes (Fig. 4, bottom left). In order to extract the characteristic conductance values, sums of Gaussian curves are fitted to these histograms and for each curve the center of the peak is stated as the mean conductance. The half width at half maximum is given as measure of the error as it roughly corresponds to the standard deviation ($HWHM = 1.17741\sigma$) (*see* **Note 28**).
2. The analysis of the time domain of current recordings yields the *open probability* of a channel describing the averaged extent

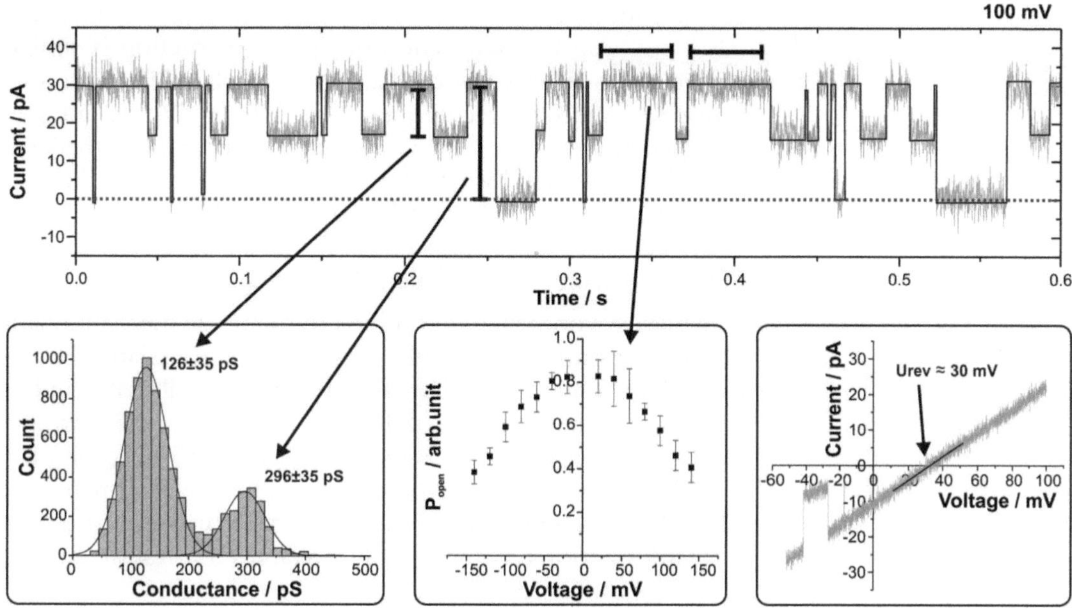

Fig. 4 Exemplified current trace with idealized representation of the gating transitions. Indicated below are representative graphs visualizing the output of the analysis methods described in Subheading 3.7. From *left* to *right*: conductance histogram, open probability and reversal potential. *arb. unit* arbitrary units, *Urev* reversal potential

to which a channel is opened at a given holding potential over a period of, e.g., 60 s. From single channel current traces at symmetric conditions, the mean current for the complete trace as well as the current corresponding to the maximum open state is determined. The ratio of both gives the averaged open probability (*see* **Note 29**). These values are plotted against the applied voltage (Fig. 4, bottom middle).

3. The reversal potential, i.e., the applied transmembrane voltage that yields zero electric current when an activity gradient is established across the channel is a measure of the *selectivity* of a pore. It is determined from voltage ramps under asymmetric electrolyte conditions, by fitting the intercept of the resulting current answer with the abscissa (Fig. 4, bottom right). The Goldman–Hodgkin–Katz equation is commonly used to calculate an approximate permeability ratio for the used cation and anion, respectively [27].

4 Notes

1. The half-chambers are typically constructed of PTFE or polyetheretherketone (PEEK) according to their dielectric strength and chemical and mechanical persistence. Figure 1 depicts a self-built vertical bilayer cuvette used in our laboratory.

A horizontal setup like the self-built one displayed in Fig. 2 [30] may also be used. The principal construction is similar. However, to achieve optical access to the bilayer for high resolution microscopy techniques, the PTFE film is inserted between horizontally stacked *cis* and *trans* chambers, the latter of which is formed by a 100 μm high channel connecting two larger reservoirs. Commercial, ready to use bilayer cuvettes are available (e.g., Ionovation GmbH, Osnabrueck, Germany or Warner Instruments, Hamden, CT, USA).

2. The bilayer workbench is designed to minimize mechanical and electrical disturbance caused by vibrations and radiative electrical pickup. Thus, a Faraday cage encloses the bilayer chamber, headstage, etc. as visualized in Fig. 3. Moreover, all components mounted into the Faraday cage need to be grounded. Excellent explanations on this topic are given by Sherman-Gold [23]. A complete bilayer workbench supporting semiautomatic bilayer formation is commercially available (Ionovation GmbH, Osnabrueck, Germany).
3. Each half-chamber is equipped with a stirring bar that is rotated by a magnetic micro stirring plate to ensure consistent mixing of the chamber content.
4. The perfusion setup must be grounded to avoid bilayer rupture upon insertion of the perfusion nozzle! To this end, a silver wire is wound around the lower end of the glass tubes with one end connected to the ground bus and the other end inserted into one of the glass tubes such that it connects the internal liquid. For manual handling of the perfusion it is advantageous if the operator is able to grab both plungers with one hand such that one syringe inserts liquid into the half-chamber, while the other simultaneously extracts liquid at the same flow rate of approximately ≤10 mL/min. Alternatively, a continuous flow (push/pull) syringe pump may be used. Both syringes and connected tubing need to be filled with liquid.
5. The bilayer technique is based on a typical electrophysiology setup including a negative feedback headstage with matching input resistance (1–5 GΩ) connected to suited amplifier circuits which allow electrical recordings with low noise at reasonable temporal resolution [23]. The system is operated in "voltage-clamp configuration" and should allow compensating electrode and bilayer capacitance as well as offsets arising from electrode potentials. The output of the amplified bilayer current recordings is fed into a dynamic AD/DA converter and if required high frequency noise is eliminated by a low-pass filter. Data are stored in a desktop computer using formats accessible for data analysis.
6. GeneClamp500b is no longer commercially available; some alternatives (not exhaustive) are the HEKA EPC 10 Patch

Clamp Amplifier (HEKA Elektronik Dr. Schulze GmbH, Lambrecht/Pfalz, Germany) or the Axopatch 200B Capacitor Feedback Patch Clamp Amplifier (Molecular Devices, Sunnyvale, CA, USA).

7. Choose the settings for sampling rate and filter frequency according to the Nyquist–Shannon sampling theorem, which states that a bandlimited analog signal can be perfectly reconstructed from an infinite sequence of samples if the sampling rate exceeds $2B$ samples per second, where B is the highest frequency of the original signal. The signal-to-noise ratio denoting the ratio of the single channel current amplitude to the standard deviation of the baseline noise should exceed 7 to allow reliable analysis [26].
8. Sometimes, individual batches of purified lipids can be contaminated with pore-forming proteins. Thus careful examination of each preparation is advisable. Additional purification of the lipids is often sufficient to yield pure and reproducibly stable bilayer membranes.
9. This practical guideline represents an example for the realization of lipid bilayer experiments we typically initially use to identify new large channels ($G \geq 200$ pS). Many experimental parameters such as the lipid composition of the bilayer, the ion, as well as buffer composition and concentration of the measurement buffers and the application sequence of holding potentials may be varied according to needs. For example, as channel conductance to a certain extent scales with the activity of the electrolyte solution, the ion concentration may be varied to yield better resolution of gating transitions. For fusion and recording of voltage ramps at asymmetric conditions we generally first try a tenfold gradient. If another pH is needed try to avoid changing salt concentrations massively by titration with KOH, NaOH, HCl, and so forth. Consider also, that liquid junction potentials may arise at the interface of dissimilar conductors due to different mobilities of the ions in both solutions. While the potentials arising during the use of KCl solutions are in the range of 1 mV or smaller, the choice of other solutions may lead to junction potentials in the range of at least 2–12 mV and thus necessitate compensation prior to the experiment or correction of the results [31].
10. The single molecule bilayer technique is often complicated by the fact that trace amounts of contaminating pore-forming proteins, which when being particularly fusogenic may even elude the detection by common gel staining techniques, can lead to a false positive identification of the pore-forming component (discussed in [32]). Thus extensive care has to be taken to make sure that the utilized components such as purified proteins, detergents, lipids, buffers, and the like are free from

contamination. To this end, the electrophysiological analysis of a control reconstitution without addition of the target protein is very informative. We have excellent experiences with synthetic and highly purified (>98 %) lipids (e.g., from Larodan Fine Chemicals, Malmö, Sweden). Special attention should be paid to the purification of the target protein (discussed in [32]). Using recombinantly expressed protein, a good negative control is, for example, the expression of a different, non-pore-forming protein and its subjection to an analogous purification and reconstitution protocol.

11. The prerequisite for insertion of membrane proteins into the bilayer by osmotic fusion is the reconstitution of the respective protein into unilamellar proteoliposomes (made from phospholipids, e.g., L-α-phosphatidylcholine from egg). In many cases, this is achieved by formation of mixed micelles consisting of detergent-solubilized protein and lipids in a protein/lipid ratio of roughly 1:50–1000 (mol/mol). The final proteoliposome suspension is generated by subsequent removal of the detergent which depends on the type of detergent used and is accomplished by dialysis or hydrophobic adsorbent (Calbiosorb™, Merck KGsA, Darmstadt, Germany; BioBeads, Bio-Rad Laboratories GmbH, Munich, Germany). For excellent reviews on different reconstitution methods see [33–36].
12. AgCl-coated silver electrodes can be repeatedly used. Only if electrode calibration gives unreasonable results or when the AgCl layer turns white or detaches from the Ag wire, the electrode has to be cleaned and freshly coated.
13. The salt bridges surrounding the electrodes are used to minimize junction potentials between the Ag/AgCl electrode and the electrolyte solution [31] and should thus be prepared freshly every day and potentially also during the day, in case they are used for very extensive measurements of several hours as well as when they fall dry or air bubbles occur in the glass capillary.
14. In our system, the electrode in the *cis* chamber is connected to the ground connector of the headstage, while the electrode in the *trans* chamber is inserted into the front connector. Thus all potentials are referred to the *trans* compartment.
15. In order to avoid contamination of lipid stocks with dust, detergents, or other solvents, the utilized syringes should be thoroughly cleaned before immersion into the solution. We do this by a cascade of different solvents such as methanol/chloroform (1/1), reused *n*-decane, and finally fresh *n*-decane.
16. The quality of the bilayer can be judged optically by the ratio of bilayer to annulus area (compare Fig. 1c) as well as electrically based on its capacitance [37]. Moreover, a stable bilayer should yield a stable zero current ($R \geq 100$ GΩ) being inert

towards conductance changes (in the absence of pore-forming proteins). It should last up to several hours.

17. The following measures may be taken to achieve a more stable bilayer, if necessary. (1) Try to vary the lipid:*n*-decane ratio slightly (i.e., 60:100), (2) extend the incubation period of the lipid droplet on the aperture before bilayer formation, (3) in extreme cases try incubating the lipid on the PTFE film in the absence of buffer, or (4) chose a different phospholipid composition. To avoid more general problems, you should (1) always try to keep the amounts of lipid solution added to the bilayer chamber at a minimum, as this may hamper the formation of a large-area bilayer or the fusion of proteoliposomes; (2) avoid spilling buffer solutions anywhere in the bilayer setup as this may lead to corrosion of the metal parts as well as the formation of unfavorable electrical connections leading to stray currents.

18. Alternative methods for the incorporation of proteins into the bilayer are (1) direct fusion or (2) nystatin/ergosterol-induced fusion, which is in principle a variant of the osmotic fusion procedure that is assisted by the formation of transient pores [38, 39]. Direct fusion of proteins can be used for soluble proteins or detergent-solubilized proteins that are able to self-assemble into the lipid bilayer. Some bacterial toxins with mostly β-barrel structure, for example, possess this ability. These can be added directly to the measurement chamber at symmetric conditions in concentrations of, e.g., 100 ng/mL. The used detergent should be suitable to keep the protein in solution even at relatively small amounts while leaving the bilayer intact upon addition to the measurement chamber (some detergents destabilize the bilayer even at very low concentrations and may produce "channel-like" current bursts at higher holding potentials). We have had good experiences with predilution of the protein sample into 0.03 % lauryldimethylamine-oxide (LDAO) or 0.1 % Triton X-100 followed by dilution into the chamber in a ratio of <1:1,000. Protein insertion into the bilayer is in this case judged by a current shift, which is only observable during application of a small holding potential (−20 to +20 mV). As the bilayer is sensitive towards detergents, the usefulness of this method is often limited by the type and concentration of detergent in use.

19. If the proteoliposomes are of lower density than the surrounding solution, they will immediately start floating upwards. In this case insert the sample below the aperture such that the liposomes pass the bilayer on their way up or try it the other way round, in case they start sinking down.

20. Depending on the protein concentration, the affinity of liposomes for the bilayer, and many other factors this may take some time. Moreover, the experimental conditions needed to

achieve fusion may vary between different proteoliposome preparations. Thus, you may extend the incubation time or repeat the procedure before you try to improve the fusion conditions by varying the amount of applied proteoliposomes. If the bilayer ruptures during this process, you may immediately form a new bilayer or in case this happened quite often (>10 times) or the connection between the two chambers remained for several minutes we would advise you to establish a fresh salt gradient and start anew.

21. As a result of the convention introduced in **Note 14**, a concentration gradient with $c_{cis} > c_{trans}$ will lead to a positive current, when an anion selective channel is inserted or to a negative current upon insertion of a cation-selective channel.

22. Several publications deal with the question how to decide whether the ion channel activity inserted into a lipid bilayer results from a single channel or multiple pore-forming proteins [40, 41]. As a first approximation you may look for the existence of distinct gating events and whether they at once reduce the current to zero. If undecided, try recording the activity, as the channel number might only become obvious from the analyzed data or after comparison with multiple individual measurements.

23. If the incorporation of the protein into the bilayer was achieved by osmotic fusion, we advise you to start with the perfusion of the *trans* chamber to match the buffer composition in the *cis* chamber. This will hinder further vesicle fusion. Afterwards, the solution in the *cis* chamber is exchanged to remove residual proteoliposomes. In most cases, the 20-fold exchange of the chamber solution is sufficient for buffer exchange however more extensive perfusion might be necessary to remove hydrophobic substances or effector molecules with low effective concentration.

24. Within our lipid bilayer setup, leakage conductances are usually well below 0.02 pS. Thus, when displaying current traces, it is advisable to indicate not only scale bars but also zero currents or even better full axes as this information is crucial for the purpose of identifying the closed state of a single channel (Fig. 4).

25. Depending on the recording and analysis software used, it might be advisable to start a new recording for every voltage applied as this prevents the generation of large data files which are more difficult to handle during analysis. We usually recommend starting with low voltages (i.e., 10 or 20 mV) and then increasing them incrementally by 10 or 20 mV. Also, recording positive and negative voltages alternately is advisable. Thus, a classic measurement series could look like this: +20, −20, +40, −40, +60, −60 mV, and so forth. The maximum holding

potential depends largely on the stability of the inserted channel protein. In general, we try to record at least up to ±140 mV.

26. Since the reversal potential of the channel adds up with the applied holding potential, the range of the voltage ramp should be chosen accordingly to avoid bilayer rupture at high potentials. For example, for a channel with a reversal potential of 50 mV at the given conditions, the voltage range of −50 to +100 mV might be chosen. For an initial approximation of the reversal potential you can either record a voltage ramp across a very narrow range or try to reduce the current to zero by manually applying small holding potentials.
27. As the described procedure aims at the analysis of single pore-forming molecules, which may underlie significant molecule-to-molecule variations, the complete set of measurements described in Subheading 3 must be repeated with several independent single channels to yield a significant description of reproducibility (at least five times). Furthermore, the identification of a new pore-forming protein must not rely on a single preparation, but ideally even result from several different expression or purification approaches as proposed in ref. [32]. Always report the measured single channel characteristics together with the ionic conditions and voltages at which they were measured.
28. Instead of conductance histograms, slope conductances may also be determined from current–voltage relations, but single-point determination of the conductance gives insufficient accuracy.
29. The open probability determination method described in Subheading 3.7 provides, in the case of the simplest channel with one open and one closed state, the fraction of time the channel spends in the open state. For channels with subconductance states, this is only a rough approximation. More complex procedures based on dwell time analysis can be performed to analyze the probability of a channel to reside in a certain state.

Acknowledgments

We acknowledge the valuable contribution of David Schmedt and Dr. Alf Honigmann in the form of the development of our single channel analysis programs. Further we would like to thank Birgit Hemmis for excellent technical assistance and the Deutsche Forschungsgemeinschaft for continuous support to RW. Finally the excellent support during the design and construction of our systems from the mechanic and electric workshops of the Physics and Biology departments is gratefully acknowledged.

References

1. Hill K, Model K, Ryan MT et al (1998) Tom40 forms the hydrophilic channel of the mitochondrial import pore for preproteins. Nature 395(6701):516–521
2. Meuser D, Splitt H, Wagner R et al (1999) Exploring the open pore of the potassium channel from Streptomyces lividans. FEBS Lett 462(3):447–452
3. Neher E, Sakmann B (1976) Single-channel currents recorded from membrane of denervated frog muscle fibres. Nature 260(5554): 799–802
4. Mueller P, Tien HT, Wescott WC et al (1962) Reconstitution of excitable cell membrane structure in vitro. Circulation 26(5):1167–1171
5. Mueller P, Rudin DO (1963) Induced excitability in reconstituted cell membrane structure. J Theor Biol 4(3):268–280
6. Mueller P, Wescott WC, Rudin DO et al (1963) Methods for Formation of Single Bimolecular Lipid Membranes in Aqueous Solution. J Phys Chem 67(2):534–535
7. Wonderlin WF, Finkel A, French RJ (1990) Optimizing planar lipid bilayer single-channel recordings for high resolution with rapid voltage steps. Biophys J 58(2):289–297
8. Wonderlin WFF, Robert J, Arispe NJ (1990) Recording and analysis of currents from single ion channels. In: Boulton AA (ed) Neurophysiological techniques: basic methods and concepts. Neuromethods, vol 14. Humana Press, Clifton, NJ, pp 35–142
9. Labarca P, Latorre R (1992) Insertion of ion channels into planar lipid bilayers by vesicle fusion. In: Rudy B, Iverson LE (eds) Methods in enzymology. Ion channels, vol 207. Academic, San Diego, CA, pp 447–462
10. Montal M, Mueller P (1972) Formation of bimolecular membranes from lipid monolayers and a study of their electrical properties. Proc Natl Acad Sci USA 69(12):3561–3566
11. White SH, Petersen DC, Simon S et al (1976) Formation of planar bilayer membranes from lipid monolayers. A critique. Biophys J 16(5): 481–489
12. Bean RC, Shepherd WC, Chan H et al (1969) Discrete conductance fluctuations in lipid bilayer protein membranes. J Gen Physiol 53(6):741–757
13. Cohen FS, Niles WD (1993) Reconstituting channels into planar membranes: a conceptual framework and methods for fusing vesicles to planar bilayer phospholipid membranes. In: Düzgüneş N (ed) Membrane fusion techniques Part A. Methods in enzymology, vol 220. Academic, San Diego, CA, pp 50–68
14. Cohen FS, Akabas MH, Zimmerberg J et al (1984) Parameters affecting the fusion of unilamellar phospholipid vesicles with planar bilayer membranes. J Cell Biol 98(3): 1054–1062
15. Niles WD, Cohen FS (1987) Video fluorescence microscopy studies of phospholipid vesicle fusion with a planar phospholipid membrane. Nature of membrane-membrane interactions and detection of release of contents. J Gen Physiol 90(5):703–735
16. Woodbury DJ, Hall JE (1988) Role of channels in the fusion of vesicles with a planar bilayer. Biophys J 54(6):1053–1063
17. Cohen FS, Niles WD, Akabas MH (1989) Fusion of phospholipid vesicles with a planar membrane depends on the membrane permeability of the solute used to create the osmotic pressure. J Gen Physiol 93(2):201–210
18. Schrempf H, Schmidt O, Kummerlen R et al (1995) A prokaryotic potassium ion channel with two predicted transmembrane segments from Streptomyces lividans. EMBO J 14(21): 5170–5178
19. Xu X, Decker W, Sampson MJ et al (1999) Mouse VDAC isoforms expressed in yeast: channel properties and their roles in mitochondrial outer membrane permeability. J Membr Biol 170(2):89–102
20. Harsman A, Kruger V, Bartsch P et al (2010) Protein conducting nanopores. J Phys Condens Matter 22(45):454102
21. Miller C (1986) Ion channel reconstitution. Plenum Press, New York, NY
22. Sakmann B, Neher E (1995) Single-channel recording, 2nd edn. Plenum Press, New York, NY
23. Sherman-Gold R (1993) The Axon guide for electrophysiology & biophysics: laboratory techniques. Axon Instruments, Foster City, CA
24. Kramar P, Miklavcic D, Kotulska M et al (2010) Voltage- and current-clamp methods for determination of planar lipid bilayer properties. In: Iglič A (ed) Advances in planar lipid bilayers and liposomes, vol 11, 1st edn. Elsevier/Academic Press, Amsterdam
25. Colquhoun D (1987) Practical analysis of single channel records. In: David CO (ed) Microelectrode techniques: the Plymouth Workshop handbook. Comp. of Biologists, Cambridge, pp 101–139

26. Patlak JB (1993) Measuring kinetics of complex single ion channel data using mean-variance histograms. Biophys J 65(1):29–42
27. Hille B (2001) Ion channels of excitable membranes, 3rd edn. Sinauer, Sunderland, MA
28. Krasilnikov OV (2002) Sizing channels with neutral polymers. Structure and dynamics of confined polymers, vol 87. Kluwer Academic, Dordrecht
29. Aguilella VM, Queralt-Martin M, Aguilella-Arzo M et al (2011) Insights on the permeability of wide protein channels: measurement and interpretation of ion selectivity. Integr Biol (Camb) 3(3):159–172
30. Honigmann A, Walter C, Erdmann F et al (2010) Characterization of horizontal lipid bilayers as a model system to study lipid phase separation. Biophys J 98(12):2886–2894
31. Neher E (1992) Correction for liquid junction potentials in patch clamp experiments. In: Ion channels. Methods in enzymology, vol 207. Academic, San Diego, CA, pp 123–131
32. Harsman A, Bartsch P, Hemmis B et al (2011) Exploring protein import pores of cellular organelles at the single molecule level using the planar lipid bilayer technique. Eur J Cell Biol 90(9):721–730
33. le Maire M, Champeil P, Moller JV (2000) Interaction of membrane proteins and lipids with solubilizing detergents. Biochim Biophys Acta 1508(1–2):86–111
34. Garavito RM, Ferguson-Miller S (2001) Detergents as tools in membrane biochemistry. J Biol Chem 276(35):32403–32406
35. Rigaud JL, Levy D (2003) Reconstitution of membrane proteins into liposomes. Methods Enzymol 372:65–86
36. Seddon AM, Curnow P, Booth PJ (2004) Membrane proteins, lipids and detergents: not just a soap opera. Biochim Biophys Acta 1666(1–2):105–117
37. Kado RT (1993) Membrane area and electrical capacitance. In: Düzgüneş N (ed) Membrane fusion techniques Part B. Methods in enzymology, vol 221. Academic, San Diego, CA, pp 273–299
38. Woodbury DJ, Miller C (1990) Nystatin-induced liposome fusion. A versatile approach to ion channel reconstitution into planar bilayers. Biophys J 58(4):833–839
39. Zagnoni M, Sandison ME, Marius P et al (2007) Controlled delivery of proteins into bilayer lipid membranes on chip. Lab Chip 7(9):1176–1183
40. Fox JA (1987) Ion channel subconductance states. J Membr Biol 97(1):1–8
41. Laver DR, Peter WG (1997) Interpretation of substates in ion channels: unipores or multipores? Prog Biophys Mol Biol 67(2–3):99–140

Chapter 23

Quantification of Protein Complexes by Blue Native Electrophoresis

Juliana Heidler, Valentina Strecker, Florian Csintalan, Lea Bleier, and Ilka Wittig

Abstract

Blue native electrophoresis (BNE) is a long established method for the analysis of native protein complexes. Applications of BNE range from investigating subunit composition, stoichiometry, and assembly of single protein complexes to profiling of whole complexomes. BNE is an indispensible tool to diagnostically analyze cells and tissues from patients with mitochondrial disorders or model organisms. Since functional proteomic studies often require quantification of protein complexes, we describe here different quantification methods subsequent to protein complex separation by BNE.

Key words Blue native electrophoresis, Western blotting, Membrane protein complexes, Mitochondria, Densitometry, Quantification, In-gel activity

1 Introduction

Blue native electrophoresis (BNE) has become increasingly popular for functional proteomics and has been continuously modified or adapted since its development in 1991 [1]. It is employed for analyzing the stoichiometry, composition, and assembly of stable soluble and membrane protein–protein interactions in different eukaryotic cellular compartments, as well as in bacteria [2–9]. Lately, the method has gained interest in particular for the characterization of complex assembly and stability defects in patients with disorders affecting the oxidative phosphorylation (OXPHOS) complexes [10].

Native acrylamide gradient gels are used to separate protein complexes according to their masses. While soluble complexes are loaded directly onto blue native gels, integral membrane protein complexes are solubilized by neutral detergents. The addition of the anionic dye Coomassie Brilliant Blue G-250 imposes a charge shift on membrane protein complexes that allows high resolution electrophoretic separation [11].

Doron Rapaport and Johannes M. Herrmann (eds.), *Membrane Biogenesis: Methods and Protocols*, Methods in Molecular Biology, vol. 1033, DOI 10.1007/978-1-62703-487-6_23, © Springer Science+Business Media, LLC 2013

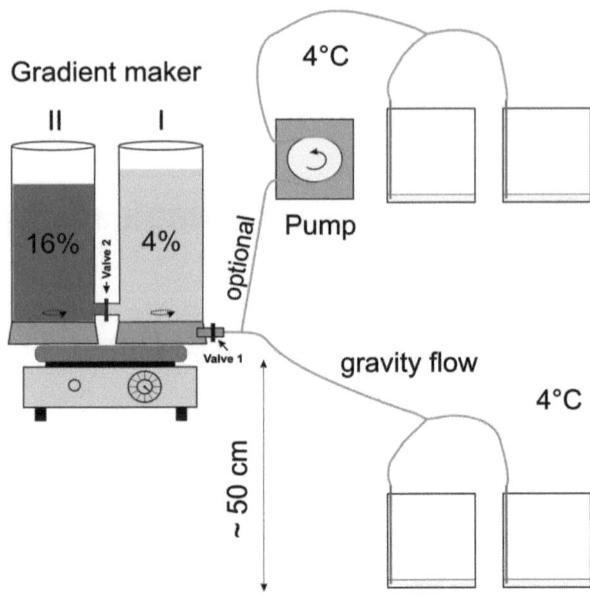

Fig. 1 Assembly of devices for casting gradient gels. Make sure to elevate gradient maker and magnetic stirrer about 0.5 m above the casting tray when using gravity flow. Optionally, incorporate a peristaltic pump for better flow control

Various modifications or variants of the original BNE have been introduced by us and other groups during the last years. A modified version of BNE, clear native electrophoresis (CNE), was developed omitting the anionic Coomassie dye and hence preventing formation of mixed micelles potentially disintegrating labile complexes. CNE is very mild and preferred for separation of very labile protein–protein interactions, but is restricted to protein complexes with an intrinsic pI<7 and suffers from low resolution [12]. The high resolution clear native electrophoresis (hrCNE) combines the advantages of BNE and CNE, separates protein complexes in high resolution comparable to BNE, allows better detection of fluorescent proteins, and enhances in-gel activity stains [13]. Recently, we further extended the applications for BNE by establishing a new method termed "Complexome profiling" [14]. Complexome profiling couples BNE analysis to highly sensitive mass spectrometric protein identification and subsequent hierarchical clustering into interaction profiles. This bottom-up approach identifies all protein–protein interactions in a sample. Furthermore the newly developed large pore native gels allow separation of megacomplexes up to 50 MDa [15] enabling isolation of oligomeric respirasomes [15] and mitochondrial nucleoids (unpublished).

In this chapter we will give instructions for casting 4–16 % acrylamide gradient mini gels (Fig. 1). This most commonly applied gradient separates protein complexes in the range of

50 kDa to 2 MDa. Casting custom-made BNE gels allows adaptation to different questions and problems. The BNE gels in mini format help saving time and precious material.

Further we describe the solubilization of mitochondrial complexes from mammalian tissues or chicken heart, which are often used as native mass ladder for estimation of native masses of membrane complexes in BNE gels [16]. We recommend solubilization of these native mass ladders with the same detergent as the sample of interest. Calibration standards for soluble proteins are commercially available.

Mitochondrial disorders are associated with defects in single subunits, assembly factors, or mitochondrial protein biosynthesis leading to impaired function of OXPHOS complexes. The functional characterization of affected complex subunits or assembly factors often involves quantification of mitochondrial complexes in tissues from model organisms, patient cells, or patient specimens from biopsies.

Hence, we also focus in this chapter on different quantification approaches for the analysis of mitochondrial complexes from human cells and various organisms, i.e., yeast (*Saccharomyces cerevisiae*, *Yarrowia lipolytica*), worm (*Caenorhabditis elegans*), fly (*Drosophila melanogaster*), zebrafish (*Danio rerio*), newt (*Notophthalmus viridescens*), chicken (*Gallus gallus*), mouse (*Mus musculus*), rat (*Rattus norvegicus*), pig (*Sus scrofa*), and bovine (*Bos taurus*) (Figs. 2, 3, 4, and 5).

Quantification of OXPHOS complexes thereby serves as an example to illustrate strategies that can be easily adapted to analyze other protein complexes of interest.

For solubilization of protein complexes from isolated mitochondria or homogenates from tissue or whole organisms the mild nonionic detergents digitonin, dodecyl-β-d-maltoside (DDM), or Triton X-100 are most commonly applied. While DDM solubilizes mitochondrial complexes as single stable OXPHOS complexes I–V, digitonin is milder and preserves higher-order assemblies of mitochondrial complexes, e.g., supercomplexes, the respirasomes, or dimeric and oligomeric ATP synthase [12, 17, 18]. For quantitative analysis of human samples (e.g., patient fibroblasts, myoblast, or biopsy samples) we favor the solubilization with digitonin since the frequently affected complex I in human mitochondria is stable only as part of respiratory chain supercomplexes [19, 20].

Total protein staining of 1D BN-gels with Coomassie is the simplest strategy to quantify protein complexes (Figs. 2, 3, and 4). However, knowing the separation pattern of the complexes or the availability of appropriate native standards (e.g., bovine heart mitochondria) is mandatory. Further, Coomassie staining is restricted to samples containing high amounts of protein complexes, like heart tissue or isolated mitochondria. A more specific approach for quantification exploits the remaining catalytic activity of native

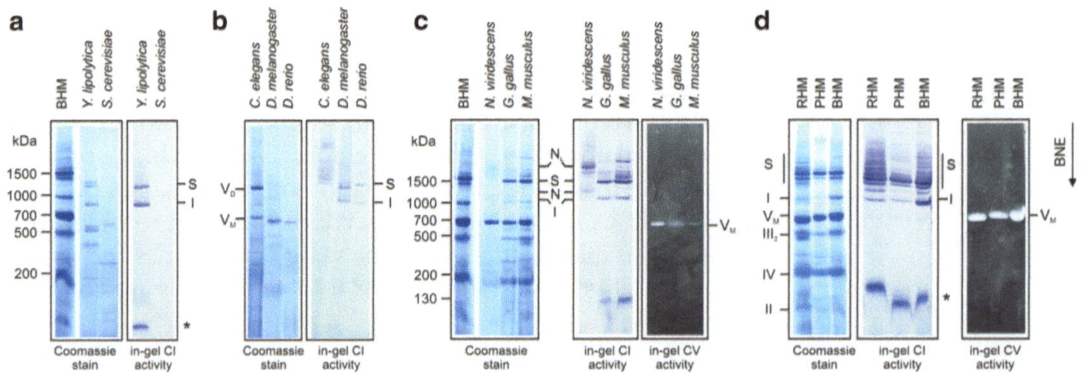

Fig. 2 Applications of blue native electrophoresis (BNE) for quantitative analysis of protein complexes from different organisms. All samples were solubilized with digitonin and separated by BNE. Separated mitochondrial complexes from (**a**) yeast mitochondria (*Y. lipolytica* and *S. cerevisiae*), (**b**) total homogenates from *C. elegans*, *D. melanogaster*, and *D. rerio*, (**c**) heart homogenates from *N. viridescens*, *G. gallus*, and *M. musculus*, and (**d**) isolated mitochondria from hearts of *R. norvegicus*, *S. scrofa*, and *B. taurus* are shown in left panels as Coomassie-stained protein complexes. (**a**, **b** *right panels*; **c**, **d** *central panels*) NADH:NTB oxidoreductase activity of complex I and complex I containing supercomplexes (S) is indicated by in-gel complex I activity stain. Interestingly the activity stain of newt homogenates reveals complex I containing complexes and supercomplexes of higher masses (*N*, N_L). (**c**, **d** *right panels*) In-gel complex V activity stain. *RHM* rat heart mitochondria, *PHM* pig heart mitochondria, *BHM* bovine heart mitochondria, *S* respiratory chain supercomplexes containing one copy of complex I, one complex III dimer, and variable copies of complex IV monomer, S_0 respiratory chain supercomplex containing one copy of complex I, one complex III dimer, and one copy of complex IV, *I* complex I, V_M monomeric complex V/ATP synthase, V_D dimeric complex V/ATP synthase, F_1 catalytic part of complex V/ATP synthase, III_2 dimeric complex III/cytochrome bc_1 complex, *IV* complex IV/cytochrome *c* oxidase, *II* complex II/succinate dehydrogenase. *Asterisks* indicate NADH activity of dehydrogenases other than respiratory chain complex I

complexes for in-gel enzyme activity stains or the presence of heme proteins in case of the heme stain (Figs. 2 and 4). Taking advantage of the high sensitivity of specific antibodies, Western blotting is an alternative strategy for quantification of low abundant protein complexes in 1D BNE gels (Figs. 3, 4, and 5). Once a set of antibodies against protein complexes of interest has been tested, cocktails of antibodies for immunodetection of multiple complexes can be applied to quantify series of complexes and to compare samples from patients, model organisms, or those where different conditions were applied (Fig. 5).

2 Materials

2.1 Special Equipment and Devices

1. Motor-driven, tightly fitting 0.5–1 mL glass/Teflon Potter-Elvehjem homogenizer or Ultraturrax mini for homogenization of tissues and cells.
2. Leak-proof electrophoresis unit for native gels in mini format (e.g., shiroGEL System).

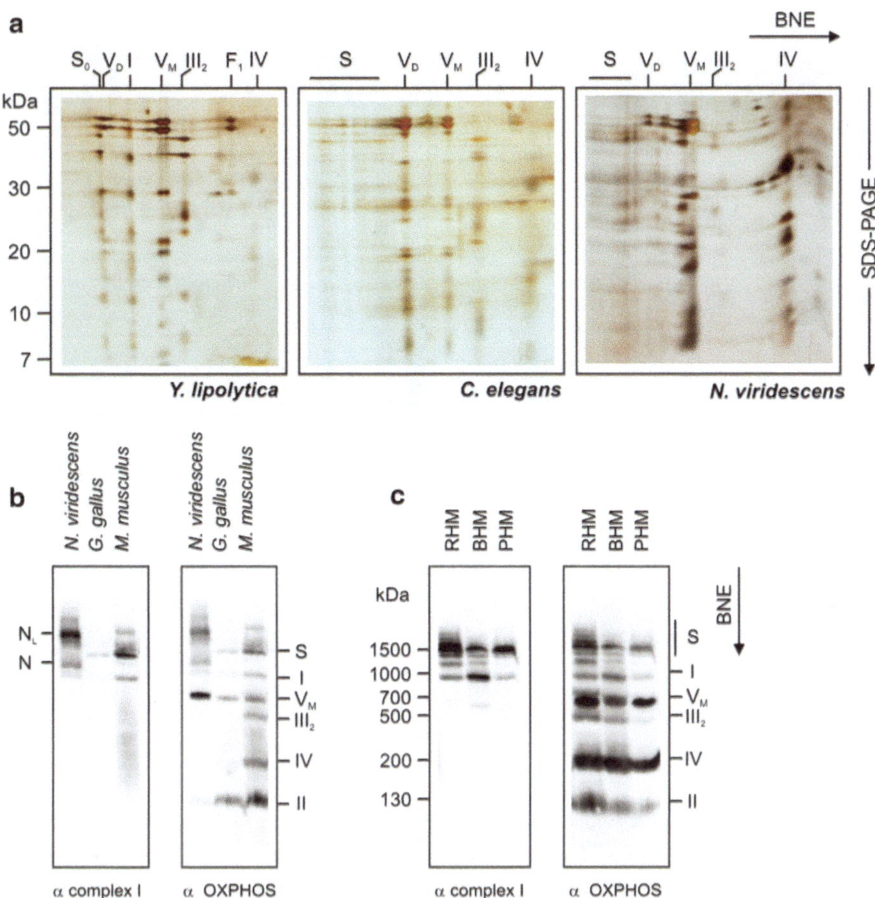

Fig. 3 Analysis of protein complexes from different organisms by silver staining and Western blotting. All samples were solubilized with digitonin and separated by BNE. (**a**) Silver-stained 2D BN/SDS-PAGE of mitochondria from *Y. lipolytica*, total homogenates from *C. elegans*, and heart homogenates of *N. viridescens*. Remarkably, all organisms show a stable dimeric ATP synthase (V_D). Separated OXPHOS complexes from (**b**) tissue homogenates of hearts of *N. viridescens*, *G. gallus*, and *M. musculus* and (**c**) rat, bovine, and PHM were detected by 1D BN-Western blotting. Blots were decorated first with antibody against complex I NDUFB8 (*left panels*) and subsequently decorated with an antibody cocktail against total OXPHOS (MitoProfile® Total OXPHOS Blue Native WB) (*right panels*). *See* legend of Fig. 2 for assignment of complexes

3. Gradient maker for gel casting (e.g., Model 385 gradient former from Bio-Rad, CBGM-20 gradient maker from C.B.S. Scientific).
4. Peristaltic pump (e.g., P1, GE Healthcare) is optional, but it helps gaining better control in gel casting as compared to simple gravity flow.
5. Cannulas (20G×2/3″ 0.9×70 mm).
6. Power supply suitable for native electrophoresis and blotting (e.g., 200 mA, 500 V).

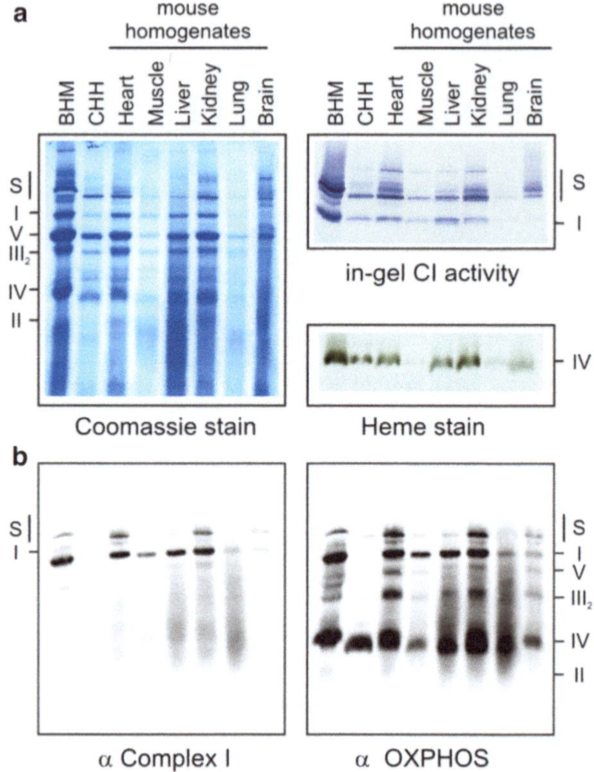

Fig. 4 Applications of BNE for quantitative analysis of protein complexes from different tissues. Homogenates of various mouse tissues were solubilized with digitonin and then separated by BNE using mini gels. For both panels same samples were used. (**a**, *left panel*) Coomassie-stain of respiratory chain complexes from heart, muscle, liver, kidney, lung, and brain homogenates. BHM and chicken heart homogenate (CHH) were used as mass calibration ladders. (**a**, *right panel*) In-gel complex I activity stain (*top*) and heme stain for complex IV (*bottom*). (**b**) 1D BN-Western blotting against complex I NDUFB8 (*left panel*), and against total OXPHOS (MitoProfile® Total OXPHOS Rodent WB Antibody Cocktail, *right panel*). *See* legend of Fig. 2 for assignment of complexes

7. ChemiDoc XRS gel documentation device (Bio-Rad) or comparable system.
8. Office flatbed scanner.
9. Quantity One (Bio-Rad) or comparable densitometry software is required to detect and quantify stained proteins.
10. Hoefer TE 77 Semi-dry blotting chamber (GE Healthcare) for blotting of gels.

2.2 Sample Preparation and Solubilization

Buffers are stored at 4 °C unless indicated otherwise.

1. Buffer for mitochondria isolation and storage of biological membranes: 250 mM sucrose, 10 mM Tris–HCl, pH 7.0, 1 mM EDTA, 5 mM 6-aminohexanoic acid, protease inhibitor cocktails.

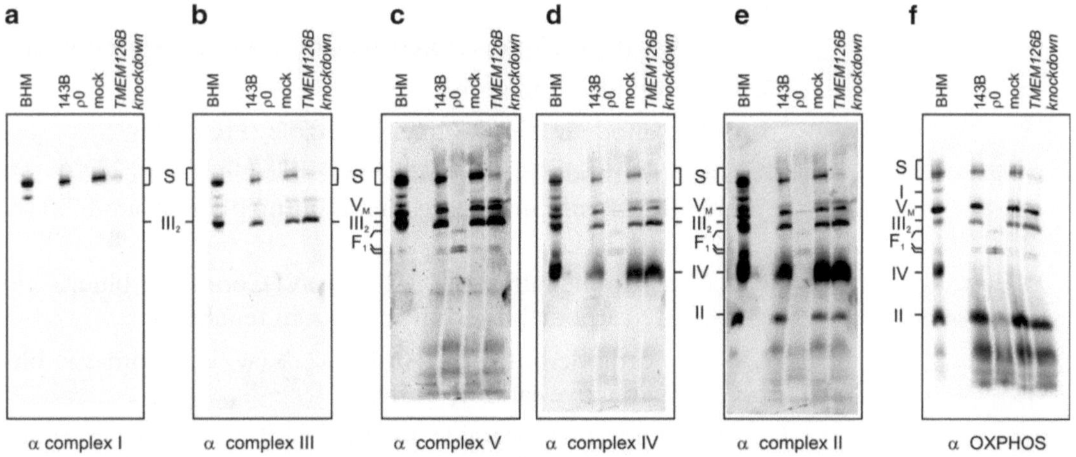

Fig. 5 Applications of BNE for quantitative analysis of human cell lines. ρ0 cells lacking mitochondrial DNA, mock-transduced and *TMEM126B* knockdown cells [14] are all derived from human osteosarcoma cells (143B, ATCC CRL-8303). Cell homogenates were solubilized with digitonin, separated by BNE, and blotted onto PVDF membranes. Sequential Western blotting against (**a**) complex I subunit NDUFB8, (**b**) core 2 subunit of complex III, (**c**) subunits α, β of complex V, (**d**) subunit COX 6a of complex IV, and finally (**e**) total complex II. (**f**) A similar blot was decorated with MitoProfile® Total OXPHOS Blue Native WB Antibody Cocktail. BHM, bovine heart mitochondria. *See* legend of Fig. 2 for assignment of complexes. F_1 subcomplexes of complex V detected in homogenates of ρ0 cells indicate deficient mitochondrial biosynthesis and are a marker for defects in *ATP6* (patients with NARP) and mitochondrial mtDNA depletion (MILS) [31]

Buffer for isolation and storage of enriched mitochondrial membranes from human cell culture (1/3 isolation buffer): 83 mM sucrose, 3.3 mM Tris–HCl, pH 7, 0.3 mM EDTA, 1.7 mM 6-aminohexanoic acid, protease inhibitor cocktails (*see* **Note 1**).

2. Solubilization buffer: 50 mM NaCl, 50 mM imidazole/HCl, pH 7.0, 5 mM 6-aminohexanoic acid, 1 mM EDTA.
3. DDM: 20 % (w/v), dissolved in water, store aliquots at −20 °C.
4. Triton X-100: 20 % (w/v), dissolved in water, store aliquots at −20 °C.
5. Digitonin: 20 % (w/v) (Serva, used without recrystallization), dissolved in water, store aliquots at −20 °C.
6. 50 % (w/v) glycerol, 0.1 % Ponceau S dye.
7. 5 % Coomassie dye: 5 % (w/v) Coomassie blue G-250 (Serva Blue G; Serva), suspension in 500 mM 6-aminohexanoic acid.

2.3 Blue Native Electrophoresis

1. Native gel buffer (3×): 75 mM imidazole/HCl, pH 7, 1.5 M 6-aminohexanoic acid.
2. Serva Blue G-250 (Serva).
3. Tetramethylethylenediamine (TEMED).

4. Ammonium persulfate (APS): 10 % (w/v), unfrozen stable only for a few days, aliquots stored at −20 °C are stable for 4 weeks.
5. Glycerol: 50 % (w/v) in water.
6. Native anode buffer: 25 mM imidazole/HCl, pH 7.
7. Native cathode buffer and native blotting buffer: 50 mM Tricine, 7.5 mM imidazole, the resulting pH is around 7.0 (do not adjust pH).
8. Native cathode buffer B: 0.02 % (w/v) Coomassie blue G-250 in native cathode buffer, store at room temperature.
9. Native cathode buffer B/10: 0.002 % (w/v) Coomassie blue G-250 in native cathode buffer.
10. Acrylamide-bisacrylamide mixture (AB-3): 48 g acrylamide (Serva) and 1.5 g bisacrylamide (Serva) per 100 mL.

2.4 Solutions and Reagents for Coomassie

1. Fixing solution: 50 % methanol, 10 % acetic acid, 10 mM ammonium acetate, store at room temperature (*see* **Note 2**).
2. Coomassie staining solution: 0.025 % (w/v) Coomassie dye in 10 % acetic acid, store at room temperature.
3. Coomassie destaining solution: 10 % acetic acid, store at room temperature.

2.5 Solutions and Reagents for Western Blotting

1. Methanol.
2. Blot staining solution: 25 % methanol, 10 % acetic acid, 0.02 % Coomassie.
3. Blot destaining solution: 25 % methanol, 10 % acetic acid.
4. PVDF blotting membranes.
5. Washing buffer: PBS, 0.1 % Tween, pH 7.4.
6. Blocking buffer: PBS, 0.5 % bovine serum albumin, 0.1 % Tween, pH 7.4.
7. Antibody against complex I (NDUFB8, Invitrogen): dilute 1:2,000 in washing buffer.
8. MitoProfile® Total OXPHOS Blue Native WB Antibody Cocktail (Abcam): dilute 1:1,000 in washing buffer.
9. MitoProfile® Total OXPHOS Rodent WB Antibody Cocktail (Abcam): dilute 1:1,000 in washing buffer.
10. Antiserum against subunits of OXPHOS complexes; raised in rabbits (our lab set, anti-ATP synthase, anti-complex III, anti-complex IV, anti-complex II), dilutions of each 1:10,000 in washing buffer.
11. Secondary antibodies conjugated with horseradish peroxidase (HRP) against mouse or rabbit: dilute 1:20,000 in washing buffer.
12. ECL chemiluminescence kit.

2.6 Solutions and Reagents for In-Gel Activity and Heme Stain

1. Incubation buffer complex I: 10 mM Tris–HCl, pH 7.5, 3 mM nitrotetrazolium blue (NTB), 120 μM NADH, freshly prepared.
2. Incubation buffer for heme stain (complex IV): 0.5 mg/mL diaminobenzidine dissolved in 50 mM sodium phosphate, pH 7.2, 50 μM cytochrome *c* (horse), freshly prepared.
3. Incubation buffer complex V: 35 mM Tris–HCl, pH 8.3, 270 mM glycine, 14 mM $MgSO_4$, 0.2 % $Pb(NO_3)_2$, 8 mM ATP, freshly prepared.
4. Stop solution for complex V stain: 50 % methanol.

2.7 Additional Reagents for 2D Tricine-SDS- PAGE

1. SDS-Gel buffer (3×): 1 M Tris, 1 M HCl, 0.3 % SDS, pH 8.45.
2. SDS-PAGE anode buffer: (10×) 1 M Tris, 0.225 M HCl, pH 8.9.
3. SDS-PAGE cathode buffer (10×): 1 M Tris, 1 M Tricine, 1 % SDS, pH ~ 8.25, adjust with Tris and Tricine only.
4. Overlay agarose: 0.5 % (w/v) agarose in 1× SDS-PAGE cathode buffer, heat the agarose in a microwave oven until melt, store at room temperature, for each use reheat for melting.

2.8 Solutions and Reagents for Silver Stain

1. Sensitization solution: 0.005 % (w/v) sodium thiosulfate ($Na_2S_2O_3$) in water, prepare always fresh.
2. Silver nitrate solution: 0.1 % (w/v) in water, prepare always fresh.
3. Developer: 0.036 % formaldehyde, 2 % sodium carbonate, prepare always fresh.
4. Stop solution: 50 mM EDTA in water, store at room temperature.

3 Methods

3.1 Casting Two BN-Minigels (Fig. 1)

Caution: acrylamide and TEMED are toxic. Wear gloves and goggles for protection.

1. Assemble glass plates and spacers (0.75 mm) and place them into casting tray according to manufacturers' instructions.
2. Prepare acrylamide mixtures 4 and 16 % without adding TEMED and APS (Table 1). Chill the mixtures on ice.
3. Connect gradient maker with tubings, use Y-piece tubing connector allowing casting two gels at once. Place cannula (20G × 2/3″ 0.9 × 70 mm) at the end of each tubing.
4. Optional: Install a peristaltic pump for better casting control.
5. Fill vessels of gradient maker with water, check flow through both ways, let water rinse in a waste beaker. The tubings should stay filled with water.

Table 1
Composition of sample gel and 4–16 % acrylamide mixtures for preparation of two acrylamide gradient mini gels

	Sample gel	Gradient separation gel	
	3.5 % acrylamide	4 % acrylamide	16 % acrylamide
AB-3	0.44 mL	0.5 mL	1.6 mL
3× native gel buffer	2 mL	2 mL	1.7 mL
50 % glycerol	–	–	1.7 mL
Water	3.56 mL	3.5 mL	–
Total volume	6 mL	6 mL	5 mL
10 % APS	50 μL	35 μL	25 μL
TEMED	5 μL	3.5 μL	2.5 μL

6. Close gradient maker valves and decant the system to remove water from both vessels. Close both valves of the gradient maker.
7. Pour 6 mL of 4 % acrylamide mixture into vessel I and 5 mL of 16 % acrylamide mixture into vessel II of gradient maker.
8. Place gradient maker on a magnetic stirrer. Add magnetic stir bars. Elevate gradient maker and magnetic stirrer if gravity flow is used for casting.
9. Place connected cannulas between glass plates.
10. Add APS and TEMED (Table 1).
11. Open first valve 1 and immediately after that valve 2 between the vessels. Check flow of both ways inspecting glass plates in casting tray. First the residual water from the tubings rinses in between the glass plates followed by the dense acrylamide solution, which therefore accumulates underlying the water (*see* **Note 3**).
12. Close valve once the level reaches the top of glass plates.
13. Pull cannulas out. Rinse the gradient maker, tubes, and cannulas immediately with water.
14. Polymerization is completed after 30 min at room temperature.
15. Decant the casting tray to remove water from the top.
16. Prepare sample gel mix, add APS and TEMED. Cast sample gel and place sample comb.
17. Polymerization is completed after 30 min at room temperature.
18. Store gels at 4 °C. Gels are stable in a sealed plastic bag for 1 week.

3.2 Isolation of Mitochondria and Enriched Mitochondrial Membranes from Tissues, Cells, and Whole Organisms

Use standard protocols as described in the literature for isolation of mitochondria from *Yarrowia lipolytica* [21, 22], *Bos taurus* [23], *Rattus norvegicus* [24, 25], and *Sus scrofa* [26].

3.2.1 Homogenization of Tissues and Whole Organisms

Carry out all procedures in a cold room at 4 °C or on ice.

1. Thaw frozen tissues and organisms:
   (a) Weigh mouse organs (liver, lung, kidney, heart, skeletal muscle, brain), newt heart, chicken heart, or flies and chill them in petri dishes on ice. Mince tissues finely with scissors. Add isolation buffer to a final concentration of 1 g tissue/10 mL.
   (b) Remove the head and fins of zebrafish with a razor blade, weigh the remaining tissue, and place in a petri dish on ice. Mince tissue finely with scissors. Add isolation buffer to a final concentration of 1 g tissue/10 mL.
   (c) Weigh *C. elegans*, homogenize 1 g tissue/10 mL isolation buffer by Ultraturrax for 5 × 10 s at 20,500 × rpm.
2. Homogenize the tissue suspension in a precooled motor-driven tightly fitting glass/Teflon Potter-Elvehjem homogenizer at 2,000 × *g* and 25 strokes.
3. Centrifuge homogenized mouse organ tissue and chicken heart tissue at 500 × *g* for 3 min at 4 °C to get a crude nuclear pellet. Transfer the supernatant into a new tube (*see* **Note 4**).
4. Centrifuge the supernatant, as well as homogenized newt heart, fly, zebrafish, and *C. elegans* at 10,000 × *g* for 10 min at 4 °C. Discard the supernatant.
5. Resuspend the pellets in isolation buffer to a concentration of 1 g original tissue weight/mL. Determine protein concentration according to Lowry [27].
6. Aliquot membranes to 400 μg protein portions and centrifuge 10 min at 10,000 × *g*, remove supernatant, shock-freeze in liquid nitrogen, and store pellets at −80 °C.

3.2.2 Homogenization of Cell Lines

1. Collect adherent cells by trypsinization or scraping in PBS into a centrifuge tube.
2. Sediment cells at 500 × *g* for 4 min and discard supernatant.
3. Determine the wet weight of the cell pellets.
4. Optional: Shock-freeze cell pellets in liquid nitrogen and store at −80 °C.
5. Suspend the cell pellets in diluted isolation buffer for cell culture (1/3 isolation buffer) to a final concentration of 100 mg/mL.

Table 2
Solubilization of mitochondrial membranes

	Yeast mitochondria	Mitochondrial membranes from tissues	Human cells
	400 μg protein pellet[a]	400 μg protein[a]	20 mg cell pellet[a]
20 % DDM	2.0 μL (1 g/g)	5 μL (2.5 g/g)	–
Or 20 % Triton X-100	4.8 μL (2.4 g/g)	6 μL (3.0 g/g)	–
Or 20 % digitonin	6.0 μg (3.0 g/g)	12 μg (6.0 g/g)	10 μL

[a]The detergent volumes are given for solubilization of 400 μg aliquots of yeast mitochondria (*Y. lipolytica* and *S. cerevisiae*), isolated mitochondria, and tissue homogenates from fly, worm, chicken, and mammals or for 20 mg (wet weight) cultured human cells, respectively. Brackets indicate the detergent/protein ratio in g/g

6. Homogenize the cell suspension in a precooled motor-driven tightly fitting glass/Teflon Potter-Elvehjem homogenizer with 40 strokes at 2,000 rpm (*see* **Note 5**).
7. Centrifuge the homogenate at 500 × *g* for 3 min at 4 °C to gain a crude nuclear pellet.
8. Aliquot the supernatant to 20 mg samples according to the wet weight and centrifuge at 10,000 × *g* for 10 min at 4 °C. Discard the supernatant.
9. Use directly or shock-freeze the samples in liquid nitrogen and store at −80 °C.

3.3 Solubilization of Mitochondrial Membranes

1. Resuspend aliquots of mitochondria, tissue, and cell homogenates in 40 μL solubilization buffer (*see* **Note 6**).
2. Choose and add the detergent according to your purposes (Table 2), vortex samples immediately after adding detergent (*see* **Note 7**).
3. Centrifuge samples at 20,000 × *g* for 20 min at 4 °C and transfer supernatant into a new tube.
4. Optional: Measure protein concentration according to Lowry [27]. Prepare tubes for equal protein loading of samples to allow quantitative analysis. Fill up with solubilization buffer to obtain equal levels, most suitable are 10 μL (*see* **Note 8**).
5. Add 1 μL 50 % glycerol, 0.1 % Ponceau S dye, and 1 μL of 5 % Coomassie solution for 10 μL solubilized protein complexes.
6. Load suspension to BN-mini gels (*see* **Note 9**).

3.4 Blue Native Electrophoresis

Carry out the procedure in a cold room or fridge at 4 °C.

1. Pour native anode buffer into the bottom chamber of the mini gel electrophoresis apparatus.
2. Pour dark blue native cathode buffer B into upper chamber. Rinse wells with cathode buffer.
3. Load samples (*see* **Note 10**).
4. Limit current to 8 mA/gel. Adjust voltage to 100 V for first 10 min, then increase up to 500 V.
5. After 30 min, replace cathode buffer B with light blue native cathode buffer B/10.
6. Continue electrophoresis until blue front reaches the bottom of the gel.
7. Electrophoresis is completed after approximately 1 h.
8. Gels can be stored at 4 °C for several days.

3.5 Coomassie Staining [28]

1. Fix BN-gel with fixing solution for 30 min.
2. Stain the gel in Coomassie staining solution for 30 min.
3. Destain BN-gel for several hours in destaining solution.
4. Store BN-gel in water.
5. Document gel by scanning with an office scanner or gel documentation system.
6. Quantify stained protein complexes using densitometry software.

3.6 Western Blotting

1. Soak eight pieces of blotting paper with native cathode buffer.
2. Place four pieces of blotting paper onto bottom electrode of the transfer unit.
3. Incubate a sheet of PVDF membrane in methanol for 1 min and wash it in cathode buffer. Place the PVDF membrane onto blotting paper sheet (*see* **Note 11**).
4. Place the BN-gel onto PVDF membrane.
5. Cover the gel with four pieces of blotting paper soaked with cathode buffer.
6. Mount the cathode of the transfer unit.
7. Place a 5 kg load on top.
8. Transfer for 2 h at 20 V and 0.5 mA/cm^2 at room temperature.
9. Stain blot in blot staining solution for 5 min.
10. Destain blot in blot destaining solution for 15 min, repeat once.
11. Wash blot in water and let it air dry before documentation using an office flatbed scanner.

12. Destain the blot completely in methanol for 3 min (*see* **Note 12**).
13. Block blotting membrane in blocking buffer for 30 min.
14. Decorate with first antibody over night at 4 °C under gentle shaking.
15. Wash blot three times with washing buffer for 10 min each.
16. Decorate blot with secondary HRP-conjugated antibody for 1 h at room temperature.
17. Wash blot three times in washing buffer 10 min each.
18. Develop blot using an ECL chemiluminescence kit and detect signals on X-ray films or by a gel documentation system.
19. Quantify detected protein complexes using densitometry software.

3.7 In-Gel Enzyme Activity Assays and Heme Stain [13, 29]

Carry out all assays at room temperature.

1. Incubate BN-gels in solutions for in-gel activity and heme stain (*see* Subheading 2.6).
2. The violet complex I activity stain develops within 1–3 h. Stop reaction by fixing solution and incubate several hours to destain residual Coomassie.
3. The lead-phosphate precipitate by complex V activity develops over night. Stop reaction in 50 % methanol and incubate several hours to destain residual Coomassie (*see* **Note 13**).
4. Complex IV heme stain is completed after several hours. Stop reaction by fixing solution und incubate several hours to destain residual Coomassie.
5. Place gels in water and document gels with an office flatbed scanner or gel documentation system.
6. Quantify detected protein complexes using densitometry software.

3.8 2D BN/SDS-PAGE [28] and Silver Staining [30]

Carry out all assays at room temperature.

1. Mount glass plates and 1 mm spacers and place them into casting tray.
2. Prepare 13 % acrylamide mixtures for Tricine-SDS-PAGE (8 mL AB-3, 10 mL 3× gel buffer SDS-PAGE, 6 mL 50 % glycerol, 6 mL water, 15 μL TEMED, 150 μL 10 % APS).
3. Cast acrylamide mixture until 0.5 cm below top level and overlay with water.
4. Polymerization is completed after 30 min.
5. Cut a slightly thinner BN-lane and let it slip on top of the SDS-gel.
6. Remove water and mount the assembly with melted Overlay agarose.

7. Add 1× SDS-PAGE cathode and anode buffers into chambers.
8. Perform SDS-PAGE and limit current to 50 mA/gel.
9. Electrophoresis is completed after 1 h.
10. Fix 2D gel in fixing solution for 30 min.
11. Wash gel twice in water for 15 min each.
12. Incubate gel 30 min in sensitization solution.
13. Incubate gel 30 min in silver nitrate solution.
14. Rinse gel for a few seconds in water.
15. Develop silver stain 1–2 min in developer.
16. Remove developer and stop reaction by 50 mM EDTA for 15 min.
17. Place gel in water and document gel by office flatbed scanner or digital camera.

4 Notes

1. Protease inhibitor cocktails are strongly recommended for cell homogenates and isolated mitochondria from yeast.
2. We recommend carbonyl-free methanol to prevent alkylation of proteins which interferes with mass spectrometric analysis.
3. When using a peristaltic pump, switch on after opening the first valve.
4. To prevent significant loss of material do not perform this low-speed centrifugation step for tiny tissue samples—fly, worm, newt tissue, zebrafish.
5. Use trypan blue and light microscopy to check that all cells are homogenized, especially when fresh cells are used.
6. Keep samples on ice during solubilization. Use prechilled centrifuge rotors. Upon solubilization supercomplexes dissociate at room temperature, except when working with membranes from thermophilic bacteria.
7. Solubilization with digitonin preserves labile protein assemblies. Use DDM or Triton X-100 for the analysis of more stable protein–protein interactions or single OXPHOS complexes.
8. This step is mandatory for quantitative analysis only. 10–20 μg protein/lane is sufficient for Coomassie stain and in-gel activity assays, lower protein contents for Western blotting. Estimation of protein content is not needed for qualitative analysis, take 10 μL/lane. 400 μg aliquots are usually sufficient for loading of four BN-mini gel lanes.
9. Solubilized complexes are stable on ice for several hours. We do not recommend freezing and storage.

10. A white plastic card behind sample wells facilitates sample loading into dark blue wells.
11. Using PVDF membrane is mandatory because other membranes are not stable in methanol used for Coomassie destaining.
12. Blot destaining in methanol is a critical step. Residual Coomassie interferes with specific antibody binding.
13. The stop solution should not contain acetic acid because it dissolves the lead-phosphate precipitates.

Acknowledgments

We are grateful for excellent technical assistance from Ilka Siebels. We thank Thilo Borchardt, Benno Jungblut, and Thomas Braun from Max Planck Institute for Heart and Lung Research (Bad Nauheim) for kindly providing newt and zebrafish samples. Many thanks to Arcangela Iuso from the Institute of Human Genetics (Helmholtz Center Munich) for wildtype flies (*D.m.*) and Stefan Dröse from the Molecular Bioenergetics group (Frankfurt) for mitochondria from *Y.l.* We thank Hermann Schägger for many helpful discussions and antibodies. The work was supported by the Bundesministerium für Bildung und Forschung (BMBF 01GM1113B; mitoNET—Deutsches Netzwerk für mitochondriale Erkrankungen) and by the Deutsche Forschungsgemeinschaft, Sonderforschungsbereich 815 (Projects Z1 (Redox-Proteomics), A02 and A10), and Excellence Initiative (EXC 115).

References

1. Schägger H, von Jagow G (1991) Blue native electrophoresis for isolation of membrane protein complexes in enzymatically active form. Anal Biochem 199:223–231
2. Wittig I, Schägger H (2009) Native electrophoretic techniques to identify protein-protein interactions. Proteomics 9:5214–5223
3. Wittig I, Schagger H (2008) Features and applications of blue-native and clear-native electrophoresis. Proteomics 8:3974–3990
4. Perales M, Eubel H, Heinemeyer J, Colaneri A, Zabaleta E, Braun HP (2005) Disruption of a nuclear gene encoding a mitochondrial gamma carbonic anhydrase reduces complex I and supercomplex I+III2 levels and alters mitochondrial physiology in Arabidopsis. J Mol Biol 350:263–277
5. Stroh A, Anderka O, Pfeiffer K, Yagi T, Finel M, Ludwig B, Schägger H (2004) Assembly of respiratory complexes I, III, and IV into NADH oxidase supercomplex stabilizes complex I in *Paracoccus denitrificans*. J Biol Chem 279: 5000–5007
6. Eubel H, Heinemeyer J, Braun HP (2004) Identification and characterization of respirasomes in potato mitochondria. Plant Physiol 134:1450–1459
7. Dudkina NV, Eubel H, Keegstra W, Boekema EJ, Braun HP (2005) Structure of a mitochondrial supercomplex formed by respiratory-chain complexes I and III. Proc Natl Acad Sci USA 102:3225–3229
8. Aufurth S, Schägger H, Müller V (2000) Identification of subunits a, b, and c1 from Acetobacterium woodii Na^{+}-F1F0 ATPase: subunits c1, c2, and c3 constitute a mixed c-oligomer. J Biol Chem 275: 33297–33301
9. Kulajta C, Thumfart JO, Haid S, Daldal F, Koch HG (2006) Multi-step assembly pathway

of the cbb(3)-type cytochrome c oxidase complex. J Mol Biol 355:989–1004

10. Calvaruso MA, Smeitink J, Nijtmans L (2008) Electrophoresis techniques to investigate defects in oxidative phosphorylation. Methods 46:281–287
11. Wittig I, Braun HP, Schägger H (2006) Blue native PAGE. Nat Protoc 1:418–428
12. Wittig I, Schägger H (2005) Advantages and limitations of clear-native PAGE. Proteomics 5:4338–4346
13. Wittig I, Karas M, Schägger H (2007) High resolution clear native electrophoresis for in-gel functional assays and fluorescence studies of membrane protein complexes. Mol Cell Proteomics 6:1215–1225
14. Heide H, Bleier L, Steger M, Ackermann J, Dröse S, Schwamb B, Zornig M, Reichert AS, Koch I, Wittig I, Brandt U (2012) Complexome profiling identifies TMEM126B as a component of the mitochondrial complex I assembly complex. Cell Metab 16:538–549
15. Strecker V, Wumaier Z, Wittig I, Schagger H (2010) Large pore gels to separate mega protein complexes larger than 10 MDa by blue native electrophoresis: isolation of putative respiratory strings or patches. Proteomics 10:3379–3387
16. Wittig I, Beckhaus T, Wumaier Z, Karas M, Schagger H (2010) Mass estimation of native proteins by blue native electrophoresis: principles and practical hints. Mol Cell Proteomics 9:2149–2161
17. Schägger H, Pfeiffer K (2000) Supercomplexes in the respiratory chains of yeast and mammalian mitochondria. EMBO J 19:1777–1783
18. Arnold I, Pfeiffer K, Neupert W, Stuart RA, Schägger H (1998) Yeast mitochondrial F_1F_0-ATP synthase exists as a dimer: identification of three dimer-specific subunits. EMBO J 17: 7170–7178
19. Acin-Perez R, Bayona-Bafaluy MP, Fernandez-Silva P, Moreno-Loshuertos R, Perez-Martos A, Bruno C, Moraes CT, Enriquez JA (2004) Respiratory complex III is required to maintain complex I in mammalian mitochondria. Mol Cell 13:805–815
20. Schägger H, de Coo R, Bauer MF, Hofmann S, Godinot C, Brandt U (2004) Significance of respirasomes for the assembly/stability of human respiratory chain complex I. J Biol Chem 279:36349–36353
21. Kashani-Poor N, Kerscher S, Zickermann V, Brandt U (2001) Efficient large scale purification of his-tagged proton translocating NADH:ubiquinone oxidoreductase (complex I) from the strictly aerobic yeast *Yarrowia lipolytica*. Biochim Biophys Acta 1504:363–370
22. Kerscher S, Okun JG, Brandt U (1999) A single external enzyme confers alternative NADH: ubiquinone oxidoreductase activity in *Yarrowia lipolytica*. J Cell Sci 112:2347–2354
23. Smith AL (1967) Preparation, properties, and conditions for assay of mitochondria: Slaughterhouse material, small-scale. Meth Enzymol 10:81–86
24. Dröse S, Brandt U, Hanley PJ (2006) K^+-independent actions of diazoxide question the role of inner membrane KATP channels in mitochondrial cytoprotective signaling. J Biol Chem 281:23733–23739
25. Jacobus WE, Saks VA (1982) Creatine-kinase of heart-mitochondria—changes in its kinetic-properties induced by coupling to oxidative-phosphorylation. Arch Biochem Biophys 219: 167–178
26. Palmer JW, Tandler B, Hoppel CL (1977) Biochemical properties of subsarcolemmal and inter-fibrillar mitochondria isolated from rat cardiac-muscle. J Biol Chem 252:8731–8739
27. Lowry OH, Rosebrough NR, Farr AL, Randall RJ (1951) Protein measurement with the folin phenol reagent. J Biol Chem 193:265–275
28. Schägger H (2006) Tricine-SDS-PAGE. Nat Protoc 1:16–22
29. Zerbetto E, Vergani L, DabbeniSala F (1997) Quantification of muscle mitochondrial oxidative phosphorylation enzymes via histochemical staining of blue native polyacrylamide gels. Electrophoresis 18:2059–2064
30. Rais I, Karas M, Schägger H (2004) Two-dimensional electrophoresis for the isolation of integral membrane proteins and mass spectrometric identification. Proteomics 4: 2567–2571
31. Carrozzo R, Wittig I, Santorelli FM, Bertini E, Hofmann S, Brandt U, Schägger H (2006) Subcomplexes of human ATP synthase mark mitochondrial biosynthesis disorders. Ann Neurol 59:265–275

Chapter 24

Optimizing *E. coli*-Based Membrane Protein Production Using Lemo21(DE3) and GFP-Fusions

Anna Hjelm, Susan Schlegel, Thomas Baumgarten, Mirjam Klepsch, David Wickström, David Drew, and Jan-Willem de Gier

Abstract

Optimizing the conditions for the overexpression of membrane proteins in *E. coli* and their subsequent purification is usually a laborious and time-consuming process. Combining the Lemo21(DE3) strain, which conveniently allows to identify the optimal expression intensity of a membrane protein using only one strain, and membrane proteins C-terminally fused to Green Fluorescent Protein (GFP) greatly facilitates the production of high-quality membrane protein material for functional and structural studies.

Key words Membrane protein, Overexpression, Purification, *E. coli*, Lemo21(DE3), Fluorescence, GFP, FSEC

1 Introduction

The natural abundance of most helical membrane proteins, hereafter referred to as membrane proteins, is usually too low for the isolation of sufficient material for functional and structural studies. The use of natural sources also excludes the possibility of genetically modifying proteins to improve their stability and to facilitate their detection and purification. Despite tremendous efforts, there are very few examples of membrane proteins that have been successfully refolded after denaturing isolation from inclusion bodies (e.g., *see* [1]). Therefore, it is preferred to overexpress membrane proteins in a membrane, from which they can be purified after detergent extraction. However, the overexpression of membrane proteins in a membrane is usually toxic to the overexpression host [1]. Optimizing the conditions for overexpression and purification of membrane proteins is usually a laborious and time-consuming process.

Doron Rapaport and Johannes M. Herrmann (eds.), *Membrane Biogenesis: Methods and Protocols*,
Methods in Molecular Biology, vol. 1033, DOI 10.1007/978-1-62703-487-6_24, © Springer Science+Business Media, LLC 2013

1.1 The Escherichia coli Lemo21(DE3) Strain

The bacterium *E. coli* is the most widely used host to overexpress both pro- and eukaryotic membrane proteins [1]. To drive expression of the membrane protein of interest in *E. coli* we utilize the widely used bacteriophage T7-based pET/T7-RNA polymerase (T7-RNAP) expression system, in which expression of the gene encoding the target protein is governed by the T7-RNAP [2]. As overexpression host we use the BL21(DE3)-derived strain Lemo21(DE3) (Fig. 1) [3, 4]. BL21(DE3) and its derivatives harbor a chromosomal copy of the T7-RNAP gene under the control of the IPTG-inducible *lac*UV5 promoter [5]. This promoter is a stronger variant of the *lac* promoter [2]. Upon addition of IPTG, repression of the *lac*UV5 promoter is released leading to expression of the gene encoding the target protein. In BL21(DE3) T7-RNAP activity is very high and fixed. In contrast, in Lemo21(DE3) the activity of the T7-RNAP can be precisely tuned by co-expression of its natural inhibitor T7 lysozyme from the pLemo plasmid [3]. This plasmid is derived from pACYC184 and the expression of the gene encoding T7 lysozyme is governed by the rhamnose promoter [3]. Notably, we used a variant of T7 lysozyme (K128Y) (LysY) that has no amidase activity but retains full inhibition of T7-RNAP [3]. The rhamnose promoter is extremely well titratable and covers a broad range of expression intensities [6]. As a consequence, Lemo21(DE3) is tunable for membrane protein overexpression and conveniently allows optimizing the overexpression of any given membrane protein using only one strain. The combination of the *lac*UV5- and the rhamnose promoters governing expression of T7-RNAP from the chromosome and T7 lysozyme from pLemo, respectively, guarantees the widest window of expression intensities possible. Therefore, the amount of membrane protein produced can easily be harmonized with the membrane protein biogenesis capacity of the cell in Lemo21(DE3) [3, 7]. The harmonization of membrane protein production with membrane protein biogenesis capacity alleviates the toxic effects of membrane protein overexpression [3, 4]. This leads to the formation of more biomass resulting in increased membrane protein overexpression yields. It should be noted that for a small number of overexpressed membrane proteins we have observed

Fig. 1 (continued) GFP-fusions in the cytoplasmic membrane. Subsequently, in-gel fluorescence is used to assess the integrity of the overexpressed membrane protein GFP-fusions. The ratio of the cytoplasmic membrane-inserted to non-inserted membrane protein is monitored using an SDS-PAGE/immuno-blotting based assay. (**c**) The GFP-moiety facilitates the identification of a detergent to optimally extract the overexpressed membrane protein from the membrane and to monitor the stability of a membrane protein in a detergent using FSEC. Membrane protein GFP-fusions can also be seen by eye. Finally, membrane proteins can be recovered from the GFP-fusion using a site-specific protease and subsequently be used for functional and structural studies. Results observed for optimal expression/solubilization conditions are marked (*asterisk*)

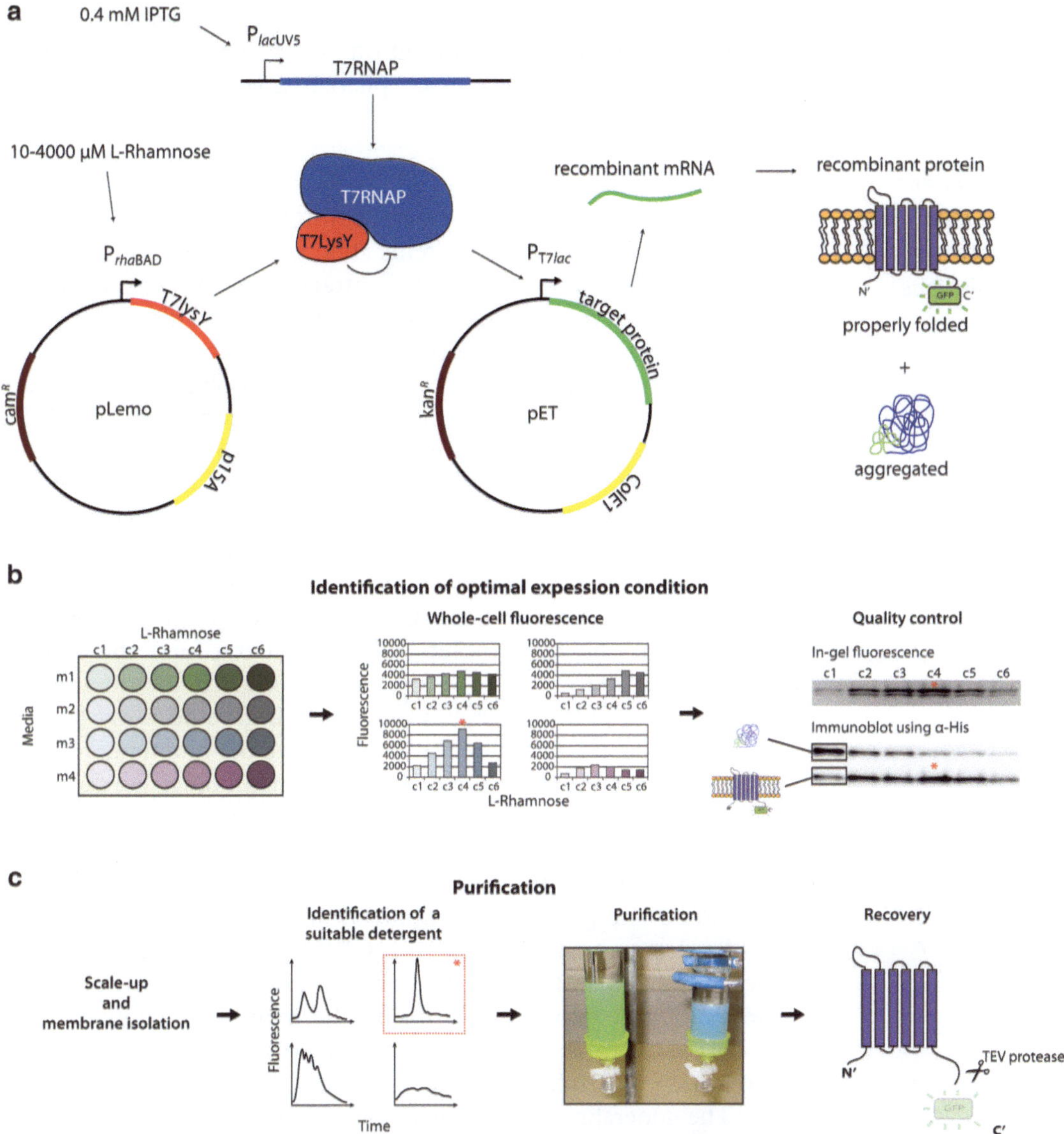

Fig. 1 Optimizing membrane protein overexpression yields in *E. coli* using Lemo21(DE3) and Green Fluorescent Protein (GFP)-fusions (**a**). Schematic representation of the Lemo21(DE3) strain. Expression of the chromosomally located gene encoding the T7 RNA polymerase is governed by the not well titratable, IPTG-inducible *lac*UV5 promoter. Expression of the gene encoding the natural inhibitor of the T7 RNA polymerase, T7 lysozyme, is governed by the exceptionally well titratable rhamnose promoter from the pLemo plasmid. The pLemo plasmid has a p15A origin of replication and contains a chloramphenicol resistance marker. The gene encoding the target membrane protein is located on a pET vector and its expression is governed by the T7*lac* promoter. The pET vector has a ColE1 origin of replication. For the overexpression of membrane proteins pET vectors with a kanamycin resistance marker are used. Membrane proteins are expressed as C-terminal GFP-fusions. The GFP-moiety only folds properly and becomes fluorescent when the membrane protein GFP-fusion is inserted in the cytoplasmic membrane. (**b**) Lemo21(DE3) cells are cultured in the presence of different concentrations of rhamnose. The expression of membrane protein GFP-fusions is induced with 0.4 mM IPTG if no autoinduction-based medium is used. Whole cell fluorescence is used to monitor the overexpression of membrane protein

that in Lemo21(DE3) the membrane protein biogenesis capacity is sufficient without any inhibition of T7-RNAP activity by T7 lysozyme [3, 4]. In these cases, plain BL21(DE3) can be used as expression host [3].

1.2 Membrane Protein Green Fluorescent Protein-Fusions

We use membrane protein Green Fluorescent Protein (GFP)-fusions to screen for optimal expression, solubilization, and purification conditions of membrane proteins (Fig. 1) [8–13]. The exceptionally stable GFP-moiety, which is attached to the C-terminus of the membrane protein, can easily be visualized. This allows monitoring both levels and integrity of the target membrane protein at any stage during the overexpression/isolation procedure. Membrane protein expression levels in the cytoplasmic membrane can be estimated by measuring fluorescence in whole cells with a detection limit as low as 10 μg GFP per liter of culture. GFP fluorescence can also be detected in standard SDS polyacrylamide gels with a detection limit of less than 5 ng of GFP [11]. This in-gel fluorescence allows rapid assessment of the integrity of membrane protein GFP-fusions and can also be used for quantification [11]. In addition, the GFP-moiety allows to determine the ratio of cytoplasmic membrane-inserted to non-inserted membrane protein using an SDS-PAGE/immuno-blotting-based assay [4, 14]. This information is very helpful to optimize the membrane protein overexpression yields in the cytoplasmic membrane. Furthermore, the GFP-moiety allows using Fluorescence-detection Size Exclusion Chromatography (FSEC) to quickly identify a detergent that is suitable for the extraction of the membrane protein from the membrane [15, 16]. Solubilization and purification efficiency can be determined by measuring GFP fluorescence in solution of the solubilized and/or purified protein [11]. After the isolation of the membrane protein GFP-fusion the membrane protein can be recovered from the fusion and subsequently used for functional and structural studies [4, 14].

The generality and simplicity of the Lemo21(DE3) "all-in-one" solution for membrane protein expression along with the use of membrane protein GFP-fusions guarantees the rapid identification of the optimal conditions for the *E. coli*-based production of membrane proteins for functional and structural studies.

2 Materials

2.1 Culturing of Cells

1. Airpore Tape sheets (to cover 24-well uniplates, *see* **item 13**).
2. Autoinduction medium: 1× M (50× M: 1.25 M K_2HPO_4, 1.25 M NaH_2PO_4, 2.5 M NH_4Cl, 0.25 M Na_2SO_4, modified after [17]), 1× Trace elements (according to [17]), 1 mM Mg_2SO_4, 0.5 % glycerol, 0.005 % glucose, 0.2 % lactose, amino acid

mixture [all amino acids except cysteine at a final concentration of 0.2 mg/mL], and 0.1 μM vitamin B_{12}.

3. Chloramphenicol, 34 mg/mL stock solution in ethanol.
4. Isopropyl β-D-1-thiogalactopyranoside (IPTG), 1 M solution, filter sterilized.
5. Kanamycin monosulfate, 50 mg/mL stock solution, filter sterilized.
6. Lysogeny Broth (aka LB medium). LB medium is usually referred to as Luria Bertani broth.
7. Lemo21(DE3) can be obtained from Xbrane Bioscience AB (www.xbrane.com, info@xbrane.com) or from New England Biolabs as competent cells (http://www.neb.com/nebecomm/products/productC2528.asp).
8. L-Rhamnose, 0.1 and 0.5 M solutions, filter sterilized.
9. Shaking incubator with temperature control.
10. Tunair 2.5 L baffled shaker flasks.
11. UV-1601 UV–VIS Spectrophotometer.
12. 50 mL Falcon tubes.
13. 24-Well uniplates (for the cover, *see* **item 1**).
14. 200 mL shaker flasks.

2.2 Monitoring Overexpression of Membrane Protein GFP-Fusions

1. Blocking buffer: 5 % milk powder in TBS-T (*see* **item 14**).
2. Coomassie staining solution: Coomassie Brilliant R250 (Fluka) 0.1 %, 40 % (v/v) methanol, 7 % (v/v) acetic acid.
3. Destaining solution: 30 % (v/v) methanol, 10 % (v/v) acetic acid.
4. ECL-Western Blotting Detection Reagents.
5. Fuji LAS-1000 charge coupled device (CCD) camera.
6. HRP-conjugated anti-His antibody.
7. Nunc 96-well optical bottom plate, black.
8. Phosphate-buffered saline (PBS): 1.44 g $Na_2HPO_4{\cdot}2H_2O$ (8.1 mM phosphate), 0.25 g KH_2PO_4 (1.9 mM phosphate), 8.00 g NaCl, 0.2 g KCl in 1,000 mL H_2O. Adjust pH to 7.4 using 1 M NaOH or 1 M HCl.
9. Polyvinylidene fluoride (PVDF) membrane (pore size 0.45 mm).
10. SDS-PAGE/Blotting set-up: your own choice.
11. Solubilization buffer (SB): 200 mM Tris–HCl, pH 8.8, 20 % glycerol, 5 mM EDTA, pH 8.0, 0.02 % bromophenol blue, make aliquots of 700 μL and keep at −20 °C. Before use, add 200 μL 20 % SDS and 100 μL 0.5 M DDT.
12. Microplate spectrofluorometer.

13. Thermomixer equipped with a thermoblock for 1.5 mL microcentrifuge tubes.
14. Tris-buffered saline 0.05 % Tween (TBS-T): 8.4 g NaCl, 3 g Tris in 1,000 mL H_2O. Adjust pH to 8.0 using 1 M NaOH or 1 M HCl and add 0.05 % Tween.
15. 5417R Eppendorf Table top centrifuge.

2.3 Membrane Isolation

1. BCA protein assay kit.
2. Beckman Optima Max XP benchtop ultracentrifuge equipped with Beckman TLA100.3 rotor.
3. Deoxyribonuclease I from bovine pancreas Type IV lyophilized powder, 1 mg/mL stock solution.
4. Disposable syringe (10 mL) with a 21-gauge needle.
5. EDTA, 0.5 M stock solution, autoclaved.
6. Emulsiflex (C3, Avestin)/French Press/Sonicator.
7. Lysozyme, 5 mg/mL stock solution.
8. $MgCl_2$, 1 M stock solution.
9. Pefabloc SC, 100 mg/mL stock solution.

2.4 FSEC and Protein Purification

1. Polyallomer 1.5 mL microcentrifuge tubes.
2. Buffer A: PBS with 0.1 % DDM (or other detergent at 5× critical micellar concentration (cmc) (for cmcs *see* Table 1).
3. Buffer B: 500 mM imidazole in Buffer A.
4. Buffer C: 20 mM Tris–HCl (pH 7.5), 5 mM EDTA (pH 8), and 100 mM NaCl.
5. 10 % CHAPS (w/v).
6. 10 % Cymal 6 (w/v).
7. 10 % Cymal 7 (w/v).
8. Imidazole, for molecular biology, minimum 99 %.
9. 10 % *N*,*N*-dimethyldodecylamine *N*-oxide (lauryldimethylamine-oxide, LDAO (w/v)).
10. 20 % *n*-decyl-β-D-maltopyranoside (DM) (w/v).
11. 20 % *n*-dodecyl-β-D-maltopyranoside (DDM) (w/v).
12. Ni-NTA Superflow.
13. 20 % *n*-octyl-β-D-glucopyranoside (w/v).
14. 20 % *n*-undecyl-β-D-maltopyranoside (UDM) (w/v).
15. Poly-Prep Chromatography Columns (Bio-Rad).
16. Superdex 200 10/300 GL Tricorn gel filtration column (GE Healthcare).
17. 20 % Triton X-100 (v/v).

Table 1
Detergents

Name	Abbreviation	Type[a]	MW	Aggregation Nr[b]	cmc (mM)[c]	cmc (% w/v)[c]	Percentage for solubilization[d]
n-Dodecyl-β-D-maltopyranoside	DDM	N	510.6	78–149	0.17	0.0087	1 %
n-Undecyl-β-D-maltopyranoside	UDM	N	496.6	74	0.59	0.029	1 %
n-Decyl-β-D-maltopyranoside	DM	N	482.6	69	1.8	0.087	1 %
Cymal 7	Cymal 7	N	522.5		0.19	0.0099	1 %
Cymal 6	Cymal 6	N	508.5	63	0.56	0.028	2 %
n-Octyl-β-D-glucopyranoside	nOG	N	292.4	78	18	0.53	2 %
N,N-dimethyldodecylamine N-oxide	LDAO	Z	229.4	76	1–2	0.023	1 %
3[(3-Cholamidopropyl) dimethylammonio] propanesulfonic acid	CHAPS	Z	614.9	10	8	0.49	1 %
Triton X-100	Triton X-100	N	647	75–165	0.23	0.015	1 % (v/v)

Suppliers like Anatrace sell different detergent grades; some crystallographers prefer the highest grade (e.g., ANAGRADE), but for most applications less expensive alternatives (e.g., SOL-GRADE) are sufficient

[a]The type of detergents: N = nonionic, Z = zwitterionic

[b]Aggregation number is the molecular weight of the micelle divided by the molecular weight of the detergent

[c]The cmc is the *c*ritical *m*icelle *c*oncentration and depends on temperature and solution conditions. For solubilizing and purification of membrane proteins one has to work always above the cmc

[d]Percentage (w/v) of detergent as used in the detergent screen (Subheading 3.5.3)

18. XK 16/20 column (GE Healthcare) or larger column.
19. ÄKTAprime or higher Äkta system (GE Healthcare).

2.5 Recovery of Membrane Proteins from GFP-Fusions

1. Centricon Centrifugal Filter Unit (Millipore): cut-off 30,000, 50,000, and 100,000 NMWL (Nominal Molecular Weight Limit) depending on size of protein and detergent.
2. Crystallization buffer of your choice (e.g., 150 mM NaCl, 20 mM Tris–HCl, pH 7.5; detergent at 3× cmc).
3. 5 mL His-Trap™ column.
4. High-grade Imidazole.
5. LAS-1000 CCD camera system.
6. TEV protease, His-tagged.
7. 0.22 μM filters.

3 Methods

3.1 Determination of Membrane Protein Topology

GFP only folds correctly and becomes fluorescent in the cytoplasm of *E. coli* whereas it does not in the periplasm [10, 18]. Therefore, using a membrane protein GFP-fusion to optimize expression/purification requires that the membrane protein of interest has a C_{in} topology. If the topology of the protein of interest is not known, use a topology predictor (e.g., MEMSAT3 [19], TOPCONS [20], or both). If the membrane protein has a C_{out} topology it can be extended with one transmembrane segment at the C-terminus (*see* Subheading 3.2). Fusing GFP to the N-terminus of N_{in}/C_{out} membrane proteins is not recommended since the GFP-moiety may interfere with the targeting of the protein to the membrane and its folding [15, 21].

3.2 Selection of Expression Vector

In our laboratories, we routinely use the pGFPd and e overexpression vectors [11]. These vectors are derived from pET28a(+) and code for a Tobacco Etch Virus (TEV) protease cleavage site between the multiple cloning site and the sequence encoding the GFP-His_8 moiety. In case a membrane protein has a C_{out} topology, pGFPd- and e-derived vectors that have the genetic information encoding the transmembrane segment of glycophorin A between the TEV protease cleavage site and the GFP-His_8 moiety can be used [22]. pGFPd and e both confer resistance to kanamycin. Kanamycin resistance is preferred to ampicillin resistance for membrane protein overexpression. This is due to the different site of action of these two antibiotic resistance markers. The antibiotic kanamycin targets the 30S subunit of the prokaryotic ribosome and the kanamycin resistance gene encodes a cytoplasmic protein. The antibiotic ampicillin interferes with the biogenesis of the peptidoglycan layer in the periplasm and is neutralized by the protein encoded by the ampicillin resistance gene, β-lactamase. To reach the periplasm, β-lactamase is translocated through the same protein-conducting channel in the cytoplasmic membrane that is also involved in mediating the biogenesis of membrane proteins into the cytoplasmic membrane, the so-called Sec-translocon. Thus, the use of the ampicillin resistance marker during membrane protein overexpression unnecessarily consumes capacity of the Sec-translocon.

3.3 Identification of the Optimal Expression Conditions

The first step to produce a membrane protein using Lemo21(DE3) is to identify the concentration of L-rhamnose that gives highest expression yields. We define the highest expression yield as the highest amount of full-length membrane protein GFP-fusion per milliliter of culture that is inserted in the cytoplasmic membrane. However, purification of the target protein may be facilitated by choosing a condition that yields the highest amount of target per

OD_{600} (per cell) instead, even if the overall yield may be lower in that case. We standard perform the initial screen at 30 °C since we have noticed that using different temperatures with Lemo21(DE3) usually does not lead to improved overexpression yields, and we routinely use Lysogeny Broth (LB) medium [3, 4]. However, Lemo21(DE3) is compatible with other culture media, including terrific broth medium and autoinduction-based media, and temperatures other than 30 °C [4, 23]. The optimal concentration of rhamnose strongly depends on the medium and the culture conditions (e.g., oxygen levels) used. Autoinduction-based overexpression facilitates overexpression screening and allows the use of 24-well plates (*see* Subheading 3.3.5). Importantly, defined autoinduction medium allows to label proteins for NMR and crystallography experiments [17]. The following steps refer to the screening of one construct, but many constructs can be screened in parallel.

3.3.1 Transforming and Culturing Lemo21(DE3)

Transform the expression vector encoding the membrane protein GFP-fusion of interest into Lemo21(DE3) (*see* **Notes 1** and **2**). Always use fresh transformants (not older than 4–5 days) for overexpression experiments. The use of glycerol stocks of transformed cells as starting material can lead to severe reduction of overexpression yields and is not recommended.

1. Set up, using a single colony from a transformation plate, an overnight (o/n) culture in a 15 mL Falcon tube containing 3 mL LB medium supplemented with 34 μg/mL chloramphenicol (for maintaining pLemo), and the appropriate antibiotic for the expression vector used (50 μg/mL kanamycin for pGFPd/e). Incubate in a shaking incubator at 30 °C, 220 rpm (*see* **Note 2**).
2. Prepare nine 50 mL Falcon tubes with 12 mL LB medium each, containing the appropriate antibiotics. Add L-rhamnose to eight of the Falcon tubes to a final concentration of 10, 50, 100, 250, 500, 1,000, 2,000, and 4,000 μM (*see* **Notes 3** and **4**). The ninth falcon tube does not contain any L-rhamnose (0 μM).
3. Inoculate each Falcon tube with a 50-fold dilution of the o/n culture. Incubate at 30 °C, 220 rpm and monitor the OD_{600} of the cultures.
4. At an OD_{600} of 0.4–0.5 (this OD_{600} will be reached approximately 2–2.5 h after inoculation) induce expression of the membrane protein GFP-fusion by adding IPTG to a final concentration of 0.4 mM.
5. Take, 4, 8, and 24 h after induction, 1 mL of culture for whole cell fluorescence measurements using a plate reader (*see* Subheading 3.3.2). Simultaneously, take 100 μL for OD_{600} measurements and approximately 500 μL for SDS-PAGE

(or immuno-blotting) (*see* Subheadings 3.3.3 and 3.3.4). Measure the OD_{600} for comparison of growth and calculations of membrane protein expression per OD_{600} (*see* **Note 5**). The whole cell and in-gel fluorescence measurements allow determining the optimal concentration of L-rhamnose for the overexpression of a membrane protein (*see* **Note 6**).

3.3.2 Whole Cell Fluorescence

1. Transfer 1 mL of culture volume to a 1.5 mL Eppendorf tube and collect the cells by centrifugation for 2 min at 15,700 × *g*. Carefully remove the supernatant.
2. Resuspend the pellet in 100 μL ice-cold PBS and leave it on ice for at least 30 min. This will allow the GFP-moiety to fold. Alternatively, wash the cell pellet once in between using 1 mL ice-cold PBS and repeat the centrifugation step.
3. Transfer the 100 μL suspension into a black Nunc 96-well optical bottom plate and measure GFP fluorescence (emission: 512 nm, excitation: 485 nm) in a microtiterplate spectrofluorometer. For maximal sensitivity select the option "bottom read" (*see* **Note 7**). To estimate membrane protein overexpression yields from whole cell fluorescence using purified GFP, follow the procedure described in subheading 3.4.

3.3.3 In-Gel Fluorescence

Measuring whole cell fluorescence does not allow discriminating between the full-length fusion protein and degradation products. GFP is an exceptionally stable molecule and remains fluorescent even if the membrane protein of interest has been degraded. The integrity of the overexpressed material can be rapidly assessed using in-gel fluorescence.

1. Harvest the cells from approximately 500 μL of culture volume by centrifugation for 2 min at 15,700 × *g*. Carefully remove the supernatant.
2. Resuspend the cell pellets in PBS to an equal OD_{600} (we dilute to a final concentration of 0.2 ODU/10 μL PBS). Add an equal volume of SB to each suspension (final concentration 0.1 ODU/10 μL solution). Ensure homogeneity of the cell suspension. Different concentrations of purified GFP of a known concentration (*see* Subheading 3.4) may be included in the analysis from here on. This will allow to accurately estimate overexpression yields and to discriminate between the full-length membrane protein GFP-fusion and degradation products (*see* **Note 8**).
3. Incubate all samples (cell suspensions and purified GFP) for 5–10 min at 37 °C (*see* **Note 9**).
4. Analyze a fraction of each sample corresponding to 0.05–0.2 ODU of sample by means of standard SDS-PAGE including

an appropriate molecular weight marker. The amount of ODU to be analyzed depends on the set-up used for SDS-PAGE (e.g., gel type, pocket size).

5. Rinse the gel with distilled water and detect in-gel fluorescence with a CCD camera system. Expose the gel to blue-light at 460 nm and capture images with increasing exposure time until the desired band intensity is reached. Fluorescence intensities can be quantified using appropriate software and expression yields estimated by comparing intensities to a GFP reference sample of known concentration (*see* **Note 10**).
6. Optionally, to control sample loading the gel can be stained for 2 h in Coomassie staining solution and subsequently destained in destaining solution.

3.3.4 Monitoring the Ratio of Membrane-Inserted to Non-inserted Membrane Protein

An assay based on SDS-PAGE/immuno-blotting of GFP-fused membrane proteins allows distinguishing between membrane proteins that are properly inserted in the cytoplasmic membrane and incorrectly folded membrane proteins, which are not inserted in the membrane [4, 14]. If a membrane protein GFP-fusion is not inserted in the cytoplasmic membrane and ends up in aggregates, its GFP-moiety does not fold properly (Fig. 1). Only if the membrane protein GFP-fusion is inserted in the cytoplasmic membrane the GFP-moiety folds properly and becomes fluorescent. Correctly folded GFP is not denatured in SDS-PAGE solubilization buffer (SB) at temperatures below 37 °C. As a consequence, a membrane protein GFP-fusion that has been inserted in the cytoplasmic membrane will migrate faster in a gel than a non-inserted fusion.

1. Follow the in-gel fluorescence protocol up until **step 4** and run an additional gel in case the one used to monitor in-gel fluorescence has been stained with Coomassie (*see* **Note 11**).
2. Transfer the proteins from the SDS-gel to a PVDF-membrane using a wet-based Western-blotting set-up of your choice.
3. For detection of the membrane protein GFP-fusions use an antibody recognizing the C-terminal His-tag of GFP-moiety (*see* **Note 12**). We routinely use an HRP-conjugated anti-His antibody. Briefly, block unspecific binding sites by incubating the membrane for 1 h in blocking buffer (5 % milk or BSA in TBS-T), then rinse the membranes three times with TBS-T, and incubate with anti-His-antibody in TBS-T for 45–60 min. Wash again with excessive amounts of TBS-T (three times 10 min) to remove non-specifically bound antibody.
4. Detect the membrane protein GFP-fusions using the detection method of your choice. We use a chemiluminescence-based assay (ECL Western blotting detection kit from Amersham) and detect the signal using a CCD camera.

3.3.5 24-Well Plate-Based Overexpression Screening

When many conditions and/or targets are included in an overexpression screen, one can use 24-well plates rather than Falcon tubes. However, the maximum culture volume of a well is 5 mL, which strongly limits sampling options. Furthermore, due to differences in growth characteristics between cultures the 24-well plate format complicates proper induction of expression. Therefore, we only use 24-well plates in combination with autoinduction-based media. We use LB medium-based o/n cultures to set up autoinduction-based cultures. Alternatively, autoinduction medium without any lactose and 0.05 % glucose can be used for the o/n cultures.

3.4 GFP-His$_8$ as a Reference for Whole Cell Fluorescence and In-Gel Fluorescence Measurements

Both whole cell fluorescence and in-gel fluorescence can be used to estimate expression yields and require purified GFP-His$_8$ as a reference [11]. Here, we briefly describe how GFP-His$_8$ is produced. For a more detailed protocol see [11].

1. Transform Lemo21(DE3) with a plasmid encoding GFP fused to a His$_8$ purification tag. We standard use pET20bGFP-His$_8$ (Amp^R) [11]. Since the overexpressed GFP-His$_8$ is soluble and located in the cytoplasm the use of the ampicillin resistance marker does not interfere with overexpression yields.
2. When using Lemo21(DE3) to express GFP-His$_8$ add L-RHAMNOSE to a final concentration of 750 μM, chloramphenicol (34 μg/mL), and ampicillin (100 μg/mL) [3].
3. Express GFP-His$_8$ for approx. 4 h as described in Subheading 3.5.1 and process the cells according to Subheading 3.5.2. Proceed with the supernatant rather than the pellet.
4. Purify GFP-His$_8$ according to the purification procedure outlined in Subheading 3.5.4. Wash the Ni-NTA column with 20 column volumes (CVs) of 10 % buffer B and elute with 50 % buffer B.
5. Pool the major GFP-His$_8$ containing fractions (as determined by fluorescence) and dialyze o/n in buffer C. As soluble GFP-His8 is expressed to very high yields and serves a reference purpose only it is not essential to retain all of it but the protein should contain as little contaminants as possible.
6. Determine the protein concentration using a BCA assay according to the instructions of the manufacturer and measure GFP fluorescence from 0.01 to 0.3 mg/mL GFP-His$_8$. Check the purity of GFP-His$_8$ by using standard SDS-PAGE followed by Coomassie staining/destaining.
7. Plot the GFP fluorescence versus the protein concentration and use the slope of the plot to convert the GFP fluorescence from any 100 μL sample to mg/mL of GFP-His$_8$.
8. Estimate expression yields by dividing the molecular weight of the expressed membrane protein GFP-fusion by 28 kDa

(MW of GFP-His_8) and multiply the obtained value with the amount of GFP-His_8 as determined in the previous step.

9. When using purified GFP-His_8 as a reference to estimate overexpression yields from whole cell fluorescence (*see* Subheading 3.3.2) and when assessing the efficiency of a detergent (*see* Subheading 3.5.3) keep in mind that GFP fluorescence is dampened by whole cells/membranes. Whole cells dampen the GFP fluorescence by a factor of approximately 1.5, membranes by a factor of 1.3 [11].

3.5 Producing Membrane Proteins for Functional and Structural Studies

After overexpression screening, the optimal condition established is used as a starting point for the production of GFP-free membrane protein material suitable for functional and structural studies. Here, we have used LB medium for scaling up the overexpression of the protein, but other media can be used as well (*see* Subheading 3.3).

3.5.1 Scaling Up of Expression

1. We use 2.5 L baffled shaker flask for scaling up the expression. However, it is also possible to use fermenters (15 L fermenters have successfully been used for scaling up expression using Lemo21(DE3)).
2. Set up an o/n culture in a 200 mL shaker flask containing 20 mL LB medium with the appropriate antibiotics (*see* Subheading 3.3.1, **step 2**). Incubate at 30 °C, 220 rpm in a shaking incubator.
3. Inoculate 1 L of LB medium with appropriate antibiotics and the optimal concentration of L-rhamnose with the o/n culture in a 2.5 L baffled shaker flask and incubate at 30 °C, 220 rpm.
4. Induce protein expression as described before, at an OD_{600} of approximately 0.4–0.5 using, 0.4 mM IPTG (final concentration) for the time determined to be optimal by the overexpression screen. Before harvesting the cells take 1 mL of culture for measuring whole cell fluorescence.

3.5.2 Isolation of Membranes

From here on, all steps should be carried out on ice or at 4 °C. Centrifugation steps are also performed at 4 °C.

1. Harvest the cells by centrifugation for 20 min at 6,200 × *g*. Discard the supernatant and carefully resuspend the pellet in 50 mL ice-cold PBS.
2. Pellet the cells according to the previous step, discard the supernatant, and resuspend the pellet in 10 mL ice-cold PBS. If needed, the pellet or the suspension can be frozen in liquid nitrogen and stored at −80 °C up to 6 months. Freezing and thawing may facilitate breaking the cells.
3. Add 1 mg/mL Pefabloc SC (or another protease inhibitor mix of your choice). Add 1 mM EDTA and 0.5 mg/mL lysozyme

(final concentration) and incubate on ice for 30–60 min (*see* **Note 13**). For improved efficiency the suspension may be stirred slowly. Add 5 μg/mL DNaseI and 2 mM $MgCl_2$ and incubate for approx. 15 min. Break the cells using an Emulsiflex (15,000 p.s.i., 3–5 cycles) or a method of your choice (e.g., French press, sonication). Most cells are broken when the suspension has turned translucent (*see* **Note 14**).

4. Clear the suspension of unbroken cells/debris by centrifugation at 24,000 × *g* for 20 min. Transfer the supernatant (containing the membranes) to a clean tube and repeat the centrifugation step.
5. To collect the membranes, centrifuge the supernatant for 45 min at 150,000 × *g*. Discard the supernatant and resuspend the membrane pellet in 10 mL ice-cold PBS using a disposable 10 mL syringe with a 21-gauge needle.
6. Fill up the centrifugation tube with ice-cold PBS and harvest the membranes once more for 45 min at 150,000 × *g*. This step will remove residual EDTA which otherwise would interfere with the immobilized metal affinity chromatography (IMAC) later on.
7. Resuspend the membrane pellet in 5 mL ice-cold PBS essentially as described before. Determine protein concentration with a standard BCA assay for **step 1** in Subheading 3.5.3. If desired, membrane suspensions may be frozen in liquid nitrogen and stored at −80 °C for up to 6 months. However, some membrane protein crystallographers avoid freezing and storing of membranes and continue with purification immediately as repeated freezing/thawing may negatively affect the material.

3.5.3 Identification of a Suitable Detergent Using FSEC

Next, a detergent has to be identified that can be used to extract the overexpressed membrane protein from the membrane optimally. Importantly, the membrane protein should remain stable in the detergent used. Each detergent has a specific cmc that depends on both the temperature and composition of the solubilization buffer used (e.g., salt content and pH). To efficiently solubilize a membrane protein, use a detergent concentration well above the cmc. In Table 1 we have listed the detergents and concentrations we use for the solubilization of membranes. None of the detergents listed in Table 1 interferes with GFP fluorescence.

1. Adjust the portion of the membrane suspension used for detergent screening to a concentration of 3.75 mg of protein/mL. This will yield a final protein concentration of 3 mg/mL in the next step. Transfer aliquots of 800 μL of membrane suspension into 1.5 mL Beckman polyallomer microcentrifuge tubes.
2. Dissolve the detergents to be screened for in PBS to 5× the final concentration.

3. Add 200 μL of each detergent to each of the tubes and incubate the samples for 1 h under mild agitation.
4. Collect non-solubilized material by centrifugation for 45 min at 100,000 × *g*.
5. Transfer the supernatant to a clean tube and measure GFP fluorescence in 100 μL of the supernatant. To determine the percentage of detergent-solubilized GFP resuspend the non-solubilized pelleted membranes in the same buffer volume and compare fluorescence to that of the detergent-solubilized membranes.
6. Using the best detergent (as established in the previous steps) determine the optimal ratio between protein and detergent by repeating **steps 1–5**. Keep the percentage of detergent constant but increase the amount of protein (starting from 3 mg/mL). The optimal protein to detergent ratio is the point at which a linear increase in protein still yields a linear increase in GFP fluorescence.
7. To assess the stability of a membrane protein GFP-fusion in a particular detergent its monodispersity in this detergent can be rapidly monitored by size exclusion column chromatography (SEC) followed by fractionation into a 96-well plate for measuring the fluorescence in each well (*see* Subheading 3.3.2, **step 3**) [12, 13, 16] (*see* **Note 15**).

3.5.4 Purification of the Membrane Protein GFP-Fusion

1. Solubilize the membranes, using the established optimal protein/detergent ratio by incubating membrane–detergent mixture for 1 h under mild agitation.
2. Collect unsolubilized material by centrifugation for 45 min at 100,000 × *g* and transfer the supernatant to a fresh tube. Take 100 μL of the supernatant to measure GFP fluorescence as described in Subheading 3.3.2, **step 5** and estimate the percentage of solubilization (*see* Subheading 3.5.1, **step 4** or Subheading 3.5.3, **step 7**). Keep another 100 μL for subsequent analysis of the purification as described below (*see* **step 8** of this section).
3. Use ~1 mL of Ni-NTA resin per mg of to be purified membrane protein GFP-fusion to pack a XK 16/20 column.
4. Equilibrate Ni-NTA column with five CVs of Buffer A.
5. Add imidazole to a final concentration of 10 mM to the solubilized membranes and load them onto the Ni-NTA column at a flow rate of approx. 0.3–0.5 mL/min.
6. Wash the column with ~20 CVs of 4 % Buffer B (*see* Subheading 2) at a flow rate of 1 mL/min to remove contaminants.
7. Wash with a gradient of 4–25 % Buffer B over 20 CVs at a flow rate of 1 mL/min and collect fractions (*see* **Note 16**).

8. Elute the membrane protein GFP-fusion with 50 % Buffer B at a flow rate of 1 mL/min and collect all fractions. Save 100 μL samples from flow-through, wash, and elution fractions (store at 4 °C). These will later be used to assess the quality and integrity of the purified material (*see* Subheadings 3.3.3 and 3.3.4).
9. Measure GFP fluorescence in the different fractions (flow-through, wash, elution) (*see* **Note 17**).
10. Determine the amount of membrane protein GFP-fusion in each fraction as described in Subheading 3.3.2, **step 3** (*see* **Note 18**) [24].

3.5.5 Recovery of a Membrane Protein from a GFP-Fusion

1. To recover the membrane protein from the membrane protein GFP-fusion incubate the membrane protein GFP-fusion with His-tagged TEV protease for 10 h or o/n at 4 °C (or another site-specific protease if a different cleavage site was chosen). Dialyze into an appropriate crystallization buffer (*see* **Note 19**).
2. Determine total protein concentrations in the different fractions (including the ones saved from the IMAC purification) using the BCA assay.
3. Add an appropriate amount of protein in a 10 μL volume to 10 μL of SB. Analyze the different fractions using SDS-PAGE followed by in-gel fluorescence and Coomassie staining/destaining. This will allow assessing the efficiency of the IMAC purification (binding/elution of the membrane protein fusion), the integrity of the fusion protein, and the efficiency of protease cleavage (*see* **Note 20**).
4. If the protease digest was complete, pass the digest through a 0.22 μM filter to remove any precipitation and then load the digested material through a 5-mL His-Trap™ column by the use of either a peristaltic pump or an appropriate sized syringe and collect the flow-through. The recovered fraction should not be fluorescent! (*see* **Note 21**).
5. Wash the column with 20 mL of crystallization buffer containing 30 mM imidizole and collect the flow-through. Repeat this step with buffer containing 250 mM imidazole. Concentrate the digest in Centricon concentrators to ~0.5 mL using an appropriate MW cut-off and measure concentration of protein in each fraction. As untagged protein can still bind to the resin we include an additional wash step to ensure that all untagged protein is recovered. Analyze the purified protein found typically in the first fraction by standard gel filtration using a Superdex 200 10/30. At this stage one symmetric peak similar to that observed by FSEC in Subheading 3.5.3, **step 7** is expected.
6. If the protein is to be used for crystallization trials concentrate to ~5–10 mg/mL in Centricon concentrators using an MW

cut-off as large as possible to minimize overconcentration of free detergent. Excessive detergent can interfere with crystallization. This is more problematic with mild detergents that have a large micelle-size, e.g., DDM with 72 kDa [25]. Typically we start with 100 kDa cut-off concentrators and only change to 50 kDa if the target protein is present in our 100 kDa concentrator flow-through fraction.

4 Notes

1. An empty expression vector may be included as control (*see* **Note 7**).
2. If "leaky" expression of the target membrane protein is toxic, the addition of L-rhamnose to the plates used for the transformation and the o/n culture will reduce the toxicity of the "leaky" expression.
3. We use these concentrations of L-rhamnose by default for the overexpression screening but they may of course be adapted.
4. If the optimal overexpression yield is reached without any rhamnose (0 μM) in Lemo21(DE3) consider to continue with plain BL21(DE3) instead.
5. OD_{600} values vary between different spectrophotometers. Make sure to measure within the linear range of the spectrophotometer.
6. As mentioned above, screening at different temperatures usually does not lead to improved overexpression yields in Lemo21(DE3). However, if severe degradation of the membrane protein GFP-fusion is observed, a switch to lower expression temperatures or a shortening of induction times may be considered. Expression at lower temperatures usually results in a different optimal L-rhamnose concentration.
7. The level of background fluorescence is usually quite low but can lead to overestimation of membrane protein expression yields especially if protein expression levels are very low. Measure whole cell fluorescence of cells harboring an empty expression vector to account for background fluorescence. If expression yields are lower than 200 μg/L, the signal to noise ratio may be improved by increasing the amount of cells analyzed. Use 5 mL of culture for fluorescence measurements in such a case.
8. Instead of adjusting the cell suspensions to the same OD_{600} they may also be adjusted to the same fluorescence levels (useful if various constructs with different expression levels as determined by whole cell fluorescence are screened simultaneously). That way, weak bands can be detected easily without interference from neighboring, stronger bands.

9. Incubation at temperatures higher than 37 °C is not recommended as this can lead to aggregation of membrane proteins and loss of GFP fluorescence. In addition, incubation at higher temperatures does not allow discriminating between the membrane-inserted and non-inserted versions of the target protein (*see* Subheading 3.3.4). If frozen cells are used add $MgCl_2$ to a final concentration of 1 mM and DNaseI (1–5 units/10 μL) to the samples and incubate for 15 min on ice before adding SB to yield a homogenous suspension. Alternatively, benzonase can be used.
10. GFP-His_8 has a molecular weight of approx. 28 kDa; however, GFP remains folded in SDS and the apparent molecular weight in SDS-gels is lower (approx. 20 kDa). For quantification, use an image without any saturated signals.
11. In order to determine the best possible ratio of membrane-inserted to non-inserted protein we advise to investigate not only the sample that yielded the highest fluorescence/mL but also some samples of adjacent rhamnose concentrations.
12. We have experienced that antisera recognizing GFP can bind differently to folded and unfolded GFP. Therefore, we use an antibody recognizing the His-tag. When evaluating the results keep in mind that the binding behavior varies between different antibodies and that binding is not necessarily linear.
13. The lysozyme version, LysY, expressed by Lemo21(DE3) is not lytic [3]. Adding lysozyme is not essential; however, it tremendously facilitates breaking the cells.
14. Devices other than the Emulsiflex can be used to break cells (e.g., the French press or a sonicator).
15. As originally outlined by Kawate and Gouaux, an in-line detector, e.g., a Prominence high performance liquid chromatography (HPLC) device (Shimadzu), can be used for higher sensitivity [15]. Notably, fluorescence-detection SEC (FSEC) has also successfully been used to screen for optimal solubilization conditions using whole cells as starting material (e.g., [15, 26, 27]). In addition, it is to be kept in mind that the addition of lipids to the buffers used for the purification of a membrane protein can considerably improve the quality of the isolated material (e.g., [28]).
16. Once the wash and elution conditions have been established, step gradients instead of continuous gradients can be used: wash for 20 CVs at 2 % less than the highest percentage of Buffer B where protein was still bound to the column.
17. If the amount of overexpressed protein has exceeded the capacities of the purification system this will be evident as a fluorescence signal in the flow-through fraction. To increase the yield of the membrane protein GFP-fusion increase the

bed-volume or, if necessary, change to a volume with a larger diameter. In case of nonspecific binding of membrane protein to Ni-NTA resin, add 5–10 mM imidazole to buffer used in batch binding.

18. The use of the BCA assay to estimate the amount of fusion protein in the fractions is not recommended as the BCA assay measures total protein and the presence of contaminants may lead to misestimating expression yields. Additionally, the imidazole used for elution cross-reacts with the assay at concentrations >50 mM [24].
19. For commonly used detergents like n-dodecyl-β-D-maltopyranoside (DDM) and Triton X-100 equimolar amounts of TEV protease typically suffice for a complete o/n digest. We generally dialyze into a crystallization buffer containing 20 mM Tris–HCl, pH 7.5, 150 mM NaCl, and the concentration of detergent at three times the cmc.
20. The different fractions obtained after purification may of course also be analyzed accordingly before proceeding with protease cleavage.
21. As some membrane proteins are sensitive to high concentrations of imidazole we prefer to use the more weakly binding Ni-NTA resin in the first IMAC step in Subheading 3.5.4, **step 3**. However, for removal of the tag and of contaminants such as AcrB [29]) the His-Trap™ column is preferred as it has better binding capacity.

Acknowledgments

Research in the laboratory of JWdG is supported by grants from the Swedish Research Council, the Carl Tryggers Stiftelse, the Marianne and Marcus Wallenberg Foundation, NIH grant 5R01GM081827-03, and the SSF supported Center for Biomembrane Research. David Drew acknowledges the support of the Royal Society (UK) through a University Research Fellowship and the Swedish Research Council.

References

1. Wagner S, Bader ML, Drew D, de Gier JW (2006) Rationalizing membrane protein overexpression. Trends Biotechnol 24:364–371
2. Samuelson JC (2011) Recent developments in difficult protein expression: a guide to *E. coli* strains, promoters, and relevant host mutations. Methods Mol Biol 705:195–209
3. Wagner S, Klepsch MM, Schlegel S, Appel A, Draheim R, Tarry M, Hogbom M, van Wijk KJ, Slotboom DJ, Persson JO, de Gier JW (2008) Tuning *Escherichia coli* for membrane protein overexpression. Proc Natl Acad Sci USA 105:14371–14376
4. Schlegel S, Lofblom J, Lee C, Hjelm A, Klepsch M, Strous M, Drew D, Slotboom DJ, de Gier JW (2012) Optimizing membrane protein overexpression in the *Escherichia coli* strain Lemo21(DE3). J Mol Biol 423(4):648–659

5. Studier FW, Moffatt BA (1986) Use of bacteriophage T7 RNA polymerase to direct selective high-level expression of cloned genes. J Mol Biol 189:113–130
6. Giacalone MJ, Gentile AM, Lovitt BT, Berkley NL, Gunderson CW, Surber MW (2006) Toxic protein expression in *Escherichia coli* using a rhamnose-based tightly regulated and tunable promoter system. Biotechniques 40:355–364
7. Wagner S, Baars L, Ytterberg AJ, Klussmeier A, Wagner CS, Nord O, Nygren PA, van Wijk KJ, de Gier JW (2007) Consequences of membrane protein overexpression in *Escherichia coli*. Mol Cell Proteomics 6:1527–1550
8. Drew DE, von Heijne G, Nordlund P, de Gier JW (2001) Green fluorescent protein as an indicator to monitor membrane protein overexpression in *Escherichia coli*. FEBS Lett 507: 220–224
9. Drew D, Slotboom DJ, Friso G, Reda T, Genevaux P, Rapp M, Meindl-Beinker NM, Lambert W, Lerch M, Daley DO, Van Wijk KJ, Hirst J, Kunji E, De Gier JW (2005) A scalable, GFP-based pipeline for membrane protein overexpression screening and purification. Protein Sci 14:2011–2017
10. Drew D, Sjostrand D, Nilsson J, Urbig T, Chin CN, de Gier JW, von Heijne G (2002) Rapid topology mapping of *Escherichia coli* inner-membrane proteins by prediction and PhoA/GFP fusion analysis. Proc Natl Acad Sci USA 99:2690–2695
11. Drew D, Lerch M, Kunji E, Slotboom DJ, de Gier JW (2006) Optimization of membrane protein overexpression and purification using GFP fusions. Nat Methods 3:303–313
12. Newstead S, Kim H, von Heijne G, Iwata S, Drew D (2007) High-throughput fluorescent-based optimization of eukaryotic membrane protein overexpression and purification in *Saccharomyces cerevisiae*. Proc Natl Acad Sci USA 104:13936–13941
13. Drew D, Newstead S, Sonoda Y, Kim H, von Heijne G, Iwata S (2008) GFP-based optimization scheme for the overexpression and purification of eukaryotic membrane proteins in *Saccharomyces cerevisiae*. Nat Protoc 3: 784–798
14. Geertsma ER, Groeneveld M, Slotboom DJ, Poolman B (2008) Quality control of overexpressed membrane proteins. Proc Natl Acad Sci USA 105:5722–5727
15. Kawate T, Gouaux E (2006) Fluorescence-detection size-exclusion chromatography for precrystallization screening of integral membrane proteins. Structure 14:673–681
16. Sonoda Y, Newstead S, Hu NJ, Alguel Y, Nji E, Beis K, Yashiro S, Lee C, Leung J, Cameron AD, Byrne B, Iwata S, Drew D (2011) Benchmarking membrane protein detergent stability for improving throughput of high-resolution X-ray structures. Structure 19:17–25
17. Studier FW (2005) Protein production by auto-induction in high density shaking cultures. Protein Expr Purif 41:207–234
18. Feilmeier BJ, Iseminger G, Schroeder D, Webber H, Phillips GJ (2000) Green fluorescent protein functions as a reporter for protein localization in *Escherichia coli*. J Bacteriol 182: 4068–4076
19. Nugent T, Jones DT (2009) Transmembrane protein topology prediction using support vector machines. BMC Bioinformatics 10:159
20. Bernsel A, Viklund H, Hennerdal A, Elofsson A (2009) TOPCONS: consensus prediction of membrane protein topology. Nucleic Acids Res 37:465–468
21. Luirink J, Yu Z, Wagner S, de Gier JW (2012) Biogenesis of inner membrane proteins in *Escherichia coli*. Biochim Biophys Acta 1817: 965–976
22. Hsieh JM, Besserer GM, Madej MG, Bui HQ, Kwon S, Abramson J (2010) Bridging the gap: à GFP-based strategy for overexpression and purification of membrane proteins with intra and extracellular C-termini. Protein Sci 19:868–880
23. Low C, Jegerschold C, Kovermann M, Moberg P, Nordlund P (2012) Optimisation of over-expression in *E. coli* and biophysical characterisation of human membrane protein synaptogyrin 1. PLoS One 7:e38244
24. http://www.piercenet.com/browse.cfm?fldID=02020101
25. Strop P, Brunger AT (2005) Refractive index-based determination of detergent concentration and its application to the study of membrane proteins. Protein Sci 14:2207–2211
26. Gonzales EB, Kawate T, Gouaux E (2009) Pore architecture and ion sites in acid-sensing ion channels and P2X receptors. Nature 460:599–604
27. Kawate T, Michel PO, Birdsong WT, Gouaux E (2009) Crystal structure of the ATP-gated P2X(4) ion channel in the closed state. Nature 460:592–598
28. Guan L, Mirza O, Verner G, Iwata S, Kaback HR (2007) Structural determination of wild-type lactose permease. Proc Natl Acad Sci USA 104:15294–15298
29. Drew D, Klepsch MM, Newstead S, Flaig R, De Gier JW, Iwata S, Beis K (2008) The structure of the efflux pump AcrB in complex with bile acid. Mol Membr Biol 25:677–682

INDEX

Doron Rapaport and Johannes M. Herrmann (eds.), *Membrane Biogenesis: Methods and Protocols*, Methods in Molecular Biology, vol. 1033, DOI 10.1007/978-1-62703-487-6, © Springer Science+Business Media, LLC 2013

Q

R

S

T

MIX
Papier aus verantwortungsvollen Quellen
Paper from responsible sources
FSC® C105338

If you have any concerns about our products,
you can contact us on
ProductSafety@springernature.com

In case Publisher is established outside the EU,
the EU authorized representative is:
Springer Nature Customer Service Center GmbH
Europaplatz 3, 69115 Heidelberg, Germany

Printed by Libri Plureos GmbH
in Hamburg, Germany